INTRODUCTION TO
General Relativity

JOHN DIRK WALECKA

College of William and Mary, USA

T0331586

World Scientific

NEW JERSEY · LONDON · SINGAPORE · BEIJING · SHANGHAI · HONG KONG · TAIPEI · CHENNAI

Published by

World Scientific Publishing Co. Pte. Ltd.

5 Toh Tuck Link, Singapore 596224

USA office: 27 Warren Street, Suite 401-402, Hackensack, NJ 07601

UK office: 57 Shelton Street, Covent Garden, London WC2H 9HE

Library of Congress Cataloging-in-Publication Data
Walecka, John Dirk, 1932–
    Introduction to general relativity / John Dirk Walecka.
        p. cm.
    Includes bibliography and index.
    ISBN-13   978-981-270-584-6
    ISBN-10   981-270-584-8
    ISBN-13   978-981-270-585-3 (pbk)
    ISBN-10   981-270-585-6 (pbk)
    1. General relativity (Physics)--Mathematics.   I. Title.
    QC173.6 .W354 2007
    530.11--dc22
                                    2007300085

British Library Cataloguing-in-Publication Data
A catalogue record for this book is available from the British Library.

First published 2007
Reprinted 2008

Printed in Singapore

For Kay

Preface

Eleven years after Einstein's theory of special relativity completely changed our understanding of the relationship between space and time [Einstein (1905)], his theory of general relativity revolutionized our understanding of how mass and energy change the underlying space-time structure of the physical universe [Einstein (1916)].

Some time ago I had to learn general relativity in connection with my research. I had great difficulty listening to the experts and understanding what they were doing. I decided I had to teach the subject to myself, and I proceeded to do so. Much later, I converted my notes to a semester course at William and Mary, which was offered three times. It was aimed at physics graduate students, but many undergraduates participated, and excelled. The students seemed to learn from it and enjoy it, and the outcome was very satisfying. I decided that I would convert these lectures into a book entitled *Introduction to General Relativity*.

General relativity is a difficult subject for two reasons. The first is that the math is unfamiliar to most physics students, and the second is that since the four-dimensional coordinate system has no intrinsic meaning, it is difficult to get at the physical interpretation and physical consequences of any result. The goals of the present text are as follows:

- The book is aimed at physics graduate students and advanced undergraduates. Only a working knowledge of classical lagrangian mechanics is assumed,[1] although an acquaintance with special relativity will make the book more meaningful. Within this framework, the material is self-contained;
- The necessary mathematics is developed within the context of a

---

[1] As presented, for example, in [Fetter and Walecka (2003)].

concrete physical problem — that of a point mass constrained to move without friction on an arbitrarily shaped two-dimensional surface;

- A strong emphasis is placed on the physical interpretation and physical consequences in all applications;
- The book is not meant to be a weighty tome for experts (indeed, I could not write one), but, in my opinion, most of the interesting applications of GR are covered;
- The final "Special Topics" chapter takes the reader up to a few areas of active research.

At the end of the course at William and Mary, students selected a topic of current interest, wrote a term paper, and then gave a talk to the class in the format of a contributed session of an American Physical Society meeting (complete with abstract). I was amazed and pleased with the level at which the students were able to perform. It is my sincere hope that this text will provide useful background for other young people and aid them in their exploration of many of the fascinating modern developments based on Einstein's theory of general relativity.

An extensive bibliography has not been attempted, although most of the comparable and more advanced texts are referenced, as are relevant websites. Since general relativity is an old subject, no effort has been made to trace the origin of the examples and problems, many of which are modified versions of those in the list of texts and monographs.

I was delighted when World Scientific Publishing Company, which had done an exceptional job with two of my previous books, showed enthusiasm for publishing this new one. I would like to thank Dr. K. K. Phua, Executive Chairman of World Scientific Publishing Company, and my editor Ms. Lakshmi Narayanan, for their help and support on this project.

I would also like to thank my colleagues Paolo Amore, Brian Serot, and J. Wallace Van Orden for their reading of the manuscript.

Williamsburg, Virginia   *John Dirk Walecka*
January 3, 2007   *Governor's Distinguished CEBAF*
                                        Professor of Physics, emeritus
                                        College of William and Mary

Contents

Chapter 1

Introduction

Newton's second law of motion, formulated over three centuries ago, forms the backbone of classical physics and continues to describe most of what we observe in the world around us. It is easily stated. One defines the primary inertial coordinate system as a system that is at rest with respect to the fixed stars. The second law then states that in this inertial frame the rate of change of momentum of a point object with inertial mass m_i and momentum $m_i \vec{v}$ is given by the applied force $\vec{F}$

$$\frac{d}{dt}(m_i \vec{v}) = \vec{F} \qquad ; \text{Newton's second law} \qquad (1.1)$$

Furthermore, any frame moving with constant velocity relative to the primary inertial coordinate system is again an inertial frame in which Newton's second law holds. In all these frames, it is assumed that there is a single, universal time t. Of course, three-dimensional space here is euclidian, obeying all of Euclid's postulates.

Newton's law of gravitation states that the force between a point object with gravitational mass m_g and another object of mass M separated by a displacement vector $\vec{r}$ is given by

$$\vec{F} = -\frac{GMm_g}{r^2}\frac{\vec{r}}{r} \qquad ; \text{Newton's law of gravitation} \qquad (1.2)$$

Here M is either a point mass, or one is outside of a spherically symmetric distributed mass with $\vec{r}$ referring to its center, and G is Newton's constant.

If these two expressions are equated, and it is assumed that $m_i = m_g$ with both being constant, then *that mass cancels* and one obtains

$$\frac{d\vec{v}}{dt} = -\frac{GM}{r^2}\frac{\vec{r}}{r} \qquad (1.3)$$

1

This expression implies that all massive objects with *any* $m_i = m_g$ move the same way in the gravitational field of the mass M, a result that accords with what we observe in the world around us and what Galileo confirmed long ago in his celebrated experiment dropping various objects from the leaning tower of Pisa.

At the beginning of the 20th century, two fundamental modifications of classical physics were discovered. The first was quantum mechanics, which describes very different behavior in the microscopic domain, yet reduces to Newton's laws in the appropriate limit. We shall have very little to say about quantum mechanics in this text. The second was Einstein's theory of special relativity [Einstein (1905)]. The Michelson-Morley experiment searched for a shift in the fringes of an interferometer moving with various velocities with respect to the primary inertial coordinate system. This experiment ultimately implied that the speed of light c is the same in any inertial frame, a most amazing and non-intuitive result at complete variance with how one adds velocities in classical physics. Lorentz had discovered an algebraic transformation that left the form of the wave equation for light invariant. It was Einstein's genius to give that transformation a physical interpretation and place the transformed coordinates in a one-to-one correspondence with what is actually observed in various inertial frames. The consequences of this association are profound: time is relative and varies from frame to frame; length is also relative; a particle's mass depends on its velocity; there is a relation between energy and mass $E = mc^2$, and so on.

The four-dimensional space $(x^1, x^2, x^3, x^4) \equiv (\vec{x}, ct)$ in which these coordinate transformations take place is no longer euclidian. If one writes an infinitesimal physical displacement in this space as $d\mathbf{s}$, then the square of this displacement, the invariant interval $(d\mathbf{s})^2 = d\mathbf{s} \cdot d\mathbf{s}$, is given by

$$(d\mathbf{s})^2 = \sum_{\mu=1}^{4} \sum_{\nu=1}^{4} g_{\mu\nu} dx^\mu dx^\nu$$

$$g_{\mu\nu} = \begin{bmatrix} 1 & 0 & 0 & 0 \\ 0 & 1 & 0 & 0 \\ 0 & 0 & 1 & 0 \\ 0 & 0 & 0 & -1 \end{bmatrix} \tag{1.4}$$

The quantity $g_{\mu\nu}$ is known as the *metric*. In chapter 6 of this book we review the basic principles of Einstein's theory of special relativity and many of its implications, which have been repeatedly confirmed experimentally over the intervening years.

It was again Einstein's genius to realize two profound implications of the above results.[1] The first is the implication of the *equivalence principle* that $m_i = m_g$

$$m_i = m_g \qquad ; \text{ equivalence principle} \qquad (1.5)$$

The first mass m_i governs the acceleration of an object with respect to the fixed stars, and the second mass m_g determines the strength of the gravitational force. Why should these things have anything to do with each other? It is a consequence of the equivalence principle that all particles follow the same trajectory in a gravitational field independent of their mass. Thus the trajectory of a particle in such a field is determined only by the *geometry* of the field. This observation provided one of the key points of departure for general relativity.

The second key insight was that the world in which we live, at least in free space and moving with uniform velocity, is not a nice four-dimensional *euclidian* space, but a rather mysterious *Minkowski* space with the indefinite metric of Eq. (1.4). Is there any reason that four-dimensional space might not have a more general structure? That it might also be *curved* rather than flat? For example, a particle can move without friction on a flat two-dimensional surface, in which case the trajectory is simply a straight line. That surface might also be curved and distorted, in which case the trajectories can become very involved. Indeed, the mechanics problem of a particle moving without friction on an arbitrarily shaped two-dimensional surface will form the paradigm for all the physics and mathematics we subsequently do in this book. We shall show that the trajectories in this case are just the *geodesics* on the surface — the curves of minimum (or stationary) physical distance. Thus they are entirely determined by the geometry of the surface!

Is it possible that the presence of mass (and energy) as a source can produce a curved four-dimensional space-time such that the equation for the geodesics in this space just reproduces Eq. (1.3) in the appropriate limit? If so, one would have a unified description of both Newton's second law and his universal law of gravitation, two cornerstones of classical physics. It is just this problem that is solved by Einstein's theory of general relativity [Einstein (1916)].

---

[1]These results are now universally presented to students at the introductory physics level. How many other such results are there that have profound implications currently beyond our grasp??

In order to formulate the theory, one has to have in his or her arsenal the mathematical tools describing a curved space in higher dimensions. This is the theory of *riemannian geometry*. In chapter 3, we first present a straightforward discussion of curvilinear coordinates in higher dimensional euclidian space, including the notion of vectors and tensors. In chapter 5, this discussion is generalized to discuss curved spaces, starting from our paradigm of a particle moving without friction on a curved two-dimensional surface. The only assumption made in this book about the reader's background is that he or she is familiar with classical lagrangian mechanics as presented, for example, in [Fetter and Walecka (2003)]. Beyond that, the reader should find the material in this text self-contained.[2]

The great power of classical lagrangian mechanics, which can be derived from Hamilton's variational principle, is that it is freed from any particular choice of coordinates. Thus in the problem of the particle moving on an arbitrary two-dimensional surface, discussed in detail in chapters 2 and 4, one can introduce *any* set of linearly independent generalized coordinates (q^1, q^2) on the surface with which to describe the particle's location and subsequent motion. We will assume that there is a unique, flat, *tangent plane* at each point on the surface. We make use of the fact that an infinitesimal physical displacement in the surface is identical to that in the tangent plane, and we then demonstrate how each choice of generalized coordinates carries with it an associated *metric* on the surface. We proceed to derive the lagrangian for particle motion on the surface and the corresponding set of Lagrange's equations for (q^1, q^2). These form a set of two coupled, second-order, non-linear differential equations in the time, with the surface entering only through the *affine connection*, a first-order, nonlinear differential form in the metric.[3] We then show that the equations of motion are identical to those for the *geodesics* on the surface. The observation that the information on the intrinsic structure of the surface must somehow be contained in the affine connection forms the basis for our development of riemannian geometry in chapter 5. In that chapter, we show how the *Riemann tensor*, a first-order, nonlinear differential form in the affine connection, characterizes the curvature of the space.

---

[2]Some familiarity with the rest of classical mechanics, as presented in [Fetter and Walecka (2003)], as well as with special relativity and other aspects of modern physics, as presented, for example, in [Ohanian (1995)], will make this book more meaningful. We do also assume the reader is familiar with vector calculus and linear algebra.

[3]We shall use the terminology "first-order, nonlinear differential form in the metric" to indicate an expression that is nonlinear in the indicated quantity and contains derivatives up through the stated order.

We subsequently, in chapter 5, proceed to prove all of the results on riemannian geometry that we use in the rest of the book.[4] Readers anxious to get to the "meat" of special and general relativity starting in chapter 6 may want to just accept the results that come after the introduction of the Riemann curvature tensor, and come back to the proof of those results at their leisure.

Einstein's theory of general relativity is introduced in chapter 7 through a set of three assumptions:

(1) We live in a four-dimensional riemannian space with a local, flat Minkowski tangent space;
(2) The structure of the space is given by the Einstein field equations

$$G^{\mu\nu} = \frac{8\pi G}{c^4} T^{\mu\nu} \qquad ; \text{Einstein equations} \qquad (1.6)$$

Here $G^{\mu\nu}$ is the Einstein tensor derived from the Riemann curvature tensor (it is a second-order, nonlinear differential form in the metric), and $T^{\mu\nu}$ is the energy-momentum tensor for the medium under consideration.
(3) Particles follow the geodesics in this space.

The Einstein field equations are solved outside of a spherically symmetric source, and it is shown how this Schwarzschild solution for the metric [Schwarzschild (1916)] then leads to Eq. (1.3) in the appropriate limit. We take this demonstration as the *basic justification of Einstein's theory*, and then proceed to investigate its further consequences.

Since coordinates, and the corresponding metric, by themselves have no meaning, it is often difficult to get at the underlying physical implications of the results obtained in general relativity. The key to our interpretation of the results lies in the *equivalence principle*. As a consequence of this principle, there is always one frame, the *local freely falling frame (LF³)*, in which one has neither gravity nor inertial forces. This is a frame that is just held and then let go in the local gravitational field. In this frame, for a short time, one has only flat Minkowski space and the laws of special relativity. All other frames, with their corresponding interpretation, can then be obtained through a coordinate transformation from this one. We show that both time dilation and radial length contraction are implied by the Schwarzschild metric, arising now from an *acceleration* of the LF^3 relative

---

[4]The theory of riemannian geometry is due to the brilliant mathematician G. F. R. Riemann (1826-1866).

to the inertial laboratory frame, rather than from the relative *velocity* of two frames in special relativity.

We proceed to investigate some of the further consequences of Einstein's theory in the subsequent chapters. In chapter 8, we show how one can generate a lagrangian for particle motion from any invariant interval, and we then construct a lagrangian $L(r, \theta, \phi, \dot{r}, \dot{\theta}, \dot{\phi})$ for particle motion in the Schwarzschild metric. From this, we calculate the precession of the perihelion of an almost circular orbit as a straightforward exercise in lagrangian mechanics. The precession of the perihelion of the planet mercury forms one of the most important confirmed predictions of general relativity. An index of refraction calculation, relegated to the problems, provides insight into the deflection of light in a gravitational field, another of the classic tests of the theory.

In chapter 9, the frequency shift of a source in the Schwarzschild metric and the redshift of light propagating out of that gravitational field are discussed. In chapter 10, the Tolman-Oppenheimer-Volkoff (TOV) equations for the structure of neutron stars are derived [Tolman (1939); Oppenheimer and Volkoff (1939)], and the solution of these equations using an equation of state derived from the relativistic mean field theory of nuclear matter is presented. Also discussed there is the transition into a black hole, where the nuclear repulsion is insufficient to overcome the gravitational attraction and the star collapses inside of its Schwarzschild radius.[5]

Chapter 11 deals with what is probably the most fascinating aspect of general relativity, at least the one that most readily captures the public's imagination, and that is cosmology — the time evolution of the universe. The Einstein field equations are solved in the case of a uniform (baryonic) matter distribution and flat, time-dependent, three-dimensional space. The resulting Robertson-Walker metric with $k = 0$ [Robertson (1935); Walker (1936)] turns out to describe well most of what is observed today. The corresponding cosmological redshift is discussed as well as the age of the universe and the role of the horizon. An expanding flat rubber sheet provides a useful two-dimensional analogy for the interpretation of the results.

Just as there is electromagnetic radiation in free space propagating with velocity c that follows from Maxwell's equations in E&M, there is gravitational radiation propagating with velocity c that follows from the Einstein

---

[5]The Schwarzschild radius is that value of the coordinate r at which the radial part of the Schwarzschild metric becomes singular.

field equations. In chapter 12, those equations are linearized in a perturbation of the Minkowski metric, and the wave equation is derived for that perturbation. A plane wave solution to the field equations is found and its properties examined. Here the simultaneous stretching and contracting of a flat, tranverse rubber sheet provides a useful analogy for the interpretation of the results.

A few special topics in general relativity are explored in chapter 13. First, the Einstein field equations in free space are derived from the Einstein-Hilbert lagrangian density and Hamilton's principle in continuum mechanics. What is varied is the metric at each point in space-time. It is tempting to just add a constant to that lagrangian density, as was done by Einstein, and the implications of that cosmological constant are explored. Once one has a lagrangian for the gravitational field, it is straightforward to augment that lagrangian with the contribution of additional matter fields, and we here explore the consequences of adding a scalar field.

We then look at the solution to the Einstein field equations for a cosmology where three-dimensional space, while homogeneous and isotropic, is also curved. The resulting metric is known as the Robertson-Walker metric with $k \neq 0$. The corresponding Friedmann equation [Friedmann (1922)] makes it appear extremely puzzling that the $k = 0$ solution should be relevant today, since with any traditional equation of state, the system diverges from that solution with time. The answer to this puzzle is given by the theory of inflation [Guth (2000)], whereby the development of a scalar field through spontaneous symmetry breaking and an initially unstable field configuration converges to the appropriate flatness, and provides as well the energy to create an associated hot plasma of particles and antiparticles.

Chapter 14 contains an extensive set of problems, some for each chapter. The goal of the problems is not to stump the reader, but to enhance and extend the coverage. Most of the problems guide the reader with steps, and give the answers. Many important concepts and applications are covered in the problems, for example:

- applications of special relativity
- geodesic equations in the Schwarzschild metric
- applications of the precession of the perihelion
- index of refraction for propagation of light in the Schwarzschild metric
- frequency and wavelength of light in the Schwarzschild metric
- numerical solution of the TOV equations
- thermodynamics of black holes

- cosmic microwave background
- time evolution from the "hot big bang"
- detection of gravitational waves
- source of gravitational waves
- solution of the Friedmann equation with various equations of state
- dynamics of a scalar field

The reader is strongly encouraged to work as many of the problems as possible in order to gain some confidence in this subject.

This book is meant as an *introduction* to general relativity. There are many good references for further study, for example, [Landau and Lifshitz (1975); Wald (1984); Shutz (1985); Hughston and Tod (1990); Kolb and Turner (1990); Linde (1990); Peebles (1993); Ohanian and Ruffini (1994); Peacock (1999); Taylor and Wheeler (2000); Hartle (2002); Foster and Nightengale (2006)]. It is hoped that the present book will allow a student to read with a deeper understanding and sense of appreciation the two classic texts [Weinberg (1972)] and [Misner, Thorne, and Wheeler (1973)].

Ours might truly be called the golden age of general relativity and cosmology. Satellite observations continue to push the boundaries of our understanding. The Cosmic Background Explorer (COBE) and Wilkinson Microwave Anisotropy Probe (WMAP) examine the isotropy and homogeneity of the cosmic microwave background left over from the big bang and give us a window back in time [COBE (2006); WMAP (2006)]. The Laser Interferometer Gravitational-Wave Observatory (LIGO) provides a powerful detector to search for gravitational radiation [LIGO (2006)]. The Chandra X-ray observatory provides important information on events yielding radiation in the high-energy end of the spectrum [Chandra (2006)]. The Hubble Space Telescope provides amazing visual images from the far outer regions of our universe [Hubble (2006)]. The Gravity Probe B experiment tests the general relativity prediction for the precession of a gyroscope in orbit around the earth [Einstein (2006)].

It is indeed a fascinating and dynamic time, where a wealth of truly impressive data continues to challenge and refine our understanding of general relativity.[6] It is important for every physicist to have a working knowledge of this subject.

---

[6]See, for example, [Sky and Telescope (2006)]. The author has found this publication to be an excellent source of information at the "semi-lay level" for keeping up with modern developments in general relativity.

Chapter 2

Particle on a Two-Dimensional Surface

Consider the motion of a point particle of mass m constrained to move without friction on a surface of arbitrary shape (Fig. 2.1).[1] There are no

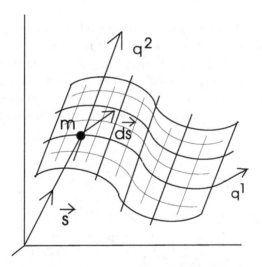

Fig. 2.1   Point particle constrained to move without friction on a two-dimensional surface of arbitrary shape. The generalized coordinates (q^1, q^2) locate the particle on the surface. The vector $d\mathbf{s}$ represents an infinitesimal displacement in the surface.

forces in the problem other than the constraint force, which remains normal to the surface. One possible physical realization of this configuration is two small magnets held together on opposite sides of a fixed smooth plastic shape in an inertial frame in space. Another is a particle moving without

---

[1]This chapter is an elaboration of problem 3.19 in [Fetter and Walecka (2003)], which provided the point of departure for the present book.

friction on the surface of the earth, where gravity provides the normal constraint force.[2] The reader can undoubtedly come up with more elegant physical realizations of this system on his or her own. We proceed to analyze the mechanics of this problem.

First, put a coordinate grid on the surface as indicated in Fig. 2.1. This can evidently be done in a wide variety of ways. All we ask of the generalized coordinates (q^1, q^2) is that they be linearly independent and uniquely specify the position of the particle. We use the superscripts $(1, 2)$ to denote these coordinates. The position of these indices, up or down, will play an important role in the subsequent developments, so we must pay careful attention to that position.[3]

2.1   Infinitesimal Displacements

It will be assumed that the surface everywhere has a unique *tangent plane*. Consider an infinitesimal displacement $d\mathbf{s}$ of the particle in the surface at the position (q^1, q^2). As long as we are talking about infinitesimal displacements, that displacement can equally well be considered as taking place in the tangent plane, as illustrated in Fig. 2.2. The tangent plane, of course, forms a nice flat euclidian surface, and here all our ordinary ideas of vector analysis apply.

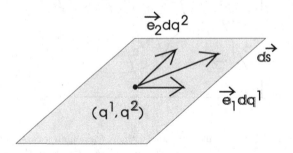

Fig. 2.2   Tangent plane to surface in Fig. 2.1 at the position of the particle (q^1, q^2), and infinitesimal displacements in this tangent plane.

Consider the following two particular infinitesimal displacements:

---

[2]Here there is an additional external force normal to the surface, but no additional forces in the surface.

[3]The quantity q raised to the power n will henceforth be denoted by $(q)^n$.

(1) Fix q^2, and let $q^1 \to q^1 + dq^1$. There will be a corresponding displacement in the tangent plane, identical to this order to that in the surface, which has some direction and magnitude. Represent this displacement by $\mathbf{e}_1\, dq^1$. This relation defines the *basis vector* $\mathbf{e}_1$;

(2) Fix q^1, and let $q^2 \to q^2 + dq^2$. There will be a corresponding displacement in the tangent plane, identical to this order to that in the surface, which has some direction and magnitude. Represent this displacement by $\mathbf{e}_2\, dq^2$. This relation defines the basis vector $\mathbf{e}_2$.

If these displacements are carried out simultaneously, the displacements in the tangent plane add as vectors to give a resultant displacement

$$d\mathbf{s} = \mathbf{e}_1 dq^1 + \mathbf{e}_2 dq^2 \qquad ; \text{ line element} \qquad (2.1)$$

The resultant is known as the *line element*, which forms a central quantity in our subsequent analysis. Note the following essential features of this relation:

- It describes the physical displacement in the tangent plane, which is identical to this order to the physical displacement in the surface;
- It defines the basis vectors $(\mathbf{e}_1, \mathbf{e}_2)$. In general they will be neither normalized, nor orthogonal;
- It is *exact to first order in infinitesimals.*

We shall subsequently rewrite Eq. (2.1) as

$$d\mathbf{s} = \sum_{i=1}^{2} \mathbf{e}_i dq^i$$
$$\equiv \mathbf{e}_i dq^i \qquad ; \text{ convention} \qquad (2.2)$$

where in the second line we adopt a shorthand *summation convention* that repeated upper and lower indices are summed over the appropriate range (here from 1 to 2). For now, we continue to retain the explicit summation signs for clarity.

Consider the square of the distance of the displacement. We shall refer to this quantity as the *interval*. It follows from ordinary vector algebra in the tangent plane that

$$(d\mathbf{s})^2 = \sum_{i=1}^{2}\sum_{j=1}^{2}(\mathbf{e}_i \cdot \mathbf{e}_j)dq^i dq^j \qquad (2.3)$$

We define the *metric* in terms of the inner product of the basis vectors according to

$$\mathbf{e}_i \cdot \mathbf{e}_j \equiv g_{ij} = g_{ji} \qquad \text{; metric matrix} \qquad (2.4)$$

The metric is a 2×2 matrix. It is explicitly symmetric in its indices $g_{ij} = g_{ji}$. The square of the physical infinitesimal displacement is then given in terms of the metric by

$$(d\mathbf{s})^2 = \sum_{i=1}^{2}\sum_{j=1}^{2} g_{ij} dq^i dq^j \qquad (2.5)$$

Note that the metric $g_{ij}(q^1, q^2) = g_{ji}(q^1, q^2)$ depends both on the choice of coordinate system and just where one is on the surface. We use the shorthand $g_{ij}(q^1, q^2) \equiv g_{ij}(q)$ to express this dependence.

2.2   Lagrange's Equations

The square of the velocity of the particle on the surface can be written in terms of the square of the infinitesimal displacement according to

$$\mathbf{v}^2 = \frac{(d\mathbf{s})^2}{(dt)^2} = \frac{d\mathbf{s}}{dt} \cdot \frac{d\mathbf{s}}{dt}$$

$$= \sum_{i=1}^{2}\sum_{j=1}^{2} g_{ij}(q) \frac{dq^i}{dt}\frac{dq^j}{dt} \qquad (2.6)$$

Hence the lagrangian for particle motion, which only contains the kinetic energy and no potential energy since there are no forces acting along the surface, is given by

$$L = T = \frac{1}{2}m\mathbf{v}^2$$

$$= \frac{1}{2}m \sum_{i=1}^{2}\sum_{j=1}^{2} g_{ij}(q) \frac{dq^i}{dt}\frac{dq^j}{dt} \qquad (2.7)$$

The motion of the particle on the surface then follows from Hamilton's principle and the resulting set of Lagrange's equations

$$\delta \int_{(1)}^{(2)} L \, dt = 0 \qquad ; \text{ Hamilton's principle}$$

fixed endpoints

$$\frac{d}{dt} \frac{\partial L}{\partial (dq^l/dt)} - \frac{\partial L}{\partial q^l} = 0 \qquad ; \text{ Lagrange's equations}$$

$$l = 1,2 \qquad (2.8)$$

Here one has $L(q^1, q^2, dq^1/dt, dq^2/dt)$, and all the other variables are kept fixed in carrying out the partial differentiations in Lagrange's equations.

Let us compute Lagrange's equations from the lagrangian in Eq. (2.7)

$$\frac{d}{dt} \left(m \sum_{j=1}^{2} g_{lj}(q) \frac{dq^j}{dt} \right) - \frac{1}{2} m \sum_{j=1}^{2} \sum_{k=1}^{2} \left[\frac{\partial}{\partial q^l} g_{jk}(q) \right] \frac{dq^j}{dt} \frac{dq^k}{dt} = 0 \quad (2.9)$$

In this expression $l = 1,2$, and the symmetry of the metric has been employed in obtaining the first term. Note the mass m now cancels. The particle trajectory on the surface is *independent of the inertial mass* (the mass that enters into Newton's second law).

The first term in Eq. (2.9) is then evaluated as

$$\frac{d}{dt} \left(\sum_{j=1}^{2} g_{lj}(q) \frac{dq^j}{dt} \right) = \sum_{j=1}^{2} \left[\frac{d}{dt} g_{lj}(q) \right] \frac{dq^j}{dt} + \sum_{j=1}^{2} g_{lj}(q) \frac{d^2 q^j}{dt^2} \quad (2.10)$$

The first term on the right-hand side (r.h.s.) is obtained by simply calculating the total time derivative of an implicit function

$$\sum_{j=1}^{2} \left[\frac{d}{dt} g_{lj}(q) \right] \frac{dq^j}{dt} = \sum_{j=1}^{2} \left\{ \sum_{k=1}^{2} \left[\frac{\partial}{\partial q^k} g_{lj}(q) \right] \frac{dq^k}{dt} \right\} \frac{dq^j}{dt}$$

$$= \frac{1}{2} \sum_{j=1}^{2} \sum_{k=1}^{2} \left[\frac{\partial g_{lj}(q)}{\partial q^k} + \frac{\partial g_{lk}(q)}{\partial q^j} \right] \frac{dq^j}{dt} \frac{dq^k}{dt} \quad (2.11)$$

A change of dummy summation indices ($j \leftrightarrow k$) has been used in the second line to write this term as the sum of two identical contributions. A combination of these results gives Lagrange's equations in terms of the generalized coordinates of choice (q^1, q^2)

$$\sum_{j=1}^{2} g_{lj}(q) \frac{d^2 q^j}{dt^2} + \frac{1}{2} \sum_{j=1}^{2} \sum_{k=1}^{2} \left[\frac{\partial g_{lk}(q)}{\partial q^j} + \frac{\partial g_{lj}(q)}{\partial q^k} - \frac{\partial g_{jk}(q)}{\partial q^l} \right] \frac{dq^j}{dt} \frac{dq^k}{dt} = 0$$

$$; l = 1, 2 \qquad (2.12)$$

This is a set of two coupled, nonlinear, second-order differential equations in the time whose coefficients involve the metric and spatial derivatives of the metric. It forms a truly remarkable result. The dynamics of just where the particle is on the surface after any time t is given in terms of the initial position and velocity $[q^1(0), q^2(0), \dot{q}^1(0), \dot{q}^2(0)]$ by the solution to Lagrange's equations $[q^1(t), q^2(t)]$ expressed in terms of the chosen set of generalized coordinates — and the lagrangian for the problem follows simply from a knowledge of the square of an infinitesimal physical displacement in the tangent plane as expressed through the metric! The dynamical equation is ultimately just Newton's second law; however, as formulated here, the force of constraint never appears in the problem.

2.3   Reciprocal Basis

The basis vectors $(\mathbf{e}_1, \mathbf{e}_2)$ depend on the choice of linearly independent generalized coordinates in the surface, and, in general, are neither normalized nor orthogonal. It is convenient to have some sort of orthonormality relation on the basis vectors in the tangent plane. To this end, we introduce a second set of basis vectors $(\mathbf{e}^1, \mathbf{e}^2)$, the *reciprocal basis*, satisfying the orthonormality relation

$$\mathbf{e}^i \cdot \mathbf{e}_j = \delta^i{}_j$$
$$= 1 \qquad ; \text{ if } i = j$$
$$= 0 \qquad \text{ if } i \neq j \qquad (2.13)$$

It is easy to see how such a basis can be constructed. Work in the subspace perpendicular to $\mathbf{e}_2$, and draw $\mathbf{e}^1$. Choose its length so that $\mathbf{e}^1 \cdot \mathbf{e}_1 = 1$. Now repeat in the process in the subspace perpendicular to $\mathbf{e}_1$. This construction serves to *define* the reciprocal basis (Fig. 2.3).

One can also uniquely decompose the line element in the reciprocal basis

$$d\mathbf{s} = \sum_{i=1}^{2} \mathbf{e}^i dq_i \qquad (2.14)$$

This relation serves to define the new infinitesimal displacements (dq_1, dq_2). As advertised, the position of the subscripts and superscripts now plays a crucial role: the superscripts on the basis vectors and the subscripts on the infinitesimal displacements now serve to denote the quantities in the reciprocal basis.

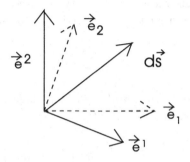

Fig. 2.3   Introduction of the reciprocal basis, which provides the useful orthonormality relation of Eq. (2.13) on the basis vectors.

Several relations involving the reciprocal basis follow immediately from its definition:

(1) The square of the physical distance of the infinitesimal displacement ds can be obtained by dotting Eq. (2.14) into Eq. (2.2) and making use of the orthonormality relation in Eq. (2.13)

$$d\mathbf{s} \cdot d\mathbf{s} = \sum_{i=1}^{2} \sum_{j=1}^{2} (\mathbf{e}^i \cdot \mathbf{e}_j)\, dq_i dq^j$$

$$= \sum_{i=1}^{2} dq_i dq^i \qquad (2.15)$$

(2) If the expressions for the line element in Eqs. (2.14) and (2.2) are dotted into $\mathbf{e}_i$ and the orthonormality relation and metric definition employed, one finds

$$dq_i = \sum_{j=1}^{2} g_{ij} dq^j \qquad (2.16)$$

Thus we observe that the metric matrix g_{ij} serves to *lower the indices*.

(3) As before, we define the metric in the reciprocal basis as

$$\mathbf{e}^i \cdot \mathbf{e}^j \equiv g^{ij} = g^{ji}$$

hence; $$ds \cdot ds = \sum_{i=1}^{2} \sum_{j=1}^{2} g^{ij} dq_i dq_j \qquad (2.17)$$

This new metric matrix is also explicity symmetric in its indices.

(4) It follows, as above, that this new metric matrix *raises the indices*

$$dq^i = \sum_{j=1}^{2} g^{ij} dq_j \qquad (2.18)$$

(5) Substitution of Eq. (2.16) into Eq. (2.18), while taking care to relabel dummy summation indices, leads to the relation

$$dq^i = \sum_{l=1}^{2} g^{il} \left(\sum_{j=1}^{2} g_{lj} dq^j \right)$$

or; $$\sum_{j=1}^{2} \left(\delta^i{}_j - \sum_{l=1}^{2} g^{il} g_{lj} \right) dq^j = 0 \qquad (2.19)$$

where the second line is just a rearrangement of the first. Since the dq^j are linearly independent, their coefficients must vanish

$$\sum_{l=1}^{2} g^{il} g_{lj} = \delta^i{}_j \qquad (2.20)$$

Let us examine this last relation in a little more detail. Introduce explicit matrix notation and define[4]

$$(\underline{g})_{ij} = g_{ij}$$
$$(\underline{1})_{ij} = \delta^i{}_j$$
$$(\underline{g}^{-1})_{ij} = g^{ij} \qquad ; \text{ inverse matrix} \qquad (2.21)$$

Equation (2.20) can then be recast in the familiar form

$$\underline{g}^{-1} \underline{g} = \underline{1} \qquad (2.22)$$

This is a very useful result, for it implies that one can obtain the metric g^{ij} in the reciprocal basis by simply finding the inverse of the matrix $\underline{g}$ of the metric in the original basis.

---

[4] We use a single bar under a symbol to denote a matrix.

Let us now apply Eq. (2.20) to simplify Eq. (2.12). Take $\sum_{l=1}^{2} g^{il}$ on that relation, and the result is

$$\frac{d^2q^i}{dt^2} + \sum_{j=1}^{2}\sum_{k=1}^{2} \Gamma^i_{jk}(q)\,\frac{dq^j}{dt}\frac{dq^k}{dt} = 0 \qquad \text{; Lagrange's equations} \qquad (2.23)$$

$$i = 1,2$$

$$\Gamma^i_{jk}(q) \equiv \frac{1}{2}\sum_{l=1}^{2} g^{il}(q)\left[\frac{\partial g_{lk}(q)}{\partial q^j} + \frac{\partial g_{lj}(q)}{\partial q^k} - \frac{\partial g_{jk}(q)}{\partial q^l}\right] \qquad \text{; affine connection}$$

These are Lagrange's equations for the motion of a point mass without friction on an arbitrary surface in final form. They provide a set of two coupled, second-order, nonlinear differential equations relating the acceleration of the generalized coordinates on the surface and the product of their velocities. The last expression, providing the coupling coefficients, is known as the *affine connection*. It depends only on the metric and derivatives of the metric.[5]

If one starts the particle at point (1) on the surface in Fig. 2.1 with an initial velocity

$$\mathbf{v}_0 = \left(\frac{d\mathbf{s}}{dt}\right)_0 = \sum_{i=1}^{2}\left(\mathbf{e}_i \frac{dq^i}{dt}\right)_0 \qquad (2.24)$$

then its trajectory will take it through some second point on the surface, which we will denote by (2) as in Fig. 2.4.

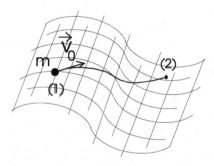

Fig. 2.4   Particle trajectory on surface in Fig. 2.1 given an initial velocity $\mathbf{v}_0$ at point (1). The particle then passes through some second point (2).

---

[5]It is worthwhile spending some time studying and remembering where the indices go in the affine connection. We will soon have many indices to contend with.

Energy is conserved, since there is no friction and the constraint force is normal to the surface and does no work ($dW = \mathbf{F}_\perp \cdot d\mathbf{s} = 0$). Thus

$$E = \frac{1}{2}m\mathbf{v}_0^2 = \frac{1}{2}m\mathbf{v}^2 = \text{constant} \tag{2.25}$$

We give a simple example of this analysis. Consider the motion of a particle of mass m in a flat plane and introduce polar coordinates $(q^1, q^2) = (r, \phi)$ as illustrated in Fig. 2.5.

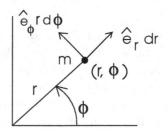

Fig. 2.5   Particle motion in a flat plane and introduction of polar coordinates.

Here $(\hat{\mathbf{e}}_r, \hat{\mathbf{e}}_\phi)$ are the familiar orthonormal *unit vectors* obtained by increasing one coordinate with the other fixed, and the previous basis vectors representing physical displacements are expressed in terms of these unit vectors as

$$\mathbf{e}_r = \hat{\mathbf{e}}_r \qquad ; \quad \mathbf{e}_\phi = r\,\hat{\mathbf{e}}_\phi \tag{2.26}$$

Thus the line element is

$$\begin{aligned} d\mathbf{s} &= \mathbf{e}_r dr + \mathbf{e}_\phi d\phi \\ &= \hat{\mathbf{e}}_r dr + \hat{\mathbf{e}}_\phi r d\phi \end{aligned} \tag{2.27}$$

The square of the displacement is

$$d\mathbf{s} \cdot d\mathbf{s} = (dr)^2 + (rd\phi)^2 \tag{2.28}$$

Hence the metric is identified as

$$g_{rr} = 1 \qquad ; \quad g_{r\phi} = g_{\phi r} = 0 \qquad ; \quad g_{\phi\phi} = r^2 \tag{2.29}$$

or in matrix form

$$\underline{g} = \begin{pmatrix} g_{rr} & g_{r\phi} \\ g_{\phi r} & g_{\phi\phi} \end{pmatrix} = \begin{pmatrix} 1 & 0 \\ 0 & r^2 \end{pmatrix} \tag{2.30}$$

The affine connection will be computed later. For now, we just observe that the lagrangian is given by[6]

$$L = \frac{1}{2}m\mathbf{v}^2 = \frac{1}{2}m\left(\frac{d\mathbf{s}}{dt}\right)^2$$

$$= \frac{1}{2}m(\dot{r}^2 + r^2\dot{\phi}^2) \qquad (2.31)$$

and Lagrange's equations immediately provide the equations of motion

$$\frac{d}{dt}(mr^2\dot{\phi}) = 0$$

$$\frac{d}{dt}(m\dot{r}) - mr\dot{\phi}^2 = 0 \qquad (2.32)$$

In terms of polar coordinates, the equations of motion are coupled and nonlinear. We know, however, that the actual particle trajectory is very simple in this case — it is just a straight line! There is an important moral here: it is not always easy to identify the physical motion from equations written in terms of an arbitrary set of generalized coordinates.

2.4 Geodesic Motion

Let us consider a different, apparently unrelated, problem. Suppose we wish to find the path of *minimum distance* between two arbitrary points (1) and (2) on the surface that we have been considering. This will be some curve C as indicated in Fig. 2.6. This curve is known as the *geodesic*.

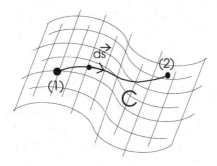

Fig. 2.6   Curve of minimum distance, the *geodesic*, between two points (1) and (2) on the surface in Fig. 2.1. Here ds is an infinitesimal displacement along C.

---

[6]We use the familiar notation $dq^j/dt \equiv \dot{q}^j$.

One can empirically determine this curve by confining a string to the surface and simply pulling it tight between the two points. The length of the string then provides the length of the geodesic.

Suppose the curve C is the geodesic. At each point, an infinitesimal displacement along C lies in the tangent plane and is given by

$$d\mathbf{s} = \hat{\mathbf{e}}_t ds \qquad ; \text{ displacement along geodesic} \qquad (2.33)$$

where $\hat{\mathbf{e}}_t$ is a unit vector tangent to the curve and ds is an infinitesimal length. The total length D of the curve C is then given by summing these infinitesimal elements

$$D = \int_{(1)}^{(2)} ds = \int_{(1)}^{(2)} (d\mathbf{s} \cdot d\mathbf{s})^{1/2} \qquad (2.34)$$

In terms of the metric and appropriate generalized coordinate displacements, this expression is given by Eq. (2.5) as

$$D = \int_{(1)}^{(2)} \left[\sum_{i=1}^{2} \sum_{j=1}^{2} g_{ij}(q^1, q^2) dq^i dq^j \right]^{1/2} \qquad (2.35)$$

It is convenient to *parameterize* the distance along the curve C by a parameter τ that runs from 0 to 1. Thus if the length of the curve C is l and the distance along C is s, then one can take

$$\tau = s/l \qquad ; \text{ parametrizes distance along C} \qquad (2.36)$$

The expression in Eq. (2.35) can then be rewritten as

$$D = \int_0^1 d\tau \left[\sum_{i=1}^{2} \sum_{j=1}^{2} g_{ij}(q^1, q^2) \frac{dq^i}{d\tau} \frac{dq^j}{d\tau} \right]^{1/2}$$
$$\equiv \int_0^1 I\left(q^1, q^2, \frac{dq^1}{d\tau}, \frac{dq^2}{d\tau} \right) d\tau \qquad (2.37)$$

where $I(q^1, q^2, dq^1/d\tau, dq^2/d\tau)$ now represents the integrand.

So far C has been the geodesic, the curve that minimizes this distance. How do we actually determine C? This is the basic problem in the *calculus of variations*.[7] One considers other curves C', which vary slightly from the geodesic, obtained by letting

$$q^i \rightarrow q^i + \delta q^i \qquad ; i = 1, 2 \qquad (2.38)$$

---

[7]See [Fetter and Walecka (2003)].

at fixed τ. This is illustrated in Fig. 2.7. The variation is required to vanish at the endpoints. The minimum distance is obtained by requiring that D be stationary under these variations[8]

$$\delta D = 0 \qquad ; \text{ fixed endpoints} \qquad (2.39)$$

The curve C that minimizes the distance, the geodesic, must then satisfy the Euler-Lagrange equations

$$\frac{d}{d\tau}\frac{\partial I}{\partial(dq^i/d\tau)} - \frac{\partial I}{\partial q^i} = 0 \qquad ; i = 1,2 \qquad (2.40)$$

These equations provide the coordinates $[q^1(\tau), q^2(\tau)]$ of the geodesic.

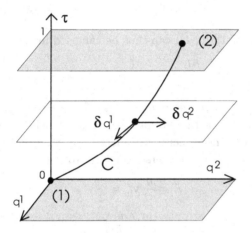

Fig. 2.7 Variations about curve of minimum distance, the geodesic, used in the calculus of variations. The picture here is in the space of the generalized coordinates (q^1, q^2) and the parameter τ. The square of the physical displacement on the surface at each point is related to small changes in the coordinates by Eq. (2.5).

Now observe that the integrand in Eq. (2.37) is simply related to the lagrangian of Eq. (2.7) expressed in terms of the parameter τ rather than the time t

$$I = \left[\frac{2}{m}L\left(q^1, q^2, \frac{dq^1}{d\tau}, \frac{dq^2}{d\tau}\right)\right]^{1/2} \qquad (2.41)$$

---

[8]We assume here, as in Fig. 2.6, that it is obvious from the physics that this procedure will find a minimum — there can be other possibilities. We will return to this.

The Euler-Lagrange equations for the geodesic then give (after canceling the factor $\sqrt{2/m}$)

$$\frac{d}{d\tau} \frac{1}{2\sqrt{L}} \frac{\partial L}{\partial(dq^i/d\tau)} - \frac{1}{2\sqrt{L}} \frac{\partial L}{\partial q^i} = 0 \quad ; \, i = 1, 2 \qquad (2.42)$$

Along the geodesic one has for this L

$$L = \frac{m}{2} \frac{d\mathbf{s}}{d\tau} \cdot \frac{d\mathbf{s}}{d\tau} = \frac{m}{2} \hat{\mathbf{e}}_t \frac{ds}{d\tau} \cdot \hat{\mathbf{e}}_t \frac{ds}{d\tau}$$

$$= \frac{m}{2} \left(\frac{ds}{d\tau} \right)^2 \qquad (2.43)$$

It then follows from our parameterization in Eq. (2.36) that

$$L = \frac{m}{2} l^2 = \text{constant} \qquad ; \text{ along geodesic} \qquad (2.44)$$

The constant factor $1/(2\sqrt{L})$ can thus be canceled in Eq. (2.42) to give

$$\frac{d}{d\tau} \frac{\partial L}{\partial(dq^i/d\tau)} - \frac{\partial L}{\partial q^i} = 0 \quad ; \, i = 1, 2 \qquad (2.45)$$

These are the *Euler-Lagrange equations for the geodesic*. They are exactly Lagrange's equations for the motion of a point particle on the surface! To make the correspondence precise, use energy conservation to determine the speed at which the particle traverses its trajectory

$$E = \frac{m}{2} \mathbf{v}_0^2 = \frac{m}{2} \mathbf{v}^2$$

$$= \frac{m}{2} \left(\frac{ds}{dt} \right)^2 \qquad (2.46)$$

where the last line follows as in Eq. (2.43). Hence, with $v_0 \equiv |\mathbf{v}_0|$

$$s = v_0 t \qquad \qquad ; \text{ along trajectory}$$

$$\tau = \frac{s}{l} = \left(\frac{v_0}{l} \right) t = \text{constant} \times t \qquad (2.47)$$

The particle covers equal distances in equal times along its orbit. Thus the geometric curve parameter τ and the time for particle motion t are equal to within a multiplicative constant. Hence one can equally well change variables in Eqs. (2.45) to the time t. The conclusion is that *Lagrange's equations for particle motion on the surface and the Euler-Lagrange equations for the geodesic are precisely the same*. These are second-order differential equations for the generalized coordinates (q^1, q^2). If one requires that the solution pass through two given points on the surface, then that solution will be unique.[9]

If the particle starts from point (1) in Fig. 2.4 with an initial velocity $\mathbf{v}_0 = v_0(\hat{\mathbf{e}}_t)_0$ that causes it to pass through point (2), then the path that the particle follows on the surface will be a geodesic. (Recall that one must specify both *position* and *velocity* to obtain a solution to Newton's second law.)

This observation provides us with a way of obtaining an *analog solution* to the mechanics problem.[10] Take a string and confine it to the surface. Now stretch it tight between points (1) and (2). This is the path that the particle travels between these points. But now observe that this is *pure geometry*! A *priori*, it has nothing to do with mechanics and Newton's laws![11]

So far, for purposes of exposition, we have talked about simple configurations as in Fig. 2.4 where the geodesic is the curve of minimum distance; however, the stationary condition in Eq. (2.39), which leads to the Euler-Lagrange equations, merely requires that the geodesic be an *extremum* — minimum, maximum, or point of inflection. We give some examples:

1) For motion in a plane, the geodesic is just a straight line [Fig. 2.8(a)];

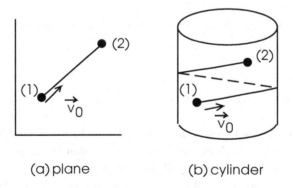

(a) plane   (b) cylinder

Fig. 2.8 Examples of geodesics between points (1) and (2): (a) motion in a plane; (b) motion on surface of a cylinder where the orbit is constrained to circle the cylinder once.

2) For motion on the surface of a cylinder, the geodesic depends on how many times the curve is constrained to wind around the cylinder. In the

[9]Or at least the solutions form a denumerable set — see later.
[10]And *vice versa*.
[11]It was Einstein, of course, who realized the deep implications of this result.

mechanics problem, the number of windings is determined by the direction of $\mathbf{v}_0$. The single winding case is illustrated in Fig. 2.8(b);

3) On the surface of the sphere, there are two possibilities. There is a second geodesic that runs on the great circle and *maximizes* the distance between the two points [Fig. 2.9(a)]. In the mechanics problem, the choice is again dictated by the direction of $\mathbf{v}_0$;

4) These considerations can be extrapolated to the case of an arbitrary surface [Fig. 2.9(b)]. Readers are urged to consider different and more complex surfaces for themselves.

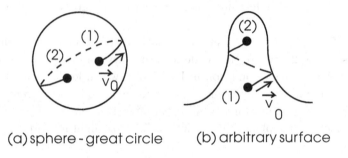

(a) sphere - great circle   (b) arbitrary surface

Fig. 2.9   Same as Fig. 2.8 for: (a) great circle on a sphere — maximum distance; (b) motion on arbitrary surface.

2.5   Role of Coordinates

So far the generalized coordinates (q^1, q^2) simply form a set of linearly independent quantities chosen to locate the position of the particle on the surface. One could equally as well have picked a completely different set (ξ^1, ξ^2) as illustrated in Fig. 2.10. The question arises, what is the relation between these two descriptions? Both are describing the same physics, but the equations of motion can look completely different.

For example, suppose we were to describe the planar motion of the particle in Fig. 2.5, which starts at some position with velocity $\mathbf{v}_0$, in terms of cartesian coordinates (x, y) rather than polar coordinates. Instead of the equations of motion in Eqs. (2.32) one would simply have

$$m\ddot{x} = 0 \qquad\qquad ; \; m\ddot{y} = 0 \qquad\qquad (2.48)$$

The equations look completely different, and the physical motion itself is just a straight line.

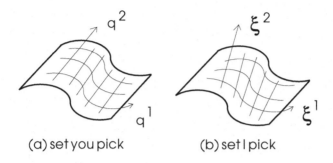

(a) set you pick   (b) set I pick

Fig. 2.10   Two sets of generalized coordinates on the surface.

The coordinates evidently have no physical meaning. The corresponding metric, since it depends on the particular choice of coordinates, by itself has no meaning. It is only the combination of metric and infinitesimal coordinate changes, related at each point through Eq. (2.5) to the square of the infinitesimal physical displacement, the *interval*, that has a direct physical interpretation.

This cannot be the whole story, however, since Lagrange's Eqs. (2.23) describe the motion on surfaces of arbitrary *shape*, and the information on the intrinsic structural properties of the surface must be contained in the affine connection, the only place the surface enters. The affine connection, in turn, is a first-order, nonlinear differential form in the metric. There must be some intrinsic structural properties of the surface that are independent of the choice of generalized coordinates and derivable from the metric. Our goal is to find these.

The problem considered so far has involved motion on a two-dimensional surface, with two-dimensional coordinate systems and two-dimensional infinitesimal displacements in the flat, euclidian tangent plane.[12] It is essential for our subsequent work to be able to extend the analysis to spaces of arbitrary dimension, and we first proceed to a discussion of this problem.

---

[12]Recall it is assumed that at each point (q^1, q^2) there is a flat, euclidian, tangent plane, and the infinitesimal displacement in the tangent plane and surface are identical to first order in infinitesimals.

Chapter 3

Curvilinear Coordinate Systems

In analyzing the frictionless motion of a particle on a surface of arbitrary shape, we have been led to a discussion of generalized coordinates in two dimensions. We have observed that for infinitesimal displacements, one can work in the tangent plane where the space is euclidian. It is essential for our subsequent development to extend this analysis to a higher number of dimensions, and we have been careful to write things in a fashion that is easy to generalize.

For the present discussion, it is assumed that we are working in that tangent space, and we thus begin with a generalization to an n-dimensional euclidian space.

3.1 Line Element

Start with the introduction of a set of n linearly independent generalized coordinates $(q^1, q^2, \cdots, q^n)$ that locate a position in the space. We will also refer to these as *curvilinear* coordinates. Now hold all the coordinates fixed except the first one and let $q^1 \to q^1 + dq^1$. There will be some physical displacement $d\mathbf{s}$ in that space

$$q^1 \to q^1 + dq^1 \qquad ; q^2, \cdots, q^n \text{ fixed}$$
$$d\mathbf{s} = \mathbf{e}_1 \, dq^1 \tag{3.1}$$

This relation serves to define the basis vector $\mathbf{e}_1$. Now repeat for the remaining $n-1$ coordinates. If the coordinate changes are carried out simul-

27

taneously, one can simply add the infinitesimal displacements as vectors

$$d\mathbf{s} = \sum_{i=1}^{n} \mathbf{e}_i \, dq^i \qquad ; \text{ line element} \qquad (3.2)$$

At this point, it is convenient to introduce the summation convention of Eq. (2.2) whereby repeated upper and lower indices are summed from 1 to n. This allows us to suppress the summation sign and simply write

$$d\mathbf{s} = \mathbf{e}_i \, dq^i \qquad ; \text{ summation convention} \qquad (3.3)$$

We say that these indices are *contracted.*. This expression for the line element is exact to first order in infinitesimals.

An example of the situation in three dimensions is illustrated in Fig. 3.1.

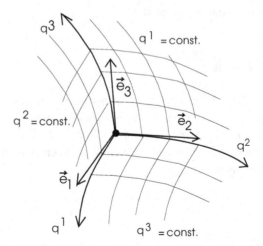

Fig. 3.1 Example of generalized coordinates and corresponding basis vectors at the point (q^1, q^2, q^3) in three dimensions.

3.2 Reciprocal Basis

We again introduce the *reciprocal basis* satisfying the orthonormality relation

$$\mathbf{e}_i \cdot \mathbf{e}^j = \delta_i{}^j$$
$$= 1 \quad ; \text{ if } i = j$$
$$= 0 \quad \text{ if } i \neq j \tag{3.4}$$

It is constructed as before. For example, in three dimensions one constructs a vector $\mathbf{e}^3$ perpendicular to the plane defined by $(\mathbf{e}_1, \mathbf{e}_2)$ and then adjusts its length so that it has unit projection on $\mathbf{e}_3$ (see Fig. 3.2). The process is then repeated for the other two cases.

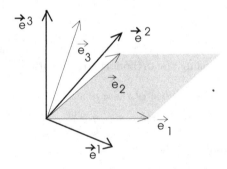

Fig. 3.2 Construction of the reciprocal basis in three dimensions.

3.3 Metric

The infinitesimal displacement $d\mathbf{s}$ is a vector, and it can be expanded in the reciprocal basis at each point as

$$d\mathbf{s} = \mathbf{e}^i \, dq_i \tag{3.5}$$

This relation serves to define the coordinate differences dq_i.

The dot product of the two expressions for $d\mathbf{s}$ gives for the interval

$$(d\mathbf{s})^2 = d\mathbf{s} \cdot d\mathbf{s}$$
$$= g_{ij} \, dq^i dq^j = g^{ij} \, dq_i dq_j \tag{3.6}$$

Here the *metric* is defined by

$$g_{ij} = \mathbf{e}_i \cdot \mathbf{e}_j = g_{ji} \qquad ; \text{ symmetric in } (i \leftrightarrow j)$$
$$g^{ij} = \mathbf{e}^i \cdot \mathbf{e}^j = g^{ji} \tag{3.7}$$

The metric is symmetric in its indices.[1] It depends on the choice of co-ordinates and on position, and we denote that dependence generically by $g(q^1, \cdots, q^n) = g(q)$.

3.4 Vectors

Since we now have two complete bases, we can expand any vector at any point in the space in either of two ways

$$\mathbf{v} = v^i \, \mathbf{e}_i = v_i \, \mathbf{e}^i \tag{3.8}$$

The components can be identified by dotting these relations into $\mathbf{e}^j$ and $\mathbf{e}_j$ respectively and then using the orthonormality relation in Eq. (3.4)

$$v^j = \mathbf{e}^j \cdot \mathbf{v} = g^{ji} v_i$$
$$v_j = \mathbf{e}_j \cdot \mathbf{v} = g_{ji} v^i \tag{3.9}$$

or, upon interchanging the labels $i \leftrightarrow j$,

$$v_i = g_{ij} v^j$$
$$v^i = g^{ij} v_j \tag{3.10}$$

The metric serves to raise and lower the indices on the components of the vector in these bases. The vector $\mathbf{v}$ in Eq. (3.8) has an intrinsic physical meaning in terms of both a magnitude and direction. The components v^i or v_i, which depend on the choice of coordinate system, by themselves do not.

The basis vectors form a particularly important vector example. One can expand a basis vector in the reciprocal basis

$$\mathbf{e}_i = c_{ij} \, \mathbf{e}^j \tag{3.11}$$

---

[1] We will henceforth feel free to make full use of the symmetry of the metric, as well as that of related quantities, without necessarily pointing it out every time it is used. The obvious relation $v^i \, \mathbf{e}_i \equiv \mathbf{e}_i \, v^i$ is used without comment. Changes of dummy summation indices, for example $a_i b^i \equiv a_j b^j$, are also frequently employed throughout.

The coefficients are obtained by dotting this into a particular $\mathbf{e}_j$

$$\mathbf{e}_j \cdot \mathbf{e}_i = c_{ij} = g_{ij} \tag{3.12}$$

Hence (the second relation is derived the same way)

$$\mathbf{e}_i = g_{ij} \, \mathbf{e}^j$$
$$\mathbf{e}^i = g^{ij} \, \mathbf{e}_j \tag{3.13}$$

The metric serves to raise and lower the indices on the basis vectors themselves.[2]

3.5 Tensors

We generalize the notion of a vector in Eq. (3.8) to a *second-rank tensor*[3]

$$\underline{\underline{T}} = T^{ij} \, \mathbf{e}_i \mathbf{e}_j$$
$$= T_{ij} \, \mathbf{e}^i \mathbf{e}^j \tag{3.14}$$

Here the second line follows from the first after the insertion of the first of Eqs. (3.13) and the identification

$$T_{ij} = g_{ii'} g_{jj'} T^{i'j'} \tag{3.15}$$

To give a familiar example, consider Hooke's law which relates stress and strain. In one-dimension Hooke's law states that for small displacements, there is a restoring force

$$\mathbf{F} = -k\mathbf{x} \tag{3.16}$$

where the constant k is a property of the material [Fig. 3.3(a)].

With a two-dimensional anisotropic membrane, if

$$\mathbf{x} = x^i \, \mathbf{e}_i \qquad ; \mathbf{F} = F_i \, \mathbf{e}^i \tag{3.17}$$

then Hooke's law states that the response is given by a *dyadic* [Fig. 3.3(b)]

$$\mathbf{F} = -\underline{\underline{K}} \cdot \mathbf{x}$$
$$\underline{\underline{K}} = K_{ij} \, \mathbf{e}^i \mathbf{e}^j \tag{3.18}$$

---

[2]This provides a particularly convenient way of determining the reciprocal basis.
[3]We use multiple bars under a symbol to denote a tensor.

Here the convention is that the last vector on the right in $\underline{K}$ gets dotted into the vector $\mathbf{x}$. Thus

$$
\begin{aligned}
\mathbf{F} &= F_i\,\mathbf{e}^i \\
&= -\underline{K}\cdot\mathbf{x} = -K_{ij}\,\mathbf{e}^i(\mathbf{e}^j\cdot\mathbf{x}) \\
&= -K_{ij}x^j\,\mathbf{e}^i
\end{aligned}
\tag{3.19}
$$

Thus in component form

$$
F_i = -K_{ij}x^j
\tag{3.20}
$$

Here the K_{ij} are properties of the membrane. This is the *tensor response* of an anisotropic membrane.

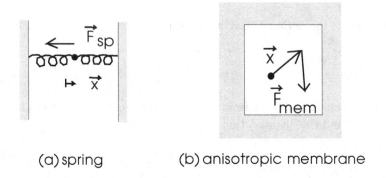

(a) spring   (b) anisotropic membrane

Fig. 3.3   Hooke's law: (a) One-dimensional spring with displacement $\mathbf{x}$ and restoring force $\mathbf{F}_{\mathrm{sp}}$; (b) Two-dimensional anisotropic membrane with displacement $\mathbf{x}$ and restoring force $\mathbf{F}_{\mathrm{mem}}$.

The dyadic in Eq. (3.18) is evidently a second-rank tensor. One particularly important quantity is the *unit dyadic*

$$
\underline{I} = g^{ij}\,\mathbf{e}_i\mathbf{e}_j = \mathbf{e}^j\mathbf{e}_j
\tag{3.21}
$$

When operating on a vector, it simply reproduces that vector

$$
\begin{aligned}
\mathbf{v} &= v_j\,\mathbf{e}^j \\
\underline{I}\cdot\mathbf{v} &= v_i\,\mathbf{e}^i = \mathbf{v}
\end{aligned}
\tag{3.22}
$$

The tensor in Eq. (3.14) is again a physical quantity. When operating on a vector, it produces another vector. Its components T^{ij}, which depend on the choice of coordinate system, by themselves are not.

The relation in Eq. (3.14) can be generalized to produce tensors of higher rank

$$\underline{\underline{T}} = T^{ijk\cdots}\, \mathbf{e}_i\mathbf{e}_j\mathbf{e}_k \cdots$$
$$= T_{ijk\cdots}\, \mathbf{e}^i\mathbf{e}^j\mathbf{e}^k \cdots \tag{3.23}$$

We say this is a tensor of first, second, third, or higher rank, respectively. The indices are again raised and lowered with the metric

$$T_{ijk\cdots} = g_{ii'}g_{jj'}g_{kk'}T^{i'j'k'\cdots} \tag{3.24}$$

Note that in a contracted expression the contracted indices can be simultaneously raised and lowered since

$$a_i b^i = a_i g^{ij} b_j = a^j b_j \tag{3.25}$$

3.6 Change of Basis

Suppose one introduces a new set of generalized curvilinear coordinates $(\xi^1, \cdots, \xi^n)$. A corresponding set of new basis vectors $(\boldsymbol{\alpha}_1, \cdots, \boldsymbol{\alpha}_n)$ can then be generated just as before (Fig. 3.4).

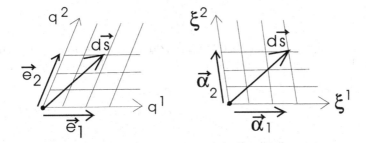

Fig. 3.4 Two different curvilinear coordinate systems describing the same physical situation. A representative value of the line element is indicated.

Now assume that there is a analytic *point transformation* between the new set of coordinates and the original set of the form

$$\xi^1 = \xi^1(q^1, \cdots, q^n)$$
$$\vdots$$
$$\xi^n = \xi^n(q^1, \cdots, q^n) \tag{3.26}$$

Assume further that this transformation is invertible

$$q^1 = q^1(\xi^1, \cdots, \xi^n)$$

$$\vdots$$

$$q^n = q^n(\xi^1, \cdots, \xi^n) \tag{3.27}$$

The total differential of Eqs. (3.26) gives

$$d\xi^i = \frac{\partial \xi^i}{\partial q^j} dq^j = \bar{a}^i{}_j \, dq^j \tag{3.28}$$

This relation defines $\bar{a}^i{}_j$

$$\bar{a}^i{}_j \equiv \frac{\partial \xi^i}{\partial q^j} \tag{3.29}$$

This is a point relation depending on position.

One similarly has from the total differential of Eqs. (3.27)

$$dq^i = \frac{\partial q^i}{\partial \xi^j} d\xi^j = a^i{}_j \, d\xi^j \tag{3.30}$$

and

$$a^i{}_j \equiv \frac{\partial q^i}{\partial \xi^j} \tag{3.31}$$

Insertion of Eq. (3.30) into Eq. (3.28) and the utilization of the linear independence of the differentials, and *vice versa*, leads to the relations

$$\bar{a}^i{}_j a^j{}_k = \delta^i{}_k$$
$$a^i{}_j \bar{a}^j{}_k = \delta^i{}_k \tag{3.32}$$

These are evidently inverse transformations.

The line element is a physical quantity representing an infinitesimal displacement in the space. We require that it be the same in either basis[4]

$$d\mathbf{s} = \mathbf{e}_i \, dq^i = \boldsymbol{\alpha}_i \, d\xi^i \tag{3.33}$$

Dotting this relation into first a specific $\mathbf{e}^i$, and then into $\boldsymbol{\alpha}^i$ leads to

$$dq^i = (\mathbf{e}^i \cdot \boldsymbol{\alpha}_j) \, d\xi^j$$
$$d\xi^i = (\boldsymbol{\alpha}^i \cdot \mathbf{e}_j) \, dq^j \tag{3.34}$$

---

[4] As before, this relation serves to define the basis vectors.

Hence

$$a^i{}_j = \frac{\partial q^i}{\partial \xi^j} = (\mathbf{e}^i \cdot \boldsymbol{\alpha}_j)$$

$$\bar{a}^i{}_j = \frac{\partial \xi^i}{\partial q^j} = (\boldsymbol{\alpha}^i \cdot \mathbf{e}_j) \tag{3.35}$$

We observe that the *partial derivatives of the analytic point relations give the dot products between the basis vectors.* It follows by inspection that[5]

$$a^i{}_j = \bar{a}_j{}^i$$

$$a_i{}^j = \bar{a}^j{}_i \tag{3.36}$$

Let us expand one basis in terms of the other

$$\boldsymbol{\alpha}_i = c_i{}^j \, \mathbf{e}_j \tag{3.37}$$

Now take the dot product with a specific $\mathbf{e}^j$

$$c_i{}^j = (\mathbf{e}^j \cdot \boldsymbol{\alpha}_i) = a^j{}_i \tag{3.38}$$

Thus

$$\boldsymbol{\alpha}_i = (\boldsymbol{\alpha}_i \cdot \mathbf{e}^j) \, \mathbf{e}_j = a^j{}_i \, \mathbf{e}_j \tag{3.39}$$

It follows in an exactly analogous fashion that

$$\mathbf{e}_i = (\mathbf{e}_i \cdot \boldsymbol{\alpha}^j) \, \boldsymbol{\alpha}_j = \bar{a}^j{}_i \, \boldsymbol{\alpha}_j \tag{3.40}$$

With the aid of Eqs. (3.36), Eqs. (3.28) and (3.39) can be rewritten as

$$\boldsymbol{\alpha}_i = \bar{a}_i{}^j \, \mathbf{e}_j$$

$$d\xi^i = \bar{a}^i{}_j \, dq^j \tag{3.41}$$

and Eqs. (3.30) and (3.40) as

$$\mathbf{e}_i = a_i{}^j \, \boldsymbol{\alpha}_j$$

$$dq^i = a^i{}_j \, d\xi^j \tag{3.42}$$

Note that the relation between the basis vectors is obtained immediately by just dotting them into the unit dyadic

$$\underline{\underline{I}} = \mathbf{e}^i \, \mathbf{e}_i = \mathbf{e}_i \, \mathbf{e}^i$$

$$= \boldsymbol{\alpha}^i \, \boldsymbol{\alpha}_i = \boldsymbol{\alpha}_i \, \boldsymbol{\alpha}^i \tag{3.43}$$

---

[5]Remember that in a, the first index goes with the first basis vector $\mathbf{e}$, and in $\bar{a}$, the first index goes with the first basis vector $\boldsymbol{\alpha}$.

Thus

$$\boldsymbol{\alpha}_i = \underline{I} \cdot \boldsymbol{\alpha}_i = \mathbf{e}_j(\mathbf{e}^j \cdot \boldsymbol{\alpha}_i) = a^j{}_i \mathbf{e}_j = \bar{a}_i{}^j \mathbf{e}_j$$
$$\mathbf{e}_i = \underline{I} \cdot \mathbf{e}_i = \boldsymbol{\alpha}_j(\boldsymbol{\alpha}^j \cdot \mathbf{c}_i) = \bar{a}^j{}_i \boldsymbol{\alpha}_j = a_i{}^j \boldsymbol{\alpha}_j \qquad (3.44)$$

This provides an easy way to remember the results.

3.7 Transformation Law

We can ask how a *vector* is described in the two coordinate systems. As noted before, a vector is a *physical quantity*. Thus it must take the same value no matter which coordinate system we use

$$\mathbf{v} = v^i \mathbf{e}_i = \bar{v}^i \boldsymbol{\alpha}_i \qquad (3.45)$$

Here the $\bar{v}^i$ are the coordinates in the new basis. Now substitute Eq. (3.39) in the last relation

$$\mathbf{v} = \bar{v}^j(a^i{}_j \mathbf{e}_i) = a^i{}_j \bar{v}^j \mathbf{e}_i \qquad (3.46)$$

The coefficients of the linearly independent $\mathbf{e}_i$ can then be equated with the result

$$v^i = a^i{}_j \bar{v}^j \qquad (3.47)$$

This is how the components of the vector must be related in the two coordinate systems. These relations can be inverted with the aid of Eqs. (3.32)

$$\bar{v}^i = \bar{a}^i{}_j v^j \qquad (3.48)$$

We say that the components of the vector with the upper index transform *contravariantly* if they transform exactly as the dq^i and $d\xi^i$ in Eqs. (3.42) and (3.41). They are the *contravariant components*.

The indices in Eqs. (3.47) and (3.48) can be lowered with the aid of the metric, and the contracted indices interchanged to give

$$v_i = a_i{}^j \bar{v}_j$$
$$\bar{v}_i = \bar{a}_i{}^j v_j \qquad (3.49)$$

We say the components with lower indices transform *covariantly* if they transform exactly as the basis vectors in Eqs. (3.42) and (3.41). They are the *covariant components*.

We can similarly ask how a *tensor* is described in the two coordinate systems. A tensor is again a physical quantity, and it must also take the same value no matter which coordinate system is used. The analysis goes exactly as above. If we now use a bar to denote the components in the new coordinate system, then it follows from the definition of a tensor in Eq. (3.23) that

$$\bar{T}^{ijk\cdots} = \bar{a}^i{}_{i'}\bar{a}^j{}_{j'}\bar{a}^k{}_{k'}\cdots T^{i'j'k'\cdots}$$

$$T^{ijk\cdots} = a^i{}_{i'}a^j{}_{j'}a^k{}_{k'}\cdots \bar{T}^{i'j'k'\cdots} \tag{3.50}$$

We again say that these are the *contravariant components* of the tensor.

Physical quantities (vectors, tensors) expressed in different coordinate systems have components that are related by these relations. If two sets of quantities satisfy these relations, they can be put into a one-to-one correspondence with a physical quantity. We observe that *physics must involve relations between physical quantities*.

We give two examples of tensors that play a central role in subsequent developments:

1) *Metric tensor.* Consider the dyadic

$$\underline{g} = g^{ij}\,\mathbf{e}_i\mathbf{e}_j \tag{3.51}$$

We claim that g^{ij} is a second-rank tensor satisfying

$$g^{ij} = a^i{}_{i'}a^j{}_{j'}\bar{g}^{i'j'} \tag{3.52}$$

The proof goes as follows. Write the r.h.s. of Eq. (3.52) as

$$\text{r.h.s.} = (\mathbf{e}^i \cdot \boldsymbol{\alpha}_{i'})(\mathbf{e}^j \cdot \boldsymbol{\alpha}_{j'})(\boldsymbol{\alpha}^{i'} \cdot \boldsymbol{\alpha}^{j'}) \tag{3.53}$$

Now identify the unit dyadic twice

$$\underline{I} = \boldsymbol{\alpha}_{i'}\boldsymbol{\alpha}^{i'} = \boldsymbol{\alpha}_{j'}\boldsymbol{\alpha}^{j'} \tag{3.54}$$

The result is that the r.h.s. reduces to

$$\text{r.h.s.} = \mathbf{e}^i \cdot \mathbf{e}^j = g^{ij} \tag{3.55}$$

This reproduces the l.h.s. of Eq. (3.52). Note that, in fact, the metric tensor *is* the unit dyadic!

$$\underline{g} = \mathbf{e}^i\mathbf{e}_i = \underline{I} = \boldsymbol{\alpha}^i\boldsymbol{\alpha}_i \tag{3.56}$$

2) *Gradient.* Consider an arbitrary scalar $S(q^1, \cdots, q^n)$ and write its total differential

$$dS = \frac{\partial S}{\partial q^i} dq^i \tag{3.57}$$

Now define the gradient as the directional derivative, so that

$$dS \equiv (\boldsymbol{\nabla} S) \cdot \boldsymbol{ds} \tag{3.58}$$

Then, with

$$\boldsymbol{\nabla} = \boldsymbol{e}^i \, \nabla_i \qquad ; \; \boldsymbol{ds} = \boldsymbol{e}_j \, dq^j \tag{3.59}$$

one has

$$
\begin{aligned}
dS &= (\boldsymbol{e}^i \, \nabla_i S) \cdot (\boldsymbol{e}_j \, dq^j) \\
&= (\nabla_i S) dq^i \tag{3.60}
\end{aligned}
$$

A comparison with Eq. (3.57) identifies the components of the gradient as

$$\nabla_i = \frac{\partial}{\partial q^i} \tag{3.61}$$

Note carefully just where the indices appear in this expression.

Now under the coordinate transformation of Eqs. (3.27), the chain rule of differentiation gives

$$
\begin{aligned}
\frac{\partial S}{\partial \xi^i} &= \frac{\partial S}{\partial q^j} \frac{\partial q^j}{\partial \xi^i} \\
&= a^j{}_i \frac{\partial S}{\partial q^j} \tag{3.62}
\end{aligned}
$$

Here Eq. (3.31) has been used in the second line. Hence

$$
\frac{\partial}{\partial \xi^i} = a^j{}_i \frac{\partial}{\partial q^j} = \bar{a}_i{}^j \frac{\partial}{\partial q^j}
$$
$$
\text{or ;} \qquad \bar{\nabla}_i = \bar{a}_i{}^j \nabla_j \tag{3.63}
$$

Thus the components of the gradient transform *covariantly* — just like the basis vectors. The covariant component of the gradient is the derivative with respect to the contravariant component of the coordinate, and *vice versa.*

3.8 Affine Connection

Consider now what happens when one moves from the point $(q^1, \cdots, q^n)$ to the neighboring point $(q^1 + dq^1, \cdots, q^n + dq^n)$ in the n-dimensional euclidian space (see Fig. 3.5).[6]

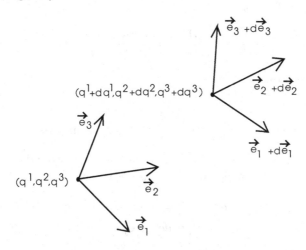

Fig. 3.5   Generalized coordinates and corresponding basis vectors in three dimensions at the point (q^1, q^2, q^3) and at the infinitesimally close point $(q^1 + dq^1, q^2 + dq^2, q^3 + dq^3)$—greatly exaggerated.

How do the *basis vectors change?* We will work to first order in infinitesimals. The change in each basis vector $d\mathbf{e}_i$ is again a vector and it can be expanded back in the original basis. Thus one can write quite generally

$$d\mathbf{e}_i = \gamma_{ij}^k \, \mathbf{e}_k dq^j \qquad (3.64)$$

Here the $\gamma_{ij}^k(q^1, \cdots, q^n)$ form a set of coefficients depending on position. We now claim that

$$\gamma_{ij}^k = \Gamma_{ij}^k \qquad ; \text{ affine connection} \qquad (3.65)$$

where Γ_{ij}^k is the affine connection of chap. 2. We proceed to give a proof of this result.

Recall that the line element is

$$d\mathbf{s} = \mathbf{e}_i \, dq^i \qquad (3.66)$$

---

[6] If we recall our problem of a particle moving on an arbitrary two-dimensional surface, it is when we start moving from point to point that the *geometry of the surface* first comes into play (see later).

Therefore

$$\frac{\partial \mathbf{s}}{\partial q^i} = \mathbf{e}_i \qquad (3.67)$$

One can interchange the order of second partial derivatives in this euclidian space,[7] and it follows that

$$\frac{\partial^2 \mathbf{s}}{\partial q^j \partial q^i} = \frac{\partial}{\partial q^j}\mathbf{e}_i = \frac{\partial^2 \mathbf{s}}{\partial q^i \partial q^j} = \frac{\partial}{\partial q^i}\mathbf{e}_j \qquad (3.68)$$

Hence from Eq. (3.64)

$$\gamma_{ij}^k \mathbf{e}_k = \gamma_{ji}^k \mathbf{e}_k$$

or ; $\qquad \gamma_{ij}^k = \gamma_{ji}^k \qquad$; symmetric in $(i \leftrightarrow j)$ $\qquad (3.69)$

Thus the coefficients must be symmetric in the lower two indices.

Now compute the change in the metric to first order in infinitesimals

$$\begin{aligned} dg_{ij} &= d\,(\mathbf{e}_i \cdot \mathbf{e}_j) = \mathbf{e}_i \cdot d\,\mathbf{e}_j + d\,\mathbf{e}_i \cdot \mathbf{e}_j \\ &= \mathbf{e}_i \cdot [\gamma_{jl}^k\,\mathbf{e}_k dq^l] + \mathbf{e}_j \cdot [\gamma_{il}^k\,\mathbf{e}_k dq^l] \\ &= \gamma_{jl}^k g_{ik} dq^l + \gamma_{il}^k g_{jk} dq^l \\ dg_{ij} &= [\gamma_{jl}^k g_{ik} + \gamma_{il}^k g_{jk}]dq^l \end{aligned} \qquad (3.70)$$

The total differential of the metric $g_{ij}(q)$ is also given by

$$dg_{ij} = \frac{\partial g_{ij}}{\partial q^l}dq^l \qquad (3.71)$$

Since the dq^l are linearly independent, one can equate their coefficients in the last two expressions

$$\frac{\partial g_{ij}}{\partial q^l} = \gamma_{jl}^k g_{ik} + \gamma_{il}^k g_{jk} \qquad (3.72)$$

We define

$$[il, j] \equiv \gamma_{il}^k g_{kj} = [li, j] \qquad ; \text{ symmetric in } (i \leftrightarrow l) \qquad (3.73)$$

This quantity is symmetric in its first pair of indices. Then

$$\frac{\partial g_{ij}}{\partial q^l} = [lj, i] + [li, j] \qquad (3.74)$$

---

[7]See Prob. 3.8.

The orthonormality relation on the metric is [this is just the unit dyadic again—see Eq. (2.20)]

$$g_{kj}g^{jm} = \delta_k{}^m \tag{3.75}$$

This allows one to invert Eq. (3.73) as

$$\gamma_{il}^m = g^{jm}[li,j] \tag{3.76}$$

Now take Eq. (3.74), subtract the result with a change of dummy labels $(l \leftrightarrow j)$, and add the result with $(i \leftrightarrow l)$

$$\frac{\partial g_{ij}}{\partial q^l} - \frac{\partial g_{il}}{\partial q^j} + \frac{\partial g_{lj}}{\partial q^i} = [lj,i] + [li,j] - [jl,i] - [ji,l] + [ij,l] + [il,j]$$

$$= 2[li,j] \tag{3.77}$$

This result can be inverted using Eq. (3.76)

$$\gamma_{il}^m = \frac{1}{2}g^{jm}\left[\frac{\partial g_{ij}}{\partial q^l} + \frac{\partial g_{lj}}{\partial q^i} - \frac{\partial g_{il}}{\partial q^j}\right] \tag{3.78}$$

Relabeling the indices $(m \to k, j \to m, l \to j)$ gives

$$\gamma_{ij}^k = \frac{1}{2}g^{km}\left[\frac{\partial g_{mi}}{\partial q^j} + \frac{\partial g_{mj}}{\partial q^i} - \frac{\partial g_{ij}}{\partial q^m}\right]$$

$$= \Gamma_{ij}^k \tag{3.79}$$

This is precisely the affine connection of Eqs. (2.23), now extended to n-dimensional euclidian space. Note that the affine connection is symmetric in its lower two indices.

$$\Gamma_{ij}^k = \Gamma_{ji}^k \qquad ; \text{ symmetric in } (i \leftrightarrow j) \tag{3.80}$$

In *summary*, to first order in infinitesimals, when one moves to a neighboring point with curvilinear coordinates in n-dimensional euclidian space, the change in basis vectors is given through the affine connection as

$$d\,\mathbf{e}_i = \Gamma_{ij}^k \, \mathbf{e}_k dq^j \tag{3.81}$$

3.9 Example: Polar Coordinates

As an example, we return to polar coordinates in a plane as presented in Eqs. (2.26)–(2.30) and illustrated in Fig. 2.5.[8] Here $(q^1, q^2) = (r, \phi)$, and

---

[8]This is a two-dimensional euclidian surface.

the line element is given by

$$d\mathbf{s} = \mathbf{e}_r \, dr + \mathbf{e}_\phi \, d\phi$$

$$\mathbf{e}_r = \hat{\mathbf{e}}_r \qquad\qquad ; \ \mathbf{e}_\phi = r \, \hat{\mathbf{e}}_\phi \tag{3.82}$$

where $(\hat{\mathbf{e}}_r, \hat{\mathbf{e}}_\phi)$ are orthogonal unit vectors. The metric and its inverse are given by

$$\underline{g} = \begin{pmatrix} g_{rr} & g_{r\phi} \\ g_{\phi r} & g_{\phi\phi} \end{pmatrix} = \begin{pmatrix} 1 & 0 \\ 0 & r^2 \end{pmatrix}$$

$$\underline{g}^{-1} = \begin{pmatrix} g^{rr} & g^{r\phi} \\ g^{\phi r} & g^{\phi\phi} \end{pmatrix} = \begin{pmatrix} 1 & 0 \\ 0 & 1/r^2 \end{pmatrix} \tag{3.83}$$

Both are diagonal. We claim that the affine connection is given by

$$\Gamma^{\phi}_{\ r\phi} = \Gamma^{\phi}_{\ \phi r} = \frac{1}{r}$$

$$\Gamma^{r}_{\ \phi\phi} = -r \qquad\qquad ; \ \text{all others vanish} \tag{3.84}$$

The change in basis vectors then follows from Eq. (3.81)

$$d\,\mathbf{e}_r = \frac{1}{r} d\phi \, \mathbf{e}_\phi$$

$$d\,\mathbf{e}_\phi = \frac{1}{r} dr \, \mathbf{e}_\phi - r d\phi \, \mathbf{e}_r \tag{3.85}$$

We derive a few of Eqs. (3.84)

$$\Gamma^{r}_{\ r\phi} = \Gamma^{r}_{\ \phi r} = \frac{1}{2} g^{rr} \left[\frac{\partial g_{r\phi}}{\partial r} + \frac{\partial g_{rr}}{\partial \phi} - \frac{\partial g_{r\phi}}{\partial r} \right] = 0$$

$$\Gamma^{r}_{\ rr} = \frac{1}{2} g^{rr} \left[\frac{\partial g_{rr}}{\partial r} + \frac{\partial g_{rr}}{\partial r} - \frac{\partial g_{rr}}{\partial r} \right] = 0$$

$$\Gamma^{\phi}_{\ \phi\phi} = \frac{1}{2} g^{\phi\phi} \left[\frac{\partial g_{\phi\phi}}{\partial \phi} + \frac{\partial g_{\phi\phi}}{\partial \phi} - \frac{\partial g_{\phi\phi}}{\partial \phi} \right] = 0$$

$$\Gamma^{r}_{\ \phi\phi} = \frac{1}{2} g^{rr} \left[\frac{\partial g_{r\phi}}{\partial \phi} + \frac{\partial g_{r\phi}}{\partial \phi} - \frac{\partial g_{\phi\phi}}{\partial r} \right] = \frac{1}{2}(1)(-2r) = -r \tag{3.86}$$

The derivation of the remaining components, as well as the direct geometrical derivation of Eqs. (3.85), are left as exercises in Prob. 3.6.

Chapter 4

Particle on a Two-Dimensional Surface—Revisited

4.1 Motion in Three-Dimensional Euclidian Space

Let us return to the problem of a particle moving without friction on an arbitrary two-dimensional surface. We now view the surface in a three-dimensional euclidian space. The first two generalized coordinates (q^1, q^2) locate the particle on the surface. Take the third coordinate q^3 to be the *distance above the surface along the normal to the tangent plane* $\mathbf{e}_3$ as illustrated in Fig. 4.1.[1] The generalized coordinates in this three-dimensional euclidian space are now (q^1, q^2, q^3) with corresponding basis vectors $(\mathbf{e}_1, \mathbf{e}_2, \mathbf{e}_3)$. The motion on the surface is obtained by imposing the constraint condition

$$q_3 = \text{constant} = 0 \qquad ; \text{ constraint} \qquad (4.1)$$

This is achieved through the application of a constraint force *normal to the surface*

$$\mathbf{F} = F^3 \mathbf{e}_3 = \mathbf{F}_\perp \qquad ; \text{ constraint force} \qquad (4.2)$$

The line element in this space is

$$d\mathbf{s} = \mathbf{e}_i \, dq^i \qquad (4.3)$$

where the contracted indices are summed from 1 to 3. The *velocity* of the particle is

$$\mathbf{v} = \frac{d\mathbf{s}}{dt} = \mathbf{e}_i \frac{dq^i}{dt} = v^i \, \mathbf{e}_i$$

$$v^i = \frac{dq^i}{dt} \qquad (4.4)$$

---

[1] We will only employ q^3 close to the surface.

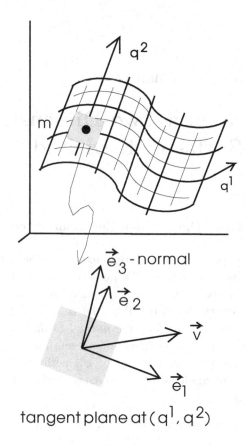

tangent plane at (q^1, q^2)

Fig. 4.1   Point particle constrained to move without friction on a two-dimensional surface of arbitrary shape. The generalized coordinates (q^1, q^2, q^3) locate the particle in three-dimensional euclidian space. The basis vector $\mathbf{e}_3$ is normal to the tangent plane, and the velocity $\mathbf{v}$ lies in it.

The velocity follows directly from the line element. To get the *acceleration*, required for Newton's second law, it is necessary to consider *changes* in the velocity.

The total differential of the velocity $\mathbf{v}$ in Eq. (4.4) is given by

$$d\mathbf{v} = dv^i \, \mathbf{e}_i + v^i \, d\mathbf{e}_i \qquad (4.5)$$

Now use the expression for the change in basis vectors when moving from point to point in a three-dimensional euclidian space as derived in Eq. (3.81)

$$d\mathbf{e}_i = \Gamma_{ij}^k dq^j \, \mathbf{e}_k \qquad (4.6)$$

here $\Gamma_{ij}^k = \Gamma_{ji}^k$ is the affine connection with these coordinates in this space; it is symmetric in the indices (ij). Substitution of Eq. (4.6) into Eq. (4.5) gives

$$d\mathbf{v} = \left(dv^k + \Gamma_{ij}^k dq^j v^i \right) \mathbf{e}_k \qquad (4.7)$$

Division by dt and the use of the last of Eqs. (4.4) then gives the *acceleration*

$$\frac{d\mathbf{v}}{dt} = \left(\frac{dv^k}{dt} + v^i v^j \Gamma_{ij}^k \right) \mathbf{e}_k \qquad (4.8)$$

Newton's second law for the motion of a particle of fixed mass m in this space is now

$$m\frac{d\mathbf{v}}{dt} = \mathbf{F}_\perp = F^3 \mathbf{e}_3 \qquad (4.9)$$

The third component of this relation yields the constraint force, which is not of interest to us here.[2] The $(1,2)$ components of this relation give, again with the aid of the last of Eqs. (4.4),

$$\frac{d^2 q^k}{dt^2} + \Gamma_{ij}^k \frac{dq^i}{dt} \frac{dq^j}{dt} = 0 \qquad ; k = 1,2 \qquad (4.10)$$

It follows from Eq. (4.1) that

$$\frac{dq^3}{dt} = 0 \qquad (4.11)$$

Thus there is no contribution from the terms with $i = 3$ or $j = 3$ in Eqs. (4.10), and *all* the indices in this equation (i,j,k) are restricted to the values $(1,2)$. The equations of motion therefore reduce to

$$\frac{d^2 q^k}{dt^2} + \Gamma_{ij}^k \frac{dq^i}{dt} \frac{dq^j}{dt} = 0 \qquad ; (i,j,k) = (1,2) \qquad (4.12)$$

Furthermore, since the vectors $\mathbf{e}_3$ and $\mathbf{e}^3$ are everywhere perpendicular to the tangent plane (Fig. 4.1), one has

$$g_{13} = g_{23} = g^{13} = g^{23} = 0 \qquad (4.13)$$

Thus the index 3 never occurs in the affine connection appearing in Eqs. (4.12).

We now make the following crucial observations:

- *These are just Lagrange's Eqs. (2.23) as derived in chap. 2;*

---

[2]See, however, Prob. 4.1.

- *The problem now sits completely in the surface, which forms a two-dimensional non-euclidian space!*

4.2   Parallel Displacement

Let us turn to the concept of *parallel displacement* (this is also sometimes referred to as *parallel transport*, and we shall use the terms interchangeably). Start first with three-dimensional euclidian space and no constraints. Consider a curve C running from points (1) to (2), and let s be the distance along the curve. The curve is then described parametrically by $q^i(s)$ with $i = (1, 2, 3)$. Let ds be an infinitesimal displacement along C and $\mathbf{u}$ be a unit vector tangent to the curve (Fig. 4.2). One then can write

$$\mathbf{u} = \frac{d\mathbf{s}}{ds} = \mathbf{e}_i \frac{dq^i(s)}{ds} \tag{4.14}$$

Evidently

$$\mathbf{u} \cdot \mathbf{u} = \frac{d\mathbf{s} \cdot d\mathbf{s}}{(ds)^2} = 1 \tag{4.15}$$

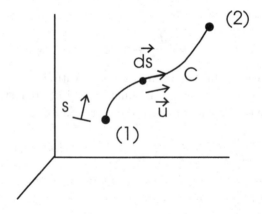

Fig. 4.2   A curve C connecting points (1) and (2) in three-dimensional euclidian space. Here s is the distance along C, $d\mathbf{s}$ is an infinitesimal displacement along C, and $\mathbf{u}$ is a unit vector tangent to the curve.

Parallel displacement is the requirement that the *tangent should be unchanged when one moves along* C. Thus

$$d\mathbf{u} = 0 \qquad ; \text{ parallel displacement} \tag{4.16}$$

It follows from this condition that

$$\frac{d\mathbf{u}}{ds} = 0 \tag{4.17}$$

A calculation exactly following Eqs. (4.3)–(4.8) then leads to the relation

$$\frac{d\mathbf{u}}{ds} = \left(\frac{d^2 q^k}{ds^2} + \Gamma^k_{ij} \frac{dq^i}{ds} \frac{dq^j}{ds} \right) \mathbf{e}_k = 0 \tag{4.18}$$

These are just the *geodesic equations* for $q^k(s)$. Parallel displacement then leads to the geodesic in three-dimensional euclidian space with no constraints. The solution to these equations in this case is, of course, trivial — it is just a straight line!

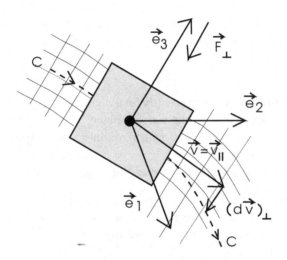

Fig. 4.3 Newton's second law for a particle moving without friction on an arbitrary surface. The tangent plane is indicated and C is the path followed by the particle. There is a normal impulse $\mathbf{F}_\perp \, dt$ which causes an infinitesimal velocity change $(d\mathbf{v})_\perp$ along the normal to the tangent plane. The velocity in the tangent plane $\mathbf{v}_\parallel$ is unchanged to lowest order.

Now let us return to the problem of the particle moving on an arbitrary surface and confine ourselves to the surface by setting $q^3 = 0$ as in Eq. (4.1). Newton's second law in Eq. (4.9) can be rearranged to read

$$m \, d\mathbf{v} = \mathbf{F}_\perp \, dt \tag{4.19}$$

The velocity changes only in the direction *normal to the tangent plane*. In the tangent plane one has

$$(d\mathbf{v})_{\parallel} = 0 \qquad ; \text{ no force in tangent plane} \qquad (4.20)$$

The component of the velocity in the tangent plane is unchanged to lowest order — there is no force in the plane. This is just the condition of parallel displacement of the velocity in the tangent plane.

$$(d\mathbf{v})_{\parallel} = 0 \qquad ; \text{ parallel displacement in tangent plane} \quad (4.21)$$

The situation is illustrated in Fig. 4.3.

In the tangent plane we have at the point (q^1, q^2)

$$\mathbf{v} = v^1\,\mathbf{e}_1 + v^2\,\mathbf{e}_2 \qquad (4.22)$$

Parallel displacement in Eq. (4.21) implies that this combination is unchanged in the tangent plane to lowest order as one moves from point to point along the trajectory.

Now

(1) Write out $(d\mathbf{v})_{\parallel} = 0$ and then $(d\mathbf{v}/dt)_{\parallel} = 0$ as in Eq. (4.8), retaining only those components in the tangent plane;
(2) Impose the condition $q^3 = 0$ which confines one to the surface; this condition implies $dq^3/dt = 0$;
(3) Use the fact $\mathbf{e}_3$ is normal to the tangent plane.

These steps lead precisely to Lagrange's Eqs. (4.12).

In the tangent plane at the point (q^1, q^2), one has $\mathbf{v} = \mathbf{v}_{\parallel}$. We know from the conservation of energy in Eq. (2.25) that

$$|\mathbf{v}| = v_0 = \text{constant} \qquad ; \text{ energy conservation} \qquad (4.23)$$

as the particle moves along the surface. The magnitude of the velocity $\mathbf{v}$ is strictly unchanged as one moves away from (q^1, q^2). There is a change $(d\mathbf{v})_{\perp}$ caused by the normal force as the particle moves along the surface (Fig. 4.3) which does imply there must be a corresponding non-zero change in the magnitude of $\mathbf{v}_{\parallel}$; however, since

$$\mathbf{v}^2 = \mathbf{v}_{\parallel}^2 + (d\mathbf{v})_{\perp}^2 = \text{constant} \qquad (4.24)$$

this change is of second order.[3]

---

[3]The normal continues to tilt in the $(\mathbf{e}_3, \mathbf{v})$ plane as one proceeds along the particle trajectory, and $|\mathbf{v}|$ is strictly constant in the new, slightly tilted, tangent plane.

To first order, the condition of parallel displacement of the velocity in Eq. (4.21), leading to Lagrange's Eqs. (4.12), can equally well be written as:

$$d\mathbf{v} = 0 \qquad ; \text{ parallel displacement in tangent plane} \qquad (4.25)$$

with the understanding that the analysis is now completely confined to the surface and its tangent plane; that is, the indices (i, j, k) are everywhere restricted to the values $(1, 2)$.[4]

We now work backwards and identify the unit vector $\mathbf{u}$ tangent to the trajectory C on the surface, and distance s along the curve, according to

$$\mathbf{u} = \frac{1}{v_0}\mathbf{v}$$

$$t = \frac{1}{v_0}s \qquad (4.26)$$

The condition of parallel displacement in the tangent plane then implies the geodesic equation for $[q^1(s), q^2(s)]$ on the surface

$$d\mathbf{u} = 0 \qquad ; \text{ parallel displacement in tangent plane}$$

$$\Rightarrow \quad \frac{d^2 q^k}{ds^2} + \Gamma^k_{ij}\frac{dq^i}{ds}\frac{dq^j}{ds} = 0 \qquad (i, j, k) = (1, 2) \qquad (4.27)$$

Once again, we have arrived at a condition and equations that have nothing to do with mechanics, but are *pure geometry*.

In *summary*, if one starts in three-dimensional euclidian space, and imposes the conditions

$$q^3 = 0 \qquad ; \text{ stay on surface}$$

$$d\mathbf{u} = 0 \qquad \text{parallel transport in tangent plane} \qquad (4.28)$$

where $\mathbf{u}$ is the tangent vector to the curve C, then one arrives at the geodesic equations for C in two-dimensional non-euclidian space. The analysis now involves quantities defined *entirely on the non-euclidean surface*.

If $\mathbf{v}$ is the velocity for particle motion, then the equivalent mechanics problem is defined by the condition

$$q^3 = 0 \qquad ; \text{ stay on surface}$$

$$d\mathbf{v} = 0 \qquad \text{parallel transport in tangent plane} \qquad (4.29)$$

---

[4]The corresponding displacement along the particle trajectory in Eq. (4.25) is $d\mathbf{s} = \mathbf{v}dt$, and along the geodesic in Eq. (4.27) is $d\mathbf{s} = \mathbf{u}\,ds$.

this leads to Lagrange's equations, and again, the analysis now involves only quantities defined *entirely on the non-euclidian surface.*

Chapter 5

Some Tensor Analysis

We proceed to first gather and extend some of the results obtained so far in n-dimensional euclidian space.

5.1 Covariant Derivative

Suppose one is given a vector field $\mathbf{v}(q^1, \cdots, q^n)$ representing some *physical quantity*.[1] This can be expanded in terms of the basis vectors at every point according to

$$\mathbf{v} = v^i(q)\,\mathbf{e}_i \tag{5.1}$$

With the aid of Eq. (3.81), the total differential of this quantity is

$$
\begin{aligned}
d\mathbf{v} &= dv^i\,\mathbf{e}_i + v^i\,d\mathbf{e}_i \\
&= \left[\frac{\partial v^i}{\partial q^j}dq^j\right]\mathbf{e}_i + v^i\left[\Gamma_{ij}^k dq^j\,\mathbf{e}_k\right] \\
&= \left(\frac{\partial v^k}{\partial q^j} + v^i\Gamma_{ij}^k\right)dq^j\,\mathbf{e}_k
\end{aligned} \tag{5.2}
$$

We define the *covariant derivative* of the vector field through the relation

$$v^k{}_{;j} \equiv \frac{\partial v^k}{\partial q^j} + v^i\Gamma_{ij}^k \tag{5.3}$$

Note the appearance of the semicolon. The total differential of $\mathbf{v}$ is thus

$$d\mathbf{v} = v^k{}_{;j}dq^j\,\mathbf{e}_k \tag{5.4}$$

---

[1] We use the terms "vector field" and "vector" interchangeably in this work.

This takes into account the fact that both components and basis vectors change as one moves away from a given point.

Recall that the reciprocal basis is defined through

$$\mathbf{e}_i \cdot \mathbf{e}^j = \delta_i{}^j \tag{5.5}$$

Write the total differential of this relation

$$d\mathbf{e}_i \cdot \mathbf{e}^j + \mathbf{e}_i \cdot d\mathbf{e}^j = 0 \tag{5.6}$$

Now use Eq. (3.81) again

$$\begin{aligned}
\mathbf{e}_i \cdot d\mathbf{e}^j &= -\mathbf{e}^j \cdot d\mathbf{e}_i = -\mathbf{e}^j \cdot \left[\Gamma_{il}^k dq^l\, \mathbf{e}_k \right] \\
&= -\Gamma_{il}^j dq^l
\end{aligned} \tag{5.7}$$

These are the components of $d\mathbf{e}^j$ in the original basis, and hence the vector itself is given by (with an appropriate change of labels)

$$d\mathbf{e}^i = -\Gamma_{kj}^i dq^j\, \mathbf{e}^k \tag{5.8}$$

This is just the expansion of the change in reciprocal basis vectors back in the reciprocal basis. It forms the analog of Eq. (3.81), but note the crucial minus sign.

A repetition of the previous calculation starting from the equivalent expansion of the vector field

$$\mathbf{v} = v_i(q)\, \mathbf{e}^i \tag{5.9}$$

then leads to

$$\begin{aligned}
d\mathbf{v} &= v_{k\,;j} dq^j\, \mathbf{e}^k \\
v_{k\,;j} &= \frac{\partial v_k}{\partial q^j} - v_i \Gamma_{kj}^i
\end{aligned} \tag{5.10}$$

It is easy to keep track of the indices here, but again note the minus sign.

Suppose one now starts from a different basis $\xi^i(q^1, \cdots, q^n)$ as in Eqs. (3.26) with corresponding basis vectors $\boldsymbol{\alpha}_i$ (see Fig. 3.4). Calculations identical to those above then lead to the relations

$$\begin{aligned}
d\mathbf{v} &= \bar{v}^k{}_{;j} d\xi^j\, \boldsymbol{\alpha}_k \\
&= \bar{v}_{k\,;j} d\xi^j\, \boldsymbol{\alpha}^k
\end{aligned} \tag{5.11}$$

Where the bar indicates the components in the new basis.

We now claim that the *covariant derivative forms a second-rank tensor.* Thus covariant differentiation takes one from a physical vector field to a physical tensor field. The proof goes as follows.

Recall the tensor transformation law from Eq. (3.50)

$$\bar{T}^{ij} = \bar{a}^i{}_l \bar{a}^j{}_m T^{lm} \tag{5.12}$$

Since one can always interchange a pair of contracted indices [Eq. (3.25)], this relation is equivalent to

$$\bar{T}_{ij} = \bar{a}_i{}^l \bar{a}_j{}^m T_{lm} \tag{5.13}$$

We proceed to show that

$$\bar{v}_{k\,;j} = \bar{a}_k{}^l \bar{a}_j{}^m \, v_{l\,;m} \tag{5.14}$$

Since $d\mathbf{v}$ is a physical quantity, one must get the same result for it in either basis

$$d\mathbf{v} = \bar{v}_{k\,;j} d\xi^j \, \boldsymbol{\alpha}^k = v_{l\,;m} dq^m \, \mathbf{e}^l \tag{5.15}$$

Now use Eqs. (3.34), (3.35), and (3.43)

$$dq^m = a^m{}_j d\xi^j = \bar{a}_j{}^m d\xi^j$$
$$\mathbf{e}^l = a^l{}_k \, \boldsymbol{\alpha}^k = \bar{a}_k{}^l \, \boldsymbol{\alpha}^k \tag{5.16}$$

Substitution in Eq. (5.15) then leads to

$$d\mathbf{v} = v_{l\,;m}[\bar{a}_j{}^m d\xi^j][\bar{a}_k{}^l \, \boldsymbol{\alpha}^k] \tag{5.17}$$

Hence

$$\bar{v}_{k\,;j} = \bar{a}_k{}^l \bar{a}_j{}^m \, v_{l\,;m} \tag{5.18}$$

which is precisely Eq. (5.14).[2]

We leave the corresponding proof of the relation

$$\bar{v}^k{}_{;j} = \bar{a}^k{}_l \bar{a}_j{}^m \, v^l{}_{;m} \tag{5.19}$$

as an exercise for the reader.

---

[2]Note that this demonstration only applies to the *combination* of terms in the second of Eqs. (5.10) and not to the individual terms themselves (see *e.g.* Prob. 5.7).

5.2 The Riemann Curvature Tensor

Let us return to the concept of parallel displacement (parallel transport). Suppose we transport a vector **v** around a closed curve in flat, two-dimensional euclidian space as illustrated in Fig. 5.1.

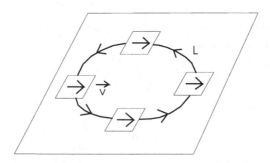

Fig. 5.1 Parallel transport of a vector **v** around a closed curve L in a flat two-dimensional euclidian space.

If **v** is kept constant in the tangent space (here identical with the whole space) from point to point along L one has

$$d\mathbf{v} = 0 \qquad \text{; in tangent space} \qquad (5.20)$$

The net change around L is then

$$\Delta\mathbf{v} = 0 \qquad \text{; around } L \qquad (5.21)$$

The vector comes back to where it started.

Suppose, on the other hand, that one is working in a *curved space*, say on the surface of a sphere (*e. g.* the earth). Now transport a vector **v** around a closed curve L keeping **v** constant in the flat tangent space as illustrated in Fig. 5.2. Here the particular L starts at the north pole, goes down to the equator where it travels one-quarter of the way around the circumference, and then returns to the pole. The condition of parallel transport is

$$d\mathbf{v} = 0 \qquad \text{; in tangent space} \qquad (5.22)$$

In this case, **v** *does not come back where it started*[3]

$$\Delta\mathbf{v} \neq 0 \qquad \text{; around } L \qquad (5.23)$$

_____

[3]In Fig. 5.2, **v** has been rotated by 90°.

The actual rotation of **v** in the tangent plane depends on the curve L, and upon its direction. We have come up with a very interesting result:

*We now have a meter, the rotation of the vector **v** in the tangent plane as we parallel transport it around a closed curve L in the non-euclidian space, with which to measure the curvature of that space.*

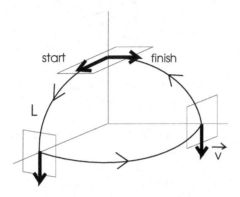

Fig. 5.2 Parallel transport of a vector **v** around a particular closed curve L on the surface of a sphere. Here L starts at the north pole, goes down to the equator where it travels one-quarter of the way around the circumference, and then returns to the pole.

Since everything has now been explicitly formulated in terms of quantities on the non-euclidian surface or its tangent plane at any point, we will now *work on the two-dimensional curved surface*. From our previous analysis, the relevant change in basis vectors as one moves an infinitesimal distance on the surface, and affine connection on the surface, are given by

$$d\,\mathbf{e}_i = \Gamma^k_{ij}\,\mathbf{e}_k dq^j \qquad\qquad ;\ d\,\mathbf{e}^i = -\Gamma^i_{kj}\,\mathbf{e}^k dq^j$$

$$\Gamma^k_{ij} = \frac{1}{2}g^{km}\left[\frac{\partial g_{mi}}{\partial q^j} + \frac{\partial g_{mj}}{\partial q^i} - \frac{\partial g_{ij}}{\partial q^m}\right] \tag{5.24}$$

where all indices are now confined to the set $(1,2)$.[4]

Now consider a closed curve L lying in the two-dimensional curved surface and surrounding the point q^i_0 as illustrated in Fig. 5.3. Let τ with $0 \leq \tau \leq 1$ parameterize the distance along L.

---

[4]Note that we have been careful to write things in a form that is immediately generalized to n-dimensional non-euclidian space — see later.

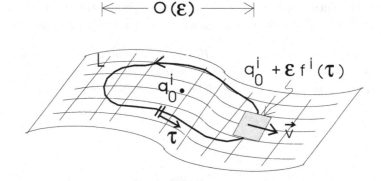

Fig. 5.3   Parallel transport of a vector **v** around a closed curve L in a two-dimensional non-euclidian space. Here ε is a small parameter.

Write the generalized coordinates q^i in the immediate neighborhood of the point q_0^i and along L as

$$q^i = q_0^i + \varepsilon f^i(\tau) \qquad ; \varepsilon \text{ a small parameter}$$
$$f^i(1) = f^i(0) \qquad\qquad \text{closed curve} \qquad\qquad (5.25)$$

where $i = (1, 2)$. We assume that ε is a small parameter.

Consider a vector **v**. At any point in the tangent space, one has

$$\mathbf{v} = v^i \, \mathbf{e}_i \qquad\qquad\qquad (5.26)$$

The total differential of this relation gives

$$d\mathbf{v} = \left(dv^k + v^i \Gamma_{ij}^k dq^j \right) \mathbf{e}_k$$
$$\equiv \left(\mathcal{D}v^k \right) \mathbf{e}_k \qquad\qquad (5.27)$$

Here the first of Eqs. (5.24) has been employed in the first line, and the *entire* change of a component in the original basis is defined as $\mathcal{D}v^k$ in the second.

We shall investigate what happens to the vector **v** under parallel transport around the curve L. Parallel transport implies that the vector is unchanged in the tangent plane

$$d\mathbf{v} = 0 \qquad ; \text{ parallel transport}$$
$$\Rightarrow \qquad \mathcal{D}v^k = 0 \qquad\qquad\qquad (5.28)$$

It follows that (with a relabeling of indices)

$$dv^i = -\Gamma^i_{jk}v^j \, dq^k$$

$$\frac{dv^i}{d\tau} = -\Gamma^i_{jk}v^j \frac{dq^k}{d\tau} \tag{5.29}$$

In parallel displacement, the components of the vector change along L because the basis vectors change from point to point. Upon the insertion of the parameterization in Eq. (5.25), this last relation becomes

$$\frac{dv^i(\tau)}{d\tau} = -\Gamma^i_{jk}(q)v^j(\tau)\varepsilon\frac{df^k(\tau)}{d\tau} \tag{5.30}$$

Now let us *expand the affine connection in the surface over the region encircled by L as a power series in ε*

$$\Gamma^i_{jk}[q_0^l + \varepsilon f^l(\tau)] = \Gamma^i_{jk}(q_0) + \varepsilon f^l(\tau)\frac{\partial}{\partial q^l}\Gamma^i_{jk}(q_0) + O(\varepsilon^2) \tag{5.31}$$

We also expand the components of the vector $\mathbf{v}$ along L as

$$v^i(\tau) = v_0^i + \varepsilon v_1^i(\tau) + \varepsilon^2 v_2^i(\tau) + \cdots \tag{5.32}$$

These two expansions can now be inserted into Eq. (5.30)[5]

$$\varepsilon\frac{dv_1^i(\tau)}{d\tau} + \varepsilon^2\frac{dv_2^i(\tau)}{d\tau} + \cdots =$$

$$-\varepsilon\frac{df^k(\tau)}{d\tau}\left[\Gamma^i_{jk}(0) + \varepsilon f^l(\tau)\frac{\partial}{\partial q^l}\Gamma^i_{jk}(0) + \cdots\right][v_0^j + \varepsilon v_1^j(\tau)\cdots] \tag{5.33}$$

Now equate coefficients of ε through second order. The first-order term gives

$$\frac{dv_1^i(\tau)}{d\tau} = -\Gamma^i_{jk}(0)v_0^j\frac{df^k(\tau)}{d\tau} \tag{5.34}$$

This relation can be integrated with respect to τ to give

$$v_1^i(\tau) = -\Gamma^i_{jk}(0)v_0^j f^k(\tau) \tag{5.35}$$

Since $v^i \to v_0^i$ when $f \to 0$, there is no additional constant of integration here. The difference between $\tau = 1$ and $\tau = 0$ tells what happens as one goes around L (Fig. 5.3), and employing Eq. (5.25) one finds

$$v_1^i(1) - v_1^i(0) = 0 \tag{5.36}$$

---

[5]We simplify the notation to $\Gamma^i_{jk}(q_0) \equiv \Gamma^i_{jk}(0)$.

The vector **v** is unchanged through $O(\varepsilon)$ under parallel transport around L.

The coefficients of the second-order terms in Eq. (5.33) are equated to give

$$\frac{dv_2^i(\tau)}{d\tau} = -\left\{ \left[\frac{\partial}{\partial q^l} \Gamma^i_{jk}(0) \right] f^l(\tau) \frac{df^k(\tau)}{d\tau} v_0^j + \Gamma^i_{jk}(0) v_1^j(\tau) \frac{df^k(\tau)}{d\tau} \right\} \quad (5.37)$$

Relabel $j \to m$ in the second term. Then substitution of Eq. (5.35) for $v_1^m(\tau)$, with k there relabeled by l, gives

$$\frac{dv_2^i(\tau)}{d\tau} = -\left[\frac{\partial}{\partial q^l} \Gamma^i_{jk}(0) - \Gamma^i_{mk}(0) \Gamma^m_{jl}(0) \right] v_0^j f^l(\tau) \frac{df^k(\tau)}{d\tau} \quad (5.38)$$

Now integrate this expression around L using

$$\oint_L f^l(\tau) \frac{df^k(\tau)}{d\tau} d\tau = \frac{1}{2} \oint_L \left[\left(f^l \frac{df^k}{d\tau} - f^k \frac{df^l}{d\tau} \right) + \frac{d}{d\tau} \left(f^l f^k \right) \right] d\tau$$

$$= \frac{1}{2} \oint_L \left(f^l \frac{df^k}{d\tau} - f^k \frac{df^l}{d\tau} \right) d\tau \quad (5.39)$$

Here the perfect differential integrates to zero around the closed loop. Now return to the coordinates as parameterized in Eq. (5.25)

$$\oint_L f^l \frac{df^k}{d\tau} d\tau = \frac{1}{2\varepsilon^2} \oint_L \left[(q^l - q_0^l) \frac{dq^k}{d\tau} - (q^k - q_0^k) \frac{dq^l}{d\tau} \right] d\tau \quad (5.40)$$

The perfect differentials again integrate to zero around L and hence one finds

$$\oint_L f^l(\tau) \frac{df^k(\tau)}{d\tau} d\tau = \frac{1}{\varepsilon^2} S^{lk}$$

$$S^{lk} \equiv \frac{1}{2} \oint_L (q^l dq^k - q^k dq^l) = -S^{kl} \quad (5.41)$$

Thus from the integration of Eq. (5.38) around L one finds

$$v_2^i(1) - v_2^i(0) = -\frac{1}{2\varepsilon^2} R^i{}_{jlk} v_0^j S^{lk}$$

$$R^i{}_{jlk} \equiv \left[\frac{\partial}{\partial q^l} \Gamma^i_{jk}(0) - \Gamma^i_{mk}(0) \Gamma^m_{jl}(0) \right] - [k \leftrightarrow l] \quad (5.42)$$

We have used the antisymmetry of $S^{lk} = -S^{kl}$ and the relations

$$T^i{}_{jlk} S^{lk} = \frac{1}{2} \left(T^i{}_{jlk} S^{lk} + T^i{}_{jkl} S^{kl} \right)$$

$$= \frac{1}{2} \left(T^i{}_{jlk} - T^i{}_{jkl} \right) S^{lk} \tag{5.43}$$

in arriving at the last form in Eqs. (5.42). We observe that there is a *nonzero second-order change* in the vector **v** under parallel transport around the loop given by

$$\Delta v^i = \varepsilon^2 \Delta v_2^i = \varepsilon^2 [v_2^i(1) - v_2^i(0)]$$

$$= -\frac{1}{2} R^i{}_{jlk} v_0^j S^{lk} \tag{5.44}$$

The quantity S^{lk} is essentially the *area of the loop* in Fig. 5.3. We show that through a simple example. Suppose the surface is, in fact, a flat plane and the closed curve L a circle of radius of radius r. Take

$$x = r \cos \theta \qquad ; \ y = r \sin \theta$$

$$\frac{1}{2} \oint (x dy - y dx) = \frac{1}{2} \int_0^{2\pi} d\theta (r^2 \cos^2 \theta + r^2 \sin^2 \theta) = \pi r^2 \tag{5.45}$$

Since the linear dimension of the loop is of $O(\varepsilon)$ in Fig. 5.3, one has

$$S^{lk} = O(\varepsilon^2) \tag{5.46}$$

The quantity $R^i{}_{jlk}$ defined in Eq. (5.42) is the *Riemann curvature tensor*[6]

$$R^i{}_{jlk} = \left(\frac{\partial}{\partial q^l} \Gamma^i_{jk} + \Gamma^i_{lm} \Gamma^m_{jk} \right) - (k \leftrightarrow l) \qquad ; \ i, j, k, l = 1, 2$$

$$= \left(\frac{\partial}{\partial q^l} \Gamma^i_{jk} + \Gamma^i_{lm} \Gamma^m_{jk} \right) - \left(\frac{\partial}{\partial q^k} \Gamma^i_{jl} + \Gamma^i_{km} \Gamma^m_{jl} \right) \tag{5.47}$$

We have arrived through this calculation at the expression in Eq. (5.44) for our "curvature meter." It is proportional to the Riemann tensor and the area of the loop around which a vector is parallel transported. The Riemann curvature tensor is a *nonlinear, second-order, differential form in the metric which varies from point to point on the surface*. It will play a central role in subsequent developments.[7] We have here achieved one of our

---

[6] Note that we have interchanged the second and fourth terms in the last of Eqs. (5.42).

[7] It is again worthwhile spending time studying the indices on this expression and remembering where they go. Notice, in particular, the role played by the third index l.

original goals of determining an intrinsic structural property of the surface derivable from the metric.

It is evident from its definition that the Riemann tensor is antisymmetric in its last two indices

$$R^i_{\ jkl} = -R^i_{\ jlk} \qquad ; \text{ antisymmetric in } (k \leftrightarrow l) \qquad (5.48)$$

5.3 Second Covariant Derivative

We remind the reader that:

> *Henceforth we are working on the two-dimensional, non-euclidian, curved surface with a non-zero Riemann tensor $R^i_{\ jlk} \neq 0$, which provides a meter in the surface to measure the curvature (Fig. 5.3). The Riemann tensor is defined in Eq. (5.47). The affine connection and change in basis vectors for an infinitesimal displacement on the surface are defined in Eqs. (5.24). All the algebra is now carried out on the surface. Vectors and tensors are defined in the local, euclidian, flat tangent plane. All indices now run over the set $(1, 2)$.*

Motivated by our initial mechanics problem, and for purposes of visualization, we are focusing on two dimensions. The analysis is developed in such a way, however, that it is readily extended to n-dimensional non-euclidian curved spaces with n-dimensional local, euclidian tangent spaces. One simply lets the indices run over the set $(1, 2, \cdots, n)$. We shall subsequently employ the extension to $n = 4$.

We first review the covariant derivative from this perspective. We expand a vector field $\mathbf{v}(q^1, q^2)$ as

$$\mathbf{v} = v^i(q)\,\mathbf{e}_i = v_i(q)\,\mathbf{e}^i \qquad (5.49)$$

The change in the basis vectors as one moves to a neighboring point on the surface is given in terms of the affine connection by Eqs. (5.24) as

$$de_i = \Gamma^k_{ij} dq^j\,\mathbf{e}_k$$
$$de^i = -\Gamma^i_{kj} dq^j\,\mathbf{e}^k \qquad (5.50)$$

Just as before, the infinitesimal change in the vector field as one moves to a neighboring point is then given by

$$d\mathbf{v} = v^k{}_{;j} \, dq^j \, \mathbf{e}_k$$

$$v^k{}_{;j} = \frac{\partial v^k}{\partial q^j} + v^i \Gamma^k_{ij} \qquad \text{; covariant derivative} \qquad (5.51)$$

If the total change in a component of the vector $\mathbf{v}$ in the original basis as one moves to the neighboring point is denoted by $\mathcal{D}v^k$, then

$$d\mathbf{v} = (\mathcal{D}v^k) \, \mathbf{e}_k$$

$$\mathcal{D}v^k = dv^k + v^i \Gamma^k_{ij} dq^j$$

$$v^k{}_{;j} = \frac{\mathcal{D}v^k}{\partial q^j} \qquad (5.52)$$

Under a coordinate transformation on the surface, the covariant derivative forms a second-rank tensor transforming as

$$\bar{v}^k{}_{;j} = \bar{a}^k{}_{k'} \bar{a}_j{}^{j'} v^{k'}{}_{;j'} \qquad (5.53)$$

We now proceed to discuss the *second* covariant derivative. From Eq. (5.53) we know that the quantity

$$\underline{\underline{T}} = v^k{}_{;j} \, \mathbf{e}_k \, \mathbf{e}^j \qquad (5.54)$$

forms a true tensor. Compute its total differential

$$d\underline{\underline{T}} = \left(dv^k{}_{;j} \right) \mathbf{e}_k \, \mathbf{e}^j + v^k{}_{;j} \, d\mathbf{e}_k \, \mathbf{e}^j + v^k{}_{;j} \, \mathbf{e}_k \, d\mathbf{e}^j$$

$$= \left[\frac{\partial v^k{}_{;j}}{\partial q^l} dq^l \right] \mathbf{e}_k \, \mathbf{e}^j + v^k{}_{;j} \left[\Gamma^m_{kl} dq^l \, \mathbf{e}_m \right] \mathbf{e}^j + v^k{}_{;j} \, \mathbf{e}_k \left[-\Gamma^j_{lm} dq^l \, \mathbf{e}^m \right] (5.55)$$

Here the total differential of the function $v^k{}_{;j}(q)$ has been written out in the first term, and Eqs. (5.50) have been used in the last two. A change in dummy summation variables $(k, m) \to (i, k)$ in the second term and $(j, m) \to (i, j)$ in the third leads to

$$d\underline{\underline{T}} = \left[\frac{\partial v^k{}_{;j}}{\partial q^l} + v^i{}_{;j} \Gamma^k_{il} - v^k{}_{;i} \Gamma^i_{lj} \right] dq^l \, \mathbf{e}_k \, \mathbf{e}^j \qquad (5.56)$$

This provides the basis for the definition of the second covariant derivative as [note the further label change $(i \to m)$]

$$v^k_{;jl} \equiv \frac{\partial v^k_{;j}}{\partial q^l} + \Gamma^k_{ml} v^m_{;j} - \Gamma^m_{jl} v^k_{;m} \qquad ; \text{2nd covariant derivative} \quad (5.57)$$

Hence the total change in the *physical quantity* $\underline{\underline{T}}$ is given by

$$\begin{aligned}
d\underline{\underline{T}} &= d\left[v^k_{;j}\, \mathbf{e}_k\, \mathbf{e}^j \right] \\
&= v^k_{;jl}\, dq^l\, \mathbf{e}_k\, \mathbf{e}^j
\end{aligned} \qquad (5.58)$$

The quantity $v^k_{;jl}$ evidently transforms as a *third-rank tensor*.

$$\bar{v}^k_{;jl} = \bar{a}^k_{k'} \bar{a}_j^{\,j'} \bar{a}_l^{\,l'}\, v^{k'}_{;j'l'} \qquad (5.59)$$

The proof is identical to that given in Eqs. (5.12)–(5.18). Covariant differentiation takes us from a physical tensor of one rank to its differential which forms a tensor of one higher rank.

We now claim that on a curved surface, taking the second covariant derivative in opposite order *does not give the same result*. In fact, the difference is given by the *Riemann curvature tensor* through the following relation

$$v^k_{;jl} - v^k_{;lj} = R^k_{ilj} v^i \qquad ; \text{2nd covariant derivative} \qquad (5.60)$$

Note that this expression holds for any vector v^i. We proceed to prove this result.

Write out the left-hand side of the above using Eqs. (5.57) and (5.51)

$$\begin{aligned}
\text{l.h.s.} = &\frac{\partial}{\partial q^l}\left(\frac{\partial v^k}{\partial q^j} + \Gamma^k_{ij} v^i \right) + \Gamma^k_{ml}\left(\frac{\partial v^m}{\partial q^j} + \Gamma^m_{ij} v^i \right) - \Gamma^m_{jl}\left(\frac{\partial v^k}{\partial q^m} + \Gamma^k_{im} v^i \right) \\
&- \frac{\partial}{\partial q^j}\left(\frac{\partial v^k}{\partial q^l} + \Gamma^k_{il} v^i \right) - \Gamma^k_{mj}\left(\frac{\partial v^m}{\partial q^l} + \Gamma^m_{il} v^i \right) + \Gamma^m_{lj}\left(\frac{\partial v^k}{\partial q^m} + \Gamma^k_{im} v^i \right)
\end{aligned} \quad (5.61)$$

After the cancelation of identical terms this becomes

$$\begin{aligned}
\text{l.h.s.} &= \left[\left(\frac{\partial}{\partial q^l}\Gamma^k_{ij} + \Gamma^k_{ml}\Gamma^m_{ij} \right) - \left(\frac{\partial}{\partial q^j}\Gamma^k_{il} + \Gamma^k_{mj}\Gamma^m_{il} \right) \right] v^i \\
&= R^k_{ilj} v^i
\end{aligned} \qquad (5.62)$$

This is the desired result.

There is another way to write this expression. Start from

$$de_i = \Gamma_{ij}^k dq^j \, \mathbf{e}_k$$

or;
$$\frac{\partial \mathbf{e}_i}{\partial q^j} = \Gamma_{ij}^k \, \mathbf{e}_k \qquad (5.63)$$

Write the total differential of this result

$$d\left(\frac{\partial \mathbf{e}_i}{\partial q^j}\right) = (d\Gamma_{ij}^k)\mathbf{e}_k + \Gamma_{ij}^m(d\mathbf{e}_m)$$

$$= \left(\frac{\partial \Gamma_{ij}^k}{\partial q^l} + \Gamma_{ij}^m \Gamma_{ml}^k\right) \mathbf{e}_k \, dq^l \qquad (5.64)$$

Hence, after an obvious rearrangement

$$\frac{\partial}{\partial q^l}\left(\frac{\partial \mathbf{e}_i}{\partial q^j}\right) = \left(\frac{\partial \Gamma_{ij}^k}{\partial q^l} + \Gamma_{lm}^k \Gamma_{ij}^m\right) \mathbf{e}_k \qquad (5.65)$$

Thus

$$\frac{\partial}{\partial q^l}\left(\frac{\partial \mathbf{e}_i}{\partial q^j}\right) - \frac{\partial}{\partial q^j}\left(\frac{\partial \mathbf{e}_i}{\partial q^l}\right) = R^k{}_{ilj}\mathbf{e}_k \qquad (5.66)$$

If one goes back to the line element $d\mathbf{s} = \mathbf{e}_i \, dq^i$, then one can write this as

$$\mathbf{e}_i = \frac{\partial \mathbf{s}}{\partial q^i}$$

$$\frac{\partial^3 \mathbf{s}}{\partial q^l \partial q^j \partial q^i} - \frac{\partial^3 \mathbf{s}}{\partial q^j \partial q^l \partial q^i} = R^k{}_{ilj}\frac{\partial \mathbf{s}}{\partial q^k} \qquad (5.67)$$

The partial derivatives do not commute on the curved surface if the order of the derivatives is high enough. This is to be contrasted with the lower order relation on the non-euclidian surface which follows here from the symmetry of the affine connection[8]

$$\frac{\partial^2 \mathbf{s}}{\partial q^j \partial q^i} - \frac{\partial^2 \mathbf{s}}{\partial q^i \partial q^j} = 0 \qquad (5.68)$$

We can make use of Eq. (5.60) to deduce the transformation properties of the Riemann curvature tensor. We first prove the following result: *the*

---

[8] In the present development, Eq. (5.68) holds more generally in the three-dimensional euclidian space in which the two-dimensional non-euclidian surface is embedded, and this fact was used to establish the symmetry of the expansion coefficients γ_{ij}^k in the indices (ij) in chap. 3.

contraction of any pair of upper and lower indices reduces the rank of a tensor by two. The proof goes as follows:

Consider the tensor transformation law

$$\bar{T}^{ij} = \bar{a}^i{}_{i'} \bar{a}^j{}_{j'} T^{i'j'}$$

therefore ; $$\bar{T}^i{}_i = \bar{a}^i{}_{i'} \bar{a}_i{}^{j'} T^{i'}{}_{j'} \qquad (5.69)$$

Now use Eqs. (3.36) and (3.32)

$$\bar{a}^i{}_{i'} \bar{a}_i{}^{j'} = a^{j'}{}_i \bar{a}^i{}_{i'} = \delta^{j'}{}_{i'} \qquad (5.70)$$

Thus Eq. (5.69) becomes

$$\bar{T}^i{}_i = T^{i'}{}_{i'} \qquad (5.71)$$

Hence the contracted indices do not transform. This establishes the result.

Since we know that the second covariant derivative forms a third-rank tensor, and a vector forms a first-rank tensor, we conclude from Eq. (5.60) that *the Riemann curvature tensor transforms as a tensor of rank four.*[9]

5.4   Covariant Differentiation

Let us proceed to a more general discussion of covariant differentiation. Suppose one has a physical tensor of arbitrary rank with some set of contravariant and covariant indices

$$\underline{\underline{T}} = T^{abc\cdots}_{lmn\cdots} \, \mathbf{e}_a \mathbf{e}_b \mathbf{e}_c \cdots \mathbf{e}^l \mathbf{e}^m \mathbf{e}^n \cdots \qquad (5.72)$$

Write out the total differential

$$d\underline{\underline{T}} = \left(dT^{abc\cdots}_{lmn\cdots} \right) \mathbf{e}_a \mathbf{e}_b \cdots \mathbf{e}^l \mathbf{e}^m \cdots + T^{a'bc\cdots}_{lmn\cdots} (d\mathbf{e}_{a'}) \cdots$$
$$+ T^{abc\cdots}_{l'mn\cdots} (d\mathbf{e}^{l'}) \cdots + \cdots \qquad (5.73)$$

Now substitute Eqs. (5.50) for the change in basis vectors

$$d\underline{\underline{T}} = \left(dT^{abc\cdots}_{lmn\cdots} \right) \mathbf{e}_a \mathbf{e}_b \cdots \mathbf{e}^l \mathbf{e}^m \cdots + T^{a'bc\cdots}_{lmn\cdots} \left(\Gamma^a{}_{a'j} dq^j \, \mathbf{e}_a \right) \cdots$$
$$+ T^{abc\cdots}_{l'mn\cdots} \left(-\Gamma^{l'}_{lj} dq^j \mathbf{e}^l \right) \cdots + \cdots \qquad (5.74)$$

---

[9]See also Prob. 3.2.

Hence, using $\mathcal{D}T^{abc\cdots}_{lmn\cdots}$ to denote the total change in the components,

$$d\underline{\underline{T}} = \left(\mathcal{D}T^{abc\cdots}_{lmn\cdots}\right) \mathbf{e}_a\mathbf{e}_b\mathbf{e}_c \cdots \mathbf{e}^l\mathbf{e}^m\mathbf{e}^n \cdots$$

$$\mathcal{D}T^{abc\cdots}_{lmn\cdots} = dT^{abc\cdots}_{lmn\cdots} + \Gamma^a_{a'j}dq^j T^{a'bc\cdots}_{lmn\cdots} + \cdots$$

$$-\Gamma^{l'}_{lj}dq^j T^{abc\cdots}_{l'mn\cdots} - \cdots \tag{5.75}$$

There is one such additional positive term in the last expression for each contravariant index and one additional negative term for each covariant index on the tensor.

Now define the *covariant derivative of the tensor* as

$$T^{abc\cdots}_{lmn\cdots;k} \equiv \frac{\mathcal{D}T^{abc\cdots}_{lmn\cdots}}{\partial q^k} \qquad ; \text{ covariant derivative} \tag{5.76}$$

Again note the semicolon. We then know from our previous discussion that the differential of $\underline{\underline{T}}$ produces a physical tensor with a rank increased by one, and

$$d\underline{\underline{T}} = T^{abc\cdots}_{lmn\cdots;k}dq^k\mathbf{e}_a\mathbf{e}_b\mathbf{e}_c \cdots \mathbf{e}^l\mathbf{e}^m\mathbf{e}^n \cdots \tag{5.77}$$

As an example, consider the metric tensor with

$$\underline{\underline{g}} = g^{ij}\,\mathbf{e}_i\mathbf{e}_j$$

$$d\underline{\underline{g}} = \left(dg^{ij} + \Gamma^i_{ml}g^{mj}dq^l + \Gamma^j_{ml}g^{im}dq^l\right)\mathbf{e}_i\mathbf{e}_j \tag{5.78}$$

Now write out the total differential of the metric using Eqs. (5.50)

$$dg^{ij} = d\left(\mathbf{e}^i \cdot \mathbf{e}^j\right) = \mathbf{e}^i \cdot d\mathbf{e}^j + d\mathbf{e}^i \cdot \mathbf{e}^j$$

$$= -\Gamma^j_{ml}g^{im}dq^l - \Gamma^i_{ml}g^{jm}dq^l \tag{5.79}$$

We conclude that the expression in parentheses in the last line of Eq. (5.78) vanishes identically, and thus

$$d\underline{\underline{g}} = \left(\mathcal{D}g^{ij}\right)\mathbf{e}_i\mathbf{e}_j = 0$$

$$g^{ij}_{;k} = 0 \tag{5.80}$$

Hence we observe that *the metric tensor is constant under covariant differentiation.*

Covariant differentiation evidently obeys a *distributive law*, just like ordinary differentiation, for suppose one has the direct product of two tensors

$$T^{abc\cdots}_{ijk\cdots} = A^{ab\cdots}_{ij\cdots} \times B^{c\cdots}_{k\cdots} \tag{5.81}$$

then by the process leading to Eq. (5.75) one finds

$$\mathcal{D}T = A(\mathcal{D}B) + (\mathcal{D}A)B \tag{5.82}$$

It follows from Eq. (5.80) that one *can always take the metric in and out of covariant derivatives.*

These observations lead us to another useful result. After lowering the index on both sides of Eq. (5.60),[10] that result can be rewritten

$$v_{i;jk} - v_{i;kj} = R_{inkj}v^n \tag{5.83}$$

In this expression, replace

$$v_i \to w_i \equiv T_{il}v^l \tag{5.84}$$

Then

$$\left(T_{il}v^l\right)_{;jk} - \left(T_{il}v^l\right)_{;kj} = R_{inkj}\left(T^n{}_l v^l\right) \tag{5.85}$$

We can now proceed as in ordinary differentiation, only keeping careful track of the order of indices. Thus the l.h.s. of the above becomes

$$\begin{aligned}
\text{l.h.s.} = {}& T_{il;jk}v^l + T_{il;j}v^l{}_{;k} + T_{il;k}v^l{}_{;j} + T_{il}v^l{}_{;jk} \\
& -T_{il;kj}v^l - T_{il;k}v^l{}_{;j} - T_{il;j}v^l{}_{;k} - T_{il}v^l{}_{;kj}
\end{aligned} \tag{5.86}$$

After the cancelation of identical terms in this expression, Eq. (5.85) becomes

$$\left(T_{il;jk} - T_{il;kj}\right)v^l + T_{il}\left(v^l{}_{;jk} - v^l{}_{;kj}\right) = R_{inkj}\left(T^n{}_l v^l\right)$$

$$\text{or ;} \qquad \left(T_{il;jk} - T_{il;kj}\right)v^l + T_{il}\left(R^l{}_{nkj}v^n\right) = R_{inkj}\left(T^n{}_l v^l\right) \tag{5.87}$$

Here Eq. (5.83) has again been employed in the second term. Now interchange summation indices ($n \leftrightarrow l$) in that term, a pair of contracted indices, and use Eq. (5.48) to rewrite it as

$$T_{il}\left(R^l{}_{nkj}v^n\right) = R_{nlkj}(T_i{}^n v^l) = -R_{nljk}(T_i{}^n v^l) \tag{5.88}$$

We then observe that since Eq. (5.87) must hold for all v^l, one can equate the coefficients of this quantity. Thus

$$T_{il;jk} - T_{il;kj} = R_{inkj}T^n{}_l + R_{nljk}T_i{}^n \quad \text{; 2nd covariant derivative} \tag{5.89}$$

---

[10]Which is now permitted since the metric can be moved in and out of the covariant differentiation.

This provides a result for the *difference in the order of second covariant derivatives of a second-rank tensor*. Note that the first term on the r.h.s. involving the sum over the first index of the tensor is just what we would expect from Eq. (5.83). There is now the additional second term involving the sum over the second index; one just has to study it to keep track of where all the indices go.

A particular second-rank tensor is formed as the first covariant derivative of a vector, $T_{il} = v_{i;l}$. Let us apply Eq. (5.89) to this case, whence it becomes (with the first index raised and a pair of contracted indices interchanged)

$$v^i{}_{;ljk} - v^i{}_{;lkj} = R^i{}_{nkj} v^n{}_{;l} + R^n{}_{ljk} v^i{}_{;n} \quad ; \text{3rd covariant derivative} \quad (5.90)$$

This relation on the third covariant derivative will prove useful in our subsequent discussion of the Bianchi identities satisfied by the Riemann tensor.

5.5 Symmetry Properties of the Riemann Tensor

The *scalar product* of two vectors will be invariant as they both rotate during parallel transport around the curve L in Fig. 5.3.[11] Thus for an infinitesimal loop with $S^{lk} \to 0$

$$\Delta(u_i v^i) = u_i \, \Delta v^i + v^i \, \Delta u_i = 0 \tag{5.91}$$

Upon employing Eqs. (5.44) and appropriately raising, lowering, and relabeling indices, this expression becomes

$$\begin{aligned}
0 &= u_i \left(\frac{-v^n}{2} R^i{}_{nlk} S^{lk} \right) + v^i \left(\frac{-u_n}{2} R_i{}^n{}_{lk} S^{lk} \right) \\
&= -\frac{1}{2} u_m v^n \left(R^m{}_{nlk} + R_n{}^m{}_{lk} \right) S^{lk} \\
&= -\frac{1}{2} u^m v^n \left(R_{mnlk} + R_{nmlk} \right) S^{lk}
\end{aligned} \tag{5.92}$$

Since this result must hold for all (u, v, S), one concludes that the Riemann tensor must also be *antisymmetric* in the first two indices

$$R_{mnlk} = -R_{nmlk} \quad ; \text{antisymmetric in } (m \leftrightarrow n) \tag{5.93}$$

---

[11]Recall the scalar product $\mathbf{v} \cdot \mathbf{v} = \mathbf{v}^2$ is constant along a geodesic in our mechanics problem.

This can be combined with our previous result in Eq. (5.48)

$$R_{mnlk} = -R_{mnkl} \qquad ; \text{ antisymmetric in } (k \leftrightarrow l) \qquad (5.94)$$

to give

$$R_{mnlk} = R_{nmkl} \qquad ; \text{ symmetric in } (m \leftrightarrow n) \text{ and } (k \leftrightarrow l) \qquad (5.95)$$

5.6 Bianchi Identities

The *first Bianchi identity* satisfied by the Riemann tensor is

$$R^k_{ilj} + R^k_{lji} + R^k_{jil} = 0 \qquad ; \text{ cyclic permutations on } (ilj) \qquad (5.96)$$

The sum of the Riemann tensors with cyclic permutations on the last three (lower) indices vanishes. The proof of this relation follows by simply writing out these quantities using their definition in Eq. (5.47) and then adding them

$$R^k_{ilj} = \left(\frac{\partial}{\partial q^l} \Gamma^k_{ij} + \Gamma^k_{ml} \Gamma^m_{ij} \right) - \left(\frac{\partial}{\partial q^j} \Gamma^k_{il} + \Gamma^k_{mj} \Gamma^m_{il} \right)$$

$$R^k_{lji} = \left(\frac{\partial}{\partial q^j} \Gamma^k_{li} + \Gamma^k_{mj} \Gamma^m_{li} \right) - \left(\frac{\partial}{\partial q^i} \Gamma^k_{jl} + \Gamma^k_{mi} \Gamma^m_{lj} \right)$$

$$R^k_{jil} = \left(\frac{\partial}{\partial q^i} \Gamma^k_{jl} + \Gamma^k_{mi} \Gamma^m_{jl} \right) - \left(\frac{\partial}{\partial q^l} \Gamma^k_{ji} + \Gamma^k_{ml} \Gamma^m_{ji} \right) \qquad (5.97)$$

If these expressions are added together, the sum of the terms on the r.h.s vanishes leading to Eq. (5.96).[12]

The *second Bianchi identity* involves the covariant derivative of the Riemann tensor and is given by

$$R^{in}_{kj;l} + R^{in}_{lk;j} + R^{in}_{jl;k} = 0 \qquad ; \text{ cyclic permutations on } (kjl) \qquad (5.98)$$

The proof goes as follows. Equation (5.83) reads

$$v_{i;jk} - v_{i;kj} = R_{inkj} v^n \qquad (5.99)$$

Take the *l*th covariant derivative of this relation

$$v_{i;jkl} - v_{i;kjl} = R_{inkj;l} v^n + R_{inkj} v^n_{;l} \qquad (5.100)$$

---

[12]We again make use of the symmetry of the affine connection in its lower two indices.

Now add the two equations with the dummy indices changed by the remaining two cyclic permutations of (jkl)

$$v_{i;klj} - v_{i;lkj} = R_{inlk;j}\, v^n + R_{inlk}\, v^n{}_{;j}$$
$$v_{i;ljk} - v_{i;jlk} = R_{injl;k}\, v^n + R_{injl}\, v^n{}_{;k} \tag{5.101}$$

Use the previous expression for the difference of third covariant derivatives in Eq. (5.90) (with the index i lowered) on the three combinations appearing on the l.h.s. Thus

$$\text{l.h.s.} = \left(R_{inlk}\, v^n{}_{;j} + R^n{}_{jkl}\, v_{i;n}\right) + \left(R_{injl}\, v^n{}_{;k} + R^n{}_{klj}\, v_{i;n}\right) +$$
$$\left(R_{inkj}\, v^n{}_{;l} + R^n{}_{ljk}\, v_{i;n}\right) \tag{5.102}$$

Now cancel the common terms on the l.h.s. and r.h.s and use the first identity in Eq. (5.96) on the l.h.s. What remains is

$$0 = \left(R_{inkj;l} + R_{inlk;j} + R_{injl;k}\right) v^n \tag{5.103}$$

Since this must hold for all v^n, the expression in parentheses must vanish. Now raise the first two indices,[13] and the result is Eq. (5.98).

5.7   The Einstein Tensor

The *Ricci tensor* is the second-rank tensor formed by contracting the first and third indices of the Riemann curvature tensor[14]

$$R_{ij} \equiv R^k{}_{ikj} \qquad ; \text{ Ricci tensor} \tag{5.104}$$

The *scalar curvature* of the space is defined through the contraction of the Ricci tensor [recall Eq. (5.95)]

$$R \equiv R^i{}_i = R^{ki}{}_{ki} \qquad ; \text{ scalar curvature} \tag{5.105}$$

The *Einstein tensor* will play a central role in the subsequent development. It is defined in terms of these two quantities and the metric as

$$G_{ij} \equiv R_{ij} - \frac{1}{2}g_{ij}R \qquad ; \text{ Einstein tensor} \tag{5.106}$$

---

[13] This simply gives an easy way to remember the identity.

[14] Recall, as previously demonstrated in Eq. (3.25), that one can always interchange a pair of contracted indices $R^i{}_{jik} = R_{ij}{}^i{}_k$, etc.

We first show that the Ricci tensor is *symmetric*, which immediately implies that the Einstein tensor is also symmetric

$$R_{jl} = R_{lj}$$
$$\Rightarrow \qquad G_{ij} = G_{ji} \qquad ; \text{ symmetric} \qquad (5.107)$$

The proof goes as follows. Start from

$$R^i_{\ ijk} = g^{mi} R_{mijk} = -g^{im} R_{imjk} = -g^{mi} R_{mijk}$$
$$\text{or} ; \qquad R^i_{\ ijk} = -R^i_{\ ijk} = 0 \qquad (5.108)$$

Here the symmetry of the metric, the antisymmetry of the Riemann tensor in its first two indices Eq. (5.93), and a relabeling of dummy indices have been used in the first line. Now use the first Bianchi identity in Eq. (5.96)

$$R^i_{\ jkl} + R^i_{\ klj} + R^i_{\ ljk} = 0 \qquad (5.109)$$

Set $i = k$ and sum, and use Eq. (5.108)

$$R^i_{\ jil} + R^i_{\ ilj} + R^i_{\ lji} = 0$$
$$\text{or} ; \qquad\qquad R^i_{\ jil} = -R^i_{\ lji} = R^i_{\ lij} \qquad (5.110)$$

Here the antisymmetry of the Riemann tensor in the last two indices in Eq. (5.94) has been used in obtaining the last result, which then gives Eqs. (5.107).

A central property of the Einstein tensor is that its *covariant divergence vanishes*[15]

$$G^{ij}_{\ ;i} = 0 \qquad\qquad ; \text{ divergenceless} \qquad (5.111)$$

The proof of this relation starts from the second Bianchi identity in Eq. (5.98)

$$R^{in}_{\ kj;l} + R^{in}_{\ jl;k} + R^{in}_{\ lk;j} = 0 \qquad (5.112)$$

Set $i = k$ and sum, use Eq. (5.94) again, and identify the Ricci tensor of Eq. (5.104)

$$R^{in}_{\ ij;l} + R^{in}_{\ jl;i} + R^{in}_{\ li;j} = 0$$
$$R^n_{\ j;l} + R^{in}_{\ jl;i} - R^n_{\ l;j} = 0 \qquad (5.113)$$

---

[15] Since the Einstein tensor is symmetric, Eq. (5.111) is equivalent to $G^{ji}_{\ ;i} = 0$.

Now set $n = l$ and sum, use Eq. (5.95), and again identify the Ricci tensor[16]

$$R^n{}_{j;n} + R^{ni}{}_{nj;i} - R^n{}_{n;j} = 0$$
$$R^n{}_{j;n} + R^i{}_{j;i} - R^n{}_{n;j} = 0 \qquad (5.114)$$

The first two terms are identical, and the last relation can therefore be rewritten as

$$R^i{}_{j;i} - \frac{1}{2} R^n{}_{n;j} = 0$$

$$R^i{}_{j;i} - \frac{1}{2} \left(R\delta^i{}_j \right)_{;i} = 0$$

$$\left(R^i{}_j - \frac{1}{2} R\delta^i{}_j \right)_{;i} = 0$$

or ; $\qquad\qquad G^i{}_{j;i} = 0 \qquad (5.115)$

The last line follows since[17]

$$G^{ij} = R^{ij} - \frac{1}{2} Rg^{ij}$$

$$G^i{}_j = R^i{}_j - \frac{1}{2} R\delta^i{}_j \qquad (5.116)$$

where we have used

$$g^i{}_j = \left(\mathbf{e}^i \cdot \mathbf{e}_j \right) = \delta^i{}_j \qquad (5.117)$$

Equation (5.115) establishes the result that the Einstein tensor is divergenceless.

There is one additional symmetry property of the Riemann tensor that follows from the first Bianchi identity in Eq. (5.96). Write it four times, simply relabeling the dummy indices each time (first index i, then first index j, etc.)

$$R_{ijkl} + R_{iklj} + R_{iljk} = 0$$
$$R_{jkli} + R_{jlik} + R_{jikl} = 0$$
$$R_{klij} + R_{kijl} + R_{kjli} = 0$$
$$R_{lijk} + R_{ljki} + R_{lkij} = 0 \qquad (5.118)$$

---

[16]Note $R^{in}{}_{jn;i} = R^{ni}{}_{nj;i}$.

[17]Note that the factor of $1/2$ in this expression arises from the presence of three terms in the Bianchi identities and has nothing to do with the *dimension* of the problem (here $D = 2$).

Now add these equations using the established symmetry properties in Eqs. (5.93)–(5.95)

$$2\left(R_{iklj} + R_{jlik}\right) = 0$$

$$\text{or ;} \qquad R_{iklj} = R_{ljik} \qquad\qquad (5.119)$$

A relabeling of indices makes this result a little more transparent

$$R_{ijkl} = R_{klij} \qquad ; \text{ symmetric in } (ij) \leftrightarrow (kl) \qquad (5.120)$$

The Riemann tensor is also symmetric in the interchange of the first and second *pair* of indices.[18]

We started with the mechanics problem of a particle moving without friction on an arbitrary two-dimensional surface and have arrived, through this tensor analysis, at *riemannian geometry*. All quantities are now defined in the surface with dimension $D = 2$, or in its flat, local tangent plane. The curvature of the surface is measured through the parallel transport of a vector around a closed loop lying in the surface, and is given by the rank-four Riemann tensor. This, in turn, is given in terms of the affine connection. The Riemann tensor is a second order, nonlinear, differential form in the metric. The divergenceless Einstein tensor is obtained through a contraction of the Riemann tensor.

The great beauty of riemannian geometry is that it can now be generalized from a curved space with dimension $D = 2$ to a space with dimension $D = n$ simply by extending the indices from the set $i = (1, 2)$ to the set $i = (1, 2, \cdots, n)$.[19] There will correspondingly be a local, n-dimensional euclidian tangent space.

We proceed to an example.

5.8   Example—Surface of a Sphere

Suppose one lives on the surface of a sphere as illustrated in Fig. 5.4.[20] Introduce the usual spherical coordinates (r, θ, ϕ), and let $(\hat{\mathbf{e}}_r, \hat{\mathbf{e}}_\theta, \hat{\mathbf{e}}_\phi)$ form the usual set of orthonormal spherical *unit vectors*. The line element in

---

[18]We leave it as a problem to show that the symmetry of the Ricci tensor follows directly from this result.

[19]Visualization, unfortunately, becomes increasingly more difficult.

[20]This does not require too much stretch of the imagination.

spherical coordinates is then given by

$$ds = dr\,\hat{\mathbf{e}}_r + rd\theta\,\hat{\mathbf{e}}_\theta + r\sin\theta d\phi\,\hat{\mathbf{e}}_\phi \qquad ; \text{ spherical coordinates} \quad (5.121)$$

We confine ourselves to the surface of the sphere with the condition $r =$ constant. The generalized coordinates on the surface of the sphere are then $(q^1, q^2) = (\theta, \phi)$. The line element on the surface of the sphere is written in terms of the *basis vectors* as

$$d\mathbf{s} = \mathbf{e}_\theta\,d\theta + \mathbf{e}_\phi\,d\phi \qquad ; \text{ surface of sphere}$$

$$\mathbf{e}_\theta = r\,\hat{\mathbf{e}}_\theta \qquad ; \mathbf{e}_\phi = r\sin\theta\,\hat{\mathbf{e}}_\phi \qquad\qquad (5.122)$$

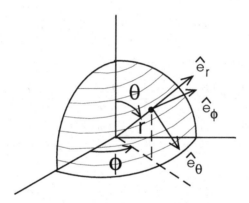

Fig. 5.4 Two-dimensional curved space on the surface of a sphere with usual spherical coordinates (r, θ, ϕ) and orthonormal spherical unit vectors $(\hat{\mathbf{e}}_r, \hat{\mathbf{e}}_\theta, \hat{\mathbf{e}}_\phi)$. The surface of the sphere is defined by $r =$ constant, and the generalized coordinates on the surface are $(q^1, q^2) = (\theta, \phi)$.

The square of the line element allows us to identify the metric and its inverse

$$(d\mathbf{s})^2 = (rd\theta)^2 + (r\sin\theta d\phi)^2$$

$$\underline{g} = \begin{bmatrix} g_{\theta\theta} & g_{\theta\phi} \\ g_{\phi\theta} & g_{\phi\phi} \end{bmatrix} = \begin{bmatrix} r^2 & 0 \\ 0 & r^2\sin^2\theta \end{bmatrix}$$

$$\underline{g}^{-1} = \begin{bmatrix} g^{\theta\theta} & g^{\theta\phi} \\ g^{\phi\theta} & g^{\phi\phi} \end{bmatrix} = \begin{bmatrix} 1/r^2 & 0 \\ 0 & 1/r^2\sin^2\theta \end{bmatrix} \qquad (5.123)$$

We first compute the affine connection given by

$$\Gamma^k_{ij} = \frac{1}{2}g^{km}\left[\frac{\partial g_{mi}}{\partial q^j} + \frac{\partial g_{mj}}{\partial q^i} - \frac{\partial g_{ij}}{\partial q^m}\right] \tag{5.124}$$

The indices run over the set (θ, ϕ). Now note

- The metric and its inverse are diagonal;
- The only functional dependence in the metric is $g_{\phi\phi}(\theta)$;
- This implies one must have two ϕ's and one θ for a non-zero result.

Thus, for example,

$$\Gamma^\theta_{\theta\theta} = \Gamma^\phi_{\phi\phi} = 0$$

$$\Gamma^\theta_{\phi\theta} = \Gamma^\theta_{\theta\phi} = \frac{1}{2}g^{\theta\theta}\left[\frac{\partial g_{\theta\phi}}{\partial \theta} + \frac{\partial g_{\theta\theta}}{\partial \phi} - \frac{\partial g_{\phi\theta}}{\partial \theta}\right] = 0$$

$$\Gamma^\phi_{\theta\theta} = \frac{1}{2}g^{\phi\phi}\left[\frac{\partial g_{\phi\theta}}{\partial \theta} + \frac{\partial g_{\phi\theta}}{\partial \theta} - \frac{\partial g_{\theta\theta}}{\partial \phi}\right] = 0 \tag{5.125}$$

The non-zero elements are computed as

$$\Gamma^\theta_{\phi\phi} = \frac{1}{2}g^{\theta\theta}\left[\frac{\partial g_{\theta\phi}}{\partial \phi} + \frac{\partial g_{\theta\phi}}{\partial \phi} - \frac{\partial g_{\phi\phi}}{\partial \theta}\right] = -\frac{1}{2r^2}\frac{\partial}{\partial \theta}(r^2\sin^2\theta)$$

$$\Gamma^\phi_{\theta\phi} = \Gamma^\phi_{\phi\theta} = \frac{1}{2}g^{\phi\phi}\left[\frac{\partial g_{\phi\theta}}{\partial \phi} + \frac{\partial g_{\phi\phi}}{\partial \theta} - \frac{\partial g_{\theta\phi}}{\partial \phi}\right] = \frac{1}{2r^2\sin^2\theta}\frac{\partial}{\partial \theta}(r^2\sin^2\theta) \tag{5.126}$$

Hence the only non-zero elements for the affine connection on the surface of the sphere are given by

$$\Gamma^\theta_{\phi\phi} = -\sin\theta\cos\theta$$

$$\Gamma^\phi_{\theta\phi} = \Gamma^\phi_{\phi\theta} = \frac{\cos\theta}{\sin\theta} \tag{5.127}$$

We next compute the Riemann tensor. Since it is antisymmetric in both the first and second pair of indices, these indices must be different to get a non-zero result.[21] Thus the only non-vanishing element of the Riemann tensor is $R_{\theta\phi\theta\phi}$ (as well as those elements obtained from it by the symmetry properties). Recall the definition of the Riemann tensor

$$R^i_{jkl} = \frac{\partial}{\partial q^k}\Gamma^i_{jl} + \Gamma^i_{mk}\Gamma^m_{jl} - (k \leftrightarrow l) \tag{5.128}$$

---

[21]Since, for example, $R_{11kl} = -R_{11kl} = 0$.

The first index is lowered with the aid of the metric, and therefore

$$R_{\theta\phi\theta\phi} = g_{\theta\theta}R^{\theta}{}_{\phi\theta\phi}$$

$$= r^2\left[\left(\frac{\partial}{\partial\theta}\Gamma^{\theta}_{\phi\phi} + \Gamma^{\theta}_{n\theta}\Gamma^{n}_{\phi\phi}\right) - \left(\frac{\partial}{\partial\phi}\Gamma^{\theta}_{\phi\theta} + \Gamma^{\theta}_{n\phi}\Gamma^{n}_{\phi\theta}\right)\right]$$

$$= r^2\left[-\frac{1}{2}\frac{\partial}{\partial\theta}\sin 2\theta - (-\sin\theta\cos\theta)\frac{\cos\theta}{\sin\theta}\right]$$

$$= r^2[-\cos 2\theta + \cos^2\theta]$$

$$= r^2\sin^2\theta \tag{5.129}$$

Thus

$$R_{\theta\phi\theta\phi} = r^2\sin^2\theta = R_{\phi\theta\phi\theta} \qquad ; \text{ surface of sphere} \tag{5.130}$$

where the last relation follows from Eq. (5.95).

The indices can be raised with the metric, and therefore

$$R^{\theta}{}_{\phi\theta\phi} = g^{\theta\theta}R_{\theta\phi\theta\phi}$$

$$= \frac{1}{r^2}r^2\sin^2\theta = \sin^2\theta$$

$$R^{\phi}{}_{\theta\phi\theta} = g^{\phi\phi}R_{\phi\theta\phi\theta}$$

$$= \frac{1}{r^2\sin^2\theta}r^2\sin^2\theta = 1 \tag{5.131}$$

The non-zero elements of the Ricci tensor follow from these relations as

$$R_{ij} = R^{k}{}_{ikj}$$

$$R_{\phi\phi} = R^{\theta}{}_{\phi\theta\phi} = \sin^2\theta$$

$$R_{\theta\theta} = R^{\phi}{}_{\theta\phi\theta} = 1 \qquad ; \text{ surface of sphere} \tag{5.132}$$

The indices are again raised with the metric

$$R^{\phi}{}_{\phi} = g^{\phi\phi}R_{\phi\phi} = \frac{1}{r^2}$$

$$R^{\theta}{}_{\theta} = g^{\theta\theta}R_{\theta\theta} = \frac{1}{r^2} \tag{5.133}$$

The scalar curvature is given by the sum of these two terms

$$R = R^{i}{}_{i} = \frac{2}{r^2} \qquad ; \text{ surface of sphere} \tag{5.134}$$

The surface of a sphere is a *space of constant curvature*, and the curvature goes to zero as the radius of the sphere goes to infinity.

It is interesting that the Einstein tensor *vanishes* on the surface of a sphere since

$$G^{\theta}{}_{\theta} = R^{\theta}{}_{\theta} - \frac{1}{2} R = 0$$

$$G^{\phi}{}_{\phi} = R^{\phi}{}_{\phi} - \frac{1}{2} R = 0 \qquad ; \text{surface of sphere} \qquad (5.135)$$

5.9   Volume element

For many subsequent developments we will require the *physical area* on the surface corresponding to the differential element $dq^1 dq^2$ in generalized coordinates (more generally, the differential volume). To lowest order in differentials, this area is *identical to that in the tangent plane.* Let us therefore again focus on the euclidian tangent space.

Two vectors **a** and **b** define an associated parallelogram as indicated in Fig. 5.5(a). The area of this parallelogram is (length of base)×(height), and therefore

$$\text{area} = ab \sin \theta = |\mathbf{a} \times \mathbf{b}| \qquad (5.136)$$

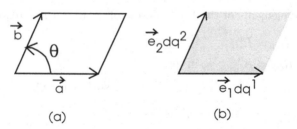

(a)   (b)

Fig. 5.5   Area element in tangent plane in two dimensions: (a) General parallelogram defined by two vectors **a** and **b**. The length of the base is a, the height is $b \sin \theta$, and the area is $ab \sin \theta$; (b) Differential physical area defined by infinitesimal displacements $\mathbf{e}_1 \, dq^1$ and $\mathbf{e}_2 \, dq^2$.

It follows that the differential area defined by the infinitesimal physical displacement vectors $\mathbf{e}_1 dq^1$ and $\mathbf{e}_2 dq^2$ in the tangent plane and shown in Fig. 5.5(b) is given by

$$dA = |\mathbf{e}_1 \times \mathbf{e}_2| \, dq^1 dq^2 \qquad (5.137)$$

The vector expression can be transformed in the following manner

$$|\mathbf{e}_1 \times \mathbf{e}_2| = [(\mathbf{e}_1 \times \mathbf{e}_2) \cdot (\mathbf{e}_1 \times \mathbf{e}_2)]^{1/2}$$
$$= \left[(\mathbf{e}_1 \cdot \mathbf{e}_1)(\mathbf{e}_2 \cdot \mathbf{e}_2) - (\mathbf{e}_1 \cdot \mathbf{e}_2)^2\right]^{1/2}$$
$$= \left[g_{11}g_{22} - g_{12}^2\right]^{1/2} \tag{5.138}$$

Here the metric has again been identified.

The metric matrix $\underline{g}$ is given by

$$\underline{g} = \begin{bmatrix} g_{11} & g_{12} \\ g_{21} & g_{22} \end{bmatrix} \qquad ; \; g_{12} = g_{21} \tag{5.139}$$

The *determinant* of this matrix is

$$\det \underline{g} = \begin{vmatrix} g_{11} & g_{12} \\ g_{21} & g_{22} \end{vmatrix} = g_{11}g_{22} - g_{12}g_{21}$$
$$= g_{11}g_{22} - g_{12}^2 \tag{5.140}$$

These results can be combined to rewrite Eq. (5.137) as

$$dA = \sqrt{g}\, dq^1 dq^2 \qquad ; \text{ physical area}$$
$$g \equiv \det \underline{g} \tag{5.141}$$

We emphasize that

- *This is the physical area in the tangent plane, identical to this order to that in the surface, defined by the infinitesimal element $dq^1 dq^2$ in generalized coordinates;*
- *This expression is exact to second order in infinitesimals.*

Let us attempt to generalize this analysis to a three-dimensional euclidian tangent space. The physical displacements $(\mathbf{e}_1 dq^1, \mathbf{e}_2 dq^2, \mathbf{e}_3 dq^3)$ define a parallelepiped in three dimensions as illustrated in Fig. 5.6. The little element of area defined by the first two displacements can be written as a *vector*, which is normal to the plane defined by these two vectors and has the correct magnitude

$$d\mathbf{A}_{12} = (\mathbf{e}_1 \times \mathbf{e}_2)\, dq^1 dq^2 \tag{5.142}$$

Since the volume of a parallelepiped is given by (area of base)×(height), the little element of volume defined by these three displacements is

$$dV = d\mathbf{A}_{12} \cdot (\mathbf{e}_3 \, dq^3)$$
$$= \mathbf{e}_3 \cdot (\mathbf{e}_1 \times \mathbf{e}_2) \, dq^1 dq^2 dq^3$$
$$= \mathbf{e}_1 \cdot (\mathbf{e}_2 \times \mathbf{e}_3) \, dq^1 dq^2 dq^3 \tag{5.143}$$

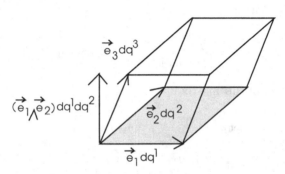

Fig. 5.6 Differential volume element in tangent space in three dimensions defined by infinitesimal displacements $(\mathbf{e}_1 \, dq^1, \mathbf{e}_2 \, dq^2, \mathbf{e}_3 \, dq^3)$.

The vector expression appearing in Eq. (5.143) can be rewritten as

$$\mathbf{e}_1 \cdot (\mathbf{e}_2 \times \mathbf{e}_3) = \sum_{i_1=1}^{3} \sum_{i_2=1}^{3} \sum_{i_3=1}^{3} \varepsilon_{i_1 i_2 i_3} (\mathbf{e}_1)_{i_1} (\mathbf{e}_2)_{i_2} (\mathbf{e}_3)_{i_3} \tag{5.144}$$

Here the *cartesian components* of the basis vectors have been introduced, and the quantity $\varepsilon_{i_1 i_2 i_3}$ is the *completely antisymmetric tensor* in three dimensions defined by

$$\varepsilon_{i_1 i_2 i_3} = +1 \qquad ; \text{ if } (i_1 i_2 i_3) \text{ is an *even* permutation of (123)}$$
$$= -1 \qquad ; \text{ if } (i_1 i_2 i_3) \text{ is an *odd* permutation of (123)}$$
$$= 0 \qquad ; \text{ otherwise} \tag{5.145}$$

Now define the *modal matrix* formed from the basis vectors as

$$e_{ij} \equiv [\mathbf{e}_j]_i \qquad ; \; \underline{e} = \begin{bmatrix} \mathbf{e}_1 & \mathbf{e}_2 & \mathbf{e}_3 \\ \downarrow & \downarrow & \downarrow \end{bmatrix} \tag{5.146}$$

Consider the matrix product

$$\left[\underline{e}^T \underline{e}\right]_{ij} = \sum_{k=1}^{3} \left[\underline{e}^T\right]_{ik} [\underline{e}]_{kj} = \sum_{k=1}^{3} [\mathbf{e}_i]_k [\mathbf{e}_j]_k$$
$$= \mathbf{e}_i \cdot \mathbf{e}_j$$
$$= g_{ij} \tag{5.147}$$

Thus we are able to write the metric matrix in three dimensions in *factored form* as

$$\underline{g} = \underline{e}^T \underline{e} \tag{5.148}$$

Use of the properties of determinants then gives

$$\det \underline{g} = \det \underline{e}^T \underline{e}$$
$$= \left(\det \underline{e}^T\right)\left(\det \underline{e}\right) = \left(\det \underline{e}\right)^2 \tag{5.149}$$

Now the expression in Eq. (5.144) is just the determinant of $\underline{e}$, and hence Eq. (5.143) becomes

$$dV = \mathbf{e}_1 \cdot (\mathbf{e}_2 \times \mathbf{e}_3)\, dq^1 dq^2 dq^3$$
$$= (\det \underline{e})\, dq^1 dq^2 dq^3 \tag{5.150}$$

Hence

$$dV = \sqrt{g}\, dq^1 dq^2 dq^3 \qquad ;\text{ physical volume}$$
$$g \equiv \det \underline{g} \tag{5.151}$$

The expressions in Eqs. (5.141) and (5.151) suggest that while the derivation is not readily generalizable, the result is

$$d^{(n)}\tau = \sqrt{g}\, dq^1 \cdots dq^n \qquad ;\ n\text{-dimensional volume element}$$
$$g \equiv \det \underline{g} \tag{5.152}$$

The factor relating the physical volume element to the volume element in the space of the generalized coordinates is the square root of the determinant of the metric matrix. The general proof of this result involves the introduction and manipulation of "p-forms" in n-dimensional euclidian space, and we are content here to give a reference to that material.[22] One can at least show that the expression in Eq. (5.152) is invariant under a change in choice of generalized coordinates, as it must be if it is to represent

---

[22]See, for example, [Abraham, Marsden, and Ratiu (1993)].

a physical volume element. The reader is guided through a proof of the fact that $d^{(n)}\tau$ is a scalar under coordinate transformations in Prob. 5.12.

Since determinants now play a crucial role, let us review some *linear algebra*. Consider a $n \times n$ matrix $\underline{m}$ of the form

$$\underline{m} = [m_{ij}] = \begin{bmatrix} m_{11} & m_{12} & \cdots & m_{1n} \\ \vdots & & & \vdots \\ m_{n1} & \cdots & & m_{nn} \end{bmatrix} \tag{5.153}$$

Here the first index i labels the row, and the second index j the column. The *cofactor* A_{ij} is the determinant formed from the $(n-1) \times (n-1)$ matrix where the ith row and jth column have been deleted

$$A_{ij} \equiv \det \begin{bmatrix} & \text{strike}\,(j) \\ & \downarrow \\ \text{strike}\,(i) & \rightarrow \end{bmatrix} \qquad ; \text{ cofactor} \tag{5.154}$$

The expansion in minors of a determinant then states that

$$\det \underline{m} = \sum_{j=1}^{n} m_{ij}(-1)^{i+j} A_{ij} \qquad ; i \text{ any row}$$

$$= \sum_{i=1}^{n} m_{ij}(-1)^{i+j} A_{ij} \qquad ; j \text{ any column} \tag{5.155}$$

The determinant of $\underline{m}$ is obtained by going along any row (or any column), taking the element m_{ij} and multiplying by the cofactor, which is a determinant of the $(n-1) \times (n-1)$ matrix obtained by striking the ith row and jth column, and then affixing a sign $(-1)^{i+j}$ to each term.

In addition to providing a systematic method for evaluating large determinants, the expansion in minors has the great advantage that it *makes the dependence of the determinant on the elements m_{ij} explicit*. A further property of determinants that we shall use is that the determinant vanishes if any two rows, or any two columns are identical.

The *inverse matrix* is defined through the relation[23]

$$\sum_{j=1}^{n} [\underline{m}^{-1}]_{ij} [\underline{m}]_{jk} = \delta_{ik} \tag{5.156}$$

---

[23]Here δ_{ik} is the usual Kronecker delta ($\delta_{ik} = 1$ if $i = k$, and $\delta_{ik} = 0$ if $i \neq k$).

It is given explicitly by

$$[\underline{m}^{-1}]_{ij} = \frac{(-1)^{i+j}A_{ji}}{\det \underline{m}} \quad ; \text{ inverse} \tag{5.157}$$

Note carefully the order of the indices in this expression. If this expression is substituted into Eq. (5.156) one has

$$\sum_{j=1}^{n}[\underline{m}^{-1}]_{ij}[\underline{m}]_{jk} = \sum_{j=1}^{n} \frac{m_{jk}(-1)^{i+j}A_{ji}}{\det \underline{m}}$$

$$= 1 \quad ; \text{ if } i = k$$

$$= 0 \qquad \text{if } i \neq k \tag{5.158}$$

The second line follows from the expansion in minors of Eq. (5.155). The last line holds since the cofactor from the ith column A_{ji} now *includes* the column m_{jk} if $i \neq k$. In this case one is computing the determinant of an $n \times n$ matrix where both the ith and kth columns are m_{jk} ($m_{ji} = m_{jk}$ for all j), and the determinant of a matrix with two identical columns vanishes as noted above.

Since Eq. (5.155) provides the functional dependence of the determinant $m \equiv \det \underline{m}$ on each of its elements

$$m \equiv \det \underline{m} = m(m_{ij}) \quad ; \ i,j = 1, \cdots, n \tag{5.159}$$

it may be differentiated to give

$$\frac{\partial m}{\partial m_{ij}} = (-1)^{i+j}A_{ij} = [\underline{m}^{-1}]_{ji}\,m \tag{5.160}$$

The *total differential* of the determinant is given by

$$dm = \sum_{i=1}^{n}\sum_{j=1}^{n} \frac{\partial m}{\partial m_{ij}}dm_{ij} \tag{5.161}$$

Substitution of Eq. (5.160) then gives for the total differential of the determinant

$$\frac{dm}{m} = \sum_{i=1}^{n}\sum_{j=1}^{n}[\underline{m}^{-1}]_{ji}\,dm_{ij} \quad ; \ m \equiv \det \underline{m} \tag{5.162}$$

Let us apply this last relation to the *metric matrix*

$$\frac{dg}{g} = \sum_{i=1}^{n} \sum_{j=1}^{n} [\underline{g}^{-1}]_{ji} dg_{ij}$$

$$= g^{ji} \, dg_{ij} \qquad\qquad ; \, g \equiv \det \underline{g} \qquad (5.163)$$

Here the relation between the inverse matrix and the metric in the reciprocal basis has been employed in the second equality [see Eq. (2.21)], and we now return to our summation convention on repeated indices. Equation (2.20) and its differential also give

$$g^{ij} g_{jk} = \delta^{i}{}_{k}$$

$$g_{jk} \, dg^{ij} + g^{ij} \, dg_{jk} = 0 \qquad (5.164)$$

Now set $k = i$ and sum

$$g^{ij} \, dg_{ji} = -g_{ji} \, dg^{ij} \qquad\qquad (5.165)$$

Hence Eq. (5.163) can also be rewritten [one can always interchange dummy summation indices $(i \leftrightarrow j)$]

$$\frac{dg}{g} = -g_{ij} \, dg^{ji} \qquad\qquad ; \, g \equiv \det \underline{g} \qquad (5.166)$$

As one consequence of these observations, let us return to the definition of the affine connection

$$\Gamma_{ij}^{k} = \frac{1}{2} g^{km} \left[\frac{\partial g_{mi}}{\partial q^{j}} + \frac{\partial g_{mj}}{\partial q^{i}} - \frac{\partial g_{ij}}{\partial q^{m}} \right] \qquad (5.167)$$

Now set $j = k$ and sum

$$\Gamma_{ik}^{k} = \frac{1}{2} g^{km} \left[\frac{\partial g_{mi}}{\partial q^{k}} + \frac{\partial g_{mk}}{\partial q^{i}} - \frac{\partial g_{ik}}{\partial q^{m}} \right] \qquad (5.168)$$

Since k and m are again just dummy summation indices, they may be interchanged and the first term just cancels the last term[24]

$$g^{km} \frac{\partial g_{mi}}{\partial q^{k}} = g^{km} \frac{\partial g_{ki}}{\partial q^{m}} \qquad (5.169)$$

---

[24]Once again we employ the symmetry of the metric, and later, of the affine connection.

Now take Eq. (5.163), relabel the dummy summation indices $(j, i) \rightarrow (k, m)$, divide by the increment dq^i, and keep all the other coordinates fixed. Substitution of the result into Eq. (5.168) then gives

$$\Gamma^k_{ik} = \frac{1}{2} g^{km} \frac{\partial g_{mk}}{\partial q^i} = \frac{1}{2g} \frac{\partial g}{\partial q^i}$$

$$\Gamma^k_{ik} = \frac{\partial}{\partial q^i} \ln \sqrt{g} \qquad \qquad \text{; affine connection} \qquad (5.170)$$

This expression for the affine connection with a pair of contracted indices will allow an elegant rewriting of the covariant divergence.

Recall the definition of the *covariant derivative* of a vector

$$v^l_{\;;m} = \frac{\partial v^l}{\partial q^m} + \Gamma^l_{mp} v^p \qquad (5.171)$$

The *covariant divergence* is then obtained by contracting the indices (l, m) and making use of Eq. (5.170)

$$
\begin{aligned}
v^l_{\;;l} &= \frac{\partial v^l}{\partial q^l} + \Gamma^l_{pl} v^p \\
&= \frac{\partial v^l}{\partial q^l} + \left(\frac{\partial}{\partial q^p} \ln \sqrt{g} \right) v^p
\end{aligned}
\qquad (5.172)
$$

This expression may be rewritten as

$$v^l_{\;;l} = \frac{1}{\sqrt{g}} \frac{\partial}{\partial q^l} \left(\sqrt{g}\, v^l \right) \qquad \text{; covariant divergence} \qquad (5.173)$$

This form of the covariant divergence suggests a simple result when integrated over the volume element in Eq. (5.152), and, indeed, we immediately deduce *Gauss' theorem* for the volume integral of the covariant divergence in curvilinear coordinates in n-dimensions

$$
\begin{aligned}
\int_{\text{Vol}} \left(v^i_{\;;i} \right) d^{(n)}\tau &= \int_{\text{Vol}} \left(v^i_{\;;i} \right) \sqrt{g}\, dq^1 \cdots dq^n \\
&= \int_{\text{Vol}} \frac{1}{\sqrt{g}} \left[\frac{\partial}{\partial q^i} \left(\sqrt{g}\, v^i \right) \right] \sqrt{g}\, dq^1 \cdots dq^n
\end{aligned}
\qquad (5.174)
$$

Notice where the factors of $\sqrt{g}$ appear, and how the one in front cancels the one in the volume element. The integration on dq^i now simply goes out to the surface, and the result is[25]

$$\int_{\text{Vol}} \left(v^i{}_{;i}\right) d^{(n)}\tau = \oint_{\text{Surface}} v^i dS_i \qquad ; \text{ Gauss' theorem}$$

$$dS_i \equiv \sqrt{g}\, dq^1 \cdots (dq^i) \cdots dq^n \qquad ; \text{ strike } dq^i \qquad (5.175)$$

As an example of the utility of these arguments, consider the familiar problem of spherical coordinates in three-dimensional euclidian space. A vector is expanded as

$$\mathbf{v} = v^r\, \mathbf{e}_r + v^\theta\, \mathbf{e}_\theta + v^\phi\, \mathbf{e}_\phi$$
$$= v^r\, \hat{\mathbf{e}}_r + \left(r v^\theta\right) \hat{\mathbf{e}}_\theta + \left(r \sin\theta\, v^\phi\right) \hat{\mathbf{e}}_\phi \qquad (5.176)$$

Here the unit vectors of Eq. (5.121) have again been introduced in the second line. The interval follows from Eq. (5.121) as

$$(d\mathbf{s})^2 = (dr)^2 + (rd\theta)^2 + (r\sin\theta\, d\phi)^2 \qquad (5.177)$$

This allows an identification of the metric

$$\underline{g} = \begin{bmatrix} g_{rr} & g_{r\theta} & g_{r\phi} \\ g_{\theta r} & g_{\theta\theta} & g_{\theta\phi} \\ g_{\phi r} & g_{\phi\theta} & g_{\phi\phi} \end{bmatrix} = \begin{bmatrix} 1 & 0 & 0 \\ 0 & r^2 & 0 \\ 0 & 0 & r^2 \sin^2\theta \end{bmatrix} \qquad (5.178)$$

The determinant of the metric is simply

$$g = \det \underline{g} = r^4 \sin^2\theta \qquad (5.179)$$

The covariant divergence is then given by Eq. (5.173) as

$$v^i{}_{;i} = \frac{1}{r^2 \sin\theta} \left[\frac{\partial}{\partial r}(r^2 \sin\theta\, v^r) + \frac{\partial}{\partial\theta}(r^2 \sin\theta\, v^\theta) + \frac{\partial}{\partial\phi}(r^2 \sin\theta\, v^\phi) \right]$$

$$= \frac{1}{r^2}\frac{\partial}{\partial r}(r^2 v^r) + \frac{1}{\sin\theta}\frac{\partial}{\partial\theta}(\sin\theta\, v^\theta) + \frac{\partial}{\partial\phi}(v^\phi) \qquad (5.180)$$

When written in terms of the components in the second of Eqs. (5.176) the covariant divergence is

$$v^i{}_{;i} = \frac{1}{r^2}\frac{\partial}{\partial r}\left[r^2 \left(v^r\right)\right] + \frac{1}{r\sin\theta}\frac{\partial}{\partial\theta}\left[\sin\theta \left(r v^\theta\right)\right] + \frac{1}{r\sin\theta}\frac{\partial}{\partial\phi}\left[\left(r\sin\theta\, v^\phi\right)\right]$$

$$\equiv \boldsymbol{\nabla} \cdot \mathbf{v} \qquad\qquad ; \text{ spherical coordinates} \qquad (5.181)$$

---

[25] As discussed, the volume element dV is a scalar under coordinate transformations. Since $dS_i = \partial V/\partial q^i$, the element of surface area transforms as the covariant component of a vector.

This is precisely the usual divergence in spherical coordinates.[26]

The volume element in spherical coordinates given by Eq. (5.151) is also the usual result

$$d^{(3)}\tau = \left(r^2 \sin\theta\right) dr d\theta d\phi \qquad ; \text{ spherical coordinates} \qquad (5.182)$$

---

[26]See, for example, [Fetter and Walecka (2003)] p. 519.

Chapter 6

Special Relativity

With this mathematical background, we proceed to review the theory of special relativity, originally due to Einstein in 1905 [Einstein (1905)].

Define the *primary inertial coordinate system* as one that is at rest with respect to the fixed stars.[1] The Michelson-Morley experiment found no shift in the interference pattern when the arms of an interferometer were moving at various velocities relative to the primary inertial coordinate system.[2] This experiment ultimately demonstrated that the speed of light is the same in all inertial frames (an amazing result), and provided the experimental foundation for the theory of special relativity, which has now been verified in countless ways.

6.1 Basic Principles

i) The basic principle of special relativity is that one must have the same laws of physics — including the fact that there is a limiting velocity on propagation of signals, namely c, the velocity of light — in any inertial frame;

ii) Any coordinate system moving with constant velocity relative to an inertial frame is inertial.

All frames moving with constant velocity relative to the primary inertial coordinate system are equivalent as far as physics is concerned.

Einstein observed that we live in a four-dimensional space-time con-

---

[1] See [Fetter and Walecka (2003)].

[2] For a discussion of this experiment, related background, and the central role played by special relativity (SR) in subsequent developments, see any good book on Modern Physics; for example, [Ohanian (1995)].

tinuum.[3]   Let us then introduce the concept of *four-vectors*, and in the following, we make full use of our previously developed framework.

6.2   Four-Vectors

Pick an inertial frame. A four-vector is written in terms of the basis vectors in four-dimensional space-time as

$$\mathbf{v} = \sum_{\mu=1}^{4} v^{\mu}\,\mathbf{e}_{\mu} = v^{\mu}\,\mathbf{e}_{\mu} \qquad (6.1)$$

The convention here is that repeated upper and lower Greek indices get summed from one to four. It is assumed that these vectors have an intrinsic physical meaning.

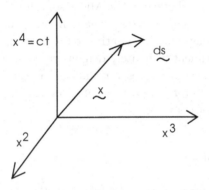

Fig. 6.1   Location of position in four-dimensional space-time and infinitesimal displacement from that position. The coordinates are $(x^1, \cdots, x^4) = (x^1, x^2, x^3, ct)$. The spatial coordinate x^1 is suppressed, and only the two spatial coordinates (x^2, x^3) are shown. Here $\mathbf{x}$ and $d\mathbf{s}$ are four-vectors (denoted in the figure with a bold-face notation).

In particular, the position vector in this inertial frame is given by (Fig. 6.1)

$$\mathbf{x} = x^{\mu}\,\mathbf{e}_{\mu}$$
$$x^{\mu} = (x^1, x^2, x^3, ct) \qquad (6.2)$$

The first three components of this four-vector are the components of the spatial location, and the fourth component is the velocity of light multiplied

---

[3]Although, as we shall see, it is *not* a euclidian space.

by the time.

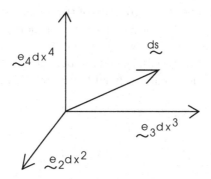

Fig. 6.2 Infinitesimal displacement in four-dimensional space-time. Again, only two spatial coordinates are shown.

The generalized coordinates are thus $x^\mu = (x^1, x^2, x^3, ct)$, and the corresponding basis vectors are $\mathbf{e}_\mu$ with $\mu = 1, \cdots, 4$. In the tangent space, which is here *identical* to the full space, the basis vectors are global and the infinitesimal displacement defining the *line element* is given by (Fig. 6.2)

$$d\mathbf{s} = \mathbf{e}_\mu \, dx^\mu \qquad \text{; line element in SR} \qquad (6.3)$$

The square of this displacement, which defines the *interval*, is given in terms of the *metric* by

$$(d\mathbf{s})^2 = g_{\mu\nu} dx^\mu dx^\nu \qquad \text{; interval in SR}$$

$$g_{\mu\nu} = \mathbf{e}_\mu \cdot \mathbf{e}_\nu \qquad\qquad\qquad\qquad (6.4)$$

While the metric depends on the choice of coordinates, the interval represents *physics* and is independent of the coordinate system. With the choice of cartesian coordinates for the spatial position, the metric (often referred to as the Lorentz metric) takes the form[4]

$$\underline{g} = \begin{bmatrix} 1 & 0 & 0 & 0 \\ 0 & 1 & 0 & 0 \\ 0 & 0 & 1 & 0 \\ 0 & 0 & 0 & -1 \end{bmatrix} \qquad \text{; Lorentz metric in SR} \qquad (6.5)$$

---

[4] Just where one puts the minus sign in the metric, that is, in either the time or the space components, and whether one defines x^4 or x^1 as ct, are matters of convention. This will be ours.

The important thing to note here is the minus sign in the fourth component of the metric, which (as we shall see) guarantees that the speed of light is the same in all inertial frames. The presence of this minus sign implies that the four-dimensional space-time continuum in *not* euclidian, but instead forms a *Minkowski space*. Fortunately, we have now developed enough experience to handle any form of metric and any type of space.

The reciprocal basis is again defined by the relation

$$\mathbf{e}_\mu \cdot \mathbf{e}^\nu = \delta_\mu{}^\nu$$
$$= 1 \qquad ; \text{ if } \mu = \nu$$
$$= 0 \qquad \text{if } \mu \neq \nu \qquad (6.6)$$

Indices are again raised and lowered with the metric. Since the matrix in Eq. (6.5) is its own inverse, one has

$$(\underline{g}^{-1})_{\mu\nu} = g^{\mu\nu} = (\underline{g})_{\mu\nu}$$
$$\text{or;} \qquad g^{\mu\nu} = g_{\mu\nu} \qquad (6.7)$$

A four-vector, a physical quantity, can be expanded in either basis

$$\mathbf{v} = v^\mu \mathbf{e}_\mu = v_\mu \mathbf{e}^\mu \qquad (6.8)$$

The square of a four-vector also represents physics and is invariant under coordinate transformations

$$\mathbf{v} \cdot \mathbf{v} = g_{\mu\nu} v^\mu v^\nu = v_\mu v^\mu \qquad ; \text{ physics} - \text{invariant} \qquad (6.9)$$

In particular, the interval takes the form[5]

$$(d\mathbf{s})^2 = g_{\mu\nu} dx^\mu dx^\nu$$
$$= d\vec{x} \cdot d\vec{x} - c^2 (dt)^2 \qquad ; \text{ interval in SR} \qquad (6.10)$$

Notice again the crucial minus sign in this relation.[6]

We may now proceed to discuss coordinate transformations in this space, and we shall subsequently discuss those coordinate transformations (Lorentz transformations) that can be put into one-to-one correspondence with transformations between inertial frames. First, however, we analyze relativistic particle motion in the given inertial frame.

---

[5]We now use a vector sign above a symbol to indicate a spatial three-dimensional vector.

[6]The theory is often said to possess an *indefinite metric*.

6.3 Relativistic Particle Motion

Consider a particle trajectory C in an inertial frame in four-dimensional space-time as illustrated in Fig. 6.3.[7] Let τ parameterize the distance along the curve with

$$\tau_1 \leq \tau \leq \tau_2 \tag{6.11}$$

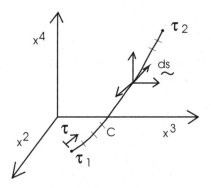

Fig. 6.3   Example of a particle trajectory C in space-time. Here $x^4 = ct$, and the spatial coordinate x^1 is suppressed. The quantity ds represents an infinitesimal displacement along the curve, and τ is the proper time.

Let ds be an infinitesimal displacement along the actual path C. Then we can define $d\tau$ through the invariant interval

$$-c^2(d\tau)^2 \equiv (ds)^2 \qquad \text{; invariant interval} \tag{6.12}$$

With the choice of metric in Eq. (6.5), the interval $(ds)^2$ in Eq. (6.10) is always negative for particle motion, since in this case

$$(d\vec{x})^2 = \vec{v}^2(dt)^2 \tag{6.13}$$

where $\vec{v}$ is the particle velocity (with $|\vec{v}| < c$), and hence

$$\begin{aligned} (ds)^2 &= (d\vec{x})^2 - c^2(dt)^2 \\ &= (\vec{v}^2 - c^2)(dt)^2 < 0 \end{aligned} \tag{6.14}$$

Thus, with the definition in Eq. (6.12), the quantity $(d\tau)^2$ is positive, and $d\tau$ is real. The parameter τ is known as the *proper time*.

---

[7]We assume here that the particle has non-zero rest mass $m > 0$ (see later).

In the tangent space, here identical to the whole space, Eqs. (6.4) and (6.12) can be divided by $(d\tau)^2$ to give

$$\frac{d\mathbf{s}}{d\tau} \cdot \frac{d\mathbf{s}}{d\tau} = \left(\frac{d\mathbf{s}}{d\tau}\right)^2 = g_{\mu\nu} \frac{dx^\mu}{d\tau} \frac{dx^\nu}{d\tau}$$
$$= -c^2 \qquad\qquad ; \text{ along } C \qquad (6.15)$$

This quantity is properly constant along the actual path C.

We define the *four-velocity* by

$$u^\mu \equiv \frac{dx^\mu}{d\tau} \qquad\qquad ; \text{ four-velocity} \qquad (6.16)$$

Since x^μ is a four-vector and $d\tau$ is an invariant, this quantity is also a four-vector. It follows from Eq. (6.15) that

$$\left(\frac{d\mathbf{s}}{d\tau}\right)^2 = g_{\mu\nu} u^\mu u^\nu$$
$$= -c^2 \qquad\qquad ; \text{ along } C \qquad (6.17)$$

With the Lorentz metric of Eq. (6.5), one has

$$(d\mathbf{s})^2 = d\vec{x} \cdot d\vec{x} - c^2 (dt)^2 = -c^2 (d\tau)^2 \qquad (6.18)$$

Upon division by the *time interval* $(dt)^2$, this leads to

$$\left(\frac{d\mathbf{s}}{dt}\right)^2 = \left(\frac{d\vec{x}}{dt}\right)^2 - c^2 = \vec{v}^2 - c^2$$
$$= -c^2 \left(\frac{d\tau}{dt}\right)^2 \qquad (6.19)$$

Thus the relation between the *time* and *proper time* is given by

$$\frac{d\tau}{dt} = (1 - \beta^2)^{1/2} \qquad\qquad ; \vec{\beta} \equiv \frac{\vec{v}}{c} \qquad (6.20)$$

Here $\vec{\beta} = \vec{v}/c$ is the ratio of the particle velocity to the speed of light.

The components of the four-velocity in Eq. (6.16) then take the form

$$u^\mu = \frac{dx^\mu}{d\tau} = \frac{dx^\mu}{dt} \frac{dt}{d\tau}$$
$$= \frac{1}{(1 - \beta^2)^{1/2}} [\vec{v}, c]$$
$$u_\mu = g_{\mu\nu} u^\nu = \frac{1}{(1 - \beta^2)^{1/2}} [\vec{v}, -c] \qquad (6.21)$$

These relations reproduce Eq. (6.17)

$$\mathbf{u}^2 = u_\mu u^\mu = \frac{\vec{v}^2 - c^2}{1 - \vec{v}^2/c^2}$$

$$= -c^2 \qquad\qquad ; \text{ invariant} \qquad (6.22)$$

We now ask the important question:

Can we find a relativistic lagrangian and relativistic form of Hamilton's principle that give the correct equations of motion?

For the relativistic lagrangian, try the following Lorentz scalar

$$\bar{L}\left(x^\mu, \frac{dx^\mu}{d\tau}; \tau\right) = mg_{\mu\nu} u^\mu u^\nu$$

$$= mg_{\mu\nu} \frac{dx^\mu}{d\tau} \frac{dx^\nu}{d\tau} \qquad ; \ m = \text{constant} \qquad (6.23)$$

and for the relativistic form of Hamilton's principle, try

$$\delta \int_{\tau_1}^{\tau_2} \bar{L}\, d\tau = 0 \qquad\qquad ; \text{ fixed endpoints} \qquad (6.24)$$

Here the quantity m, the particle's *rest mass*, is a constant. Note that Hamilton's principle involves *variations* along the path (recall Fig. 2.7), and along the *actual* path

$$(d\mathbf{s})^2 = -c^2(d\tau)^2 \qquad\qquad ; \text{ along actual path} \qquad (6.25)$$

We can now just do mechanics. The Euler-Lagrange equations are

$$\frac{d}{d\tau} \frac{\partial \bar{L}}{\partial(dx^\mu/d\tau)} - \frac{\partial \bar{L}}{\partial x^\mu} = 0 \qquad\qquad ; \ \mu = 1, \cdots, 4 \qquad (6.26)$$

Hence

$$\frac{d}{d\tau}\left(2mg_{\mu\nu} \frac{dx^\nu}{d\tau}\right) = 0 \qquad\qquad (6.27)$$

Since the metric in Eq. (6.5) is constant everywhere, one has $dg_{\mu\nu}/d\tau = 0$, and thus Eqs. (6.27) reduce to

$$\frac{d}{d\tau}(mu^\mu) = 0 \qquad\qquad ; \text{ E-L equations}$$

$$mu^\mu = m\frac{dx^\mu}{d\tau} \qquad\qquad \text{four-momentum} \qquad (6.28)$$

As written in terms of the rate of change of the four-momentum mu^μ with respect to the proper time τ, these provide the appropriate relativistic equations of motion.

Let us write Eqs. (6.28) out in terms of the time t

$$\left[\frac{d}{dt}\left(m\frac{dx^\mu}{dt}\frac{dt}{d\tau} \right) \right] \frac{dt}{d\tau} = 0$$

$$\text{or;} \qquad \frac{d}{dt}\left[\frac{m}{(1-\beta^2)^{1/2}}(\vec{v},\,c) \right] = 0 \qquad\qquad (6.29)$$

In detail, these equations state that

$$\frac{d\vec{p}}{dt} = \frac{d}{dt}\left[\frac{m\vec{v}}{(1-\beta^2)^{1/2}} \right] = 0 \qquad ; \text{ Newton's law}$$

$$\frac{d}{dt}\left(\frac{E}{c} \right) = \frac{d}{dt}\left[\frac{mc}{(1-\beta^2)^{1/2}} \right] = 0 \qquad ; \text{ energy conservation} \quad (6.30)$$

The spatial components correspond to the relativistic extension of Newton's second law with the correct relativistic momentum.[8] The fourth component yields the statement of energy conservation with the correct relativistic energy. Note that along the *actual* trajectory C, the lagrangian $\bar{L}$ is constant, with $\bar{L} = -mc^2$.

In summary, for relativistic particle motion

$$mu^\mu = \frac{m}{(1-\beta^2)^{1/2}}(\vec{v},\,c) = \left(\vec{p},\,\frac{E}{c} \right)$$

$$(m\mathbf{u})^2 = -(mc)^2 = \vec{p}^{\,2} - \left(\frac{E}{c} \right)^2$$

$$\text{or ;} \qquad E^2 = \vec{p}^{\,2}c^2 + m^2c^4 \qquad\qquad (6.31)$$

Note that for a particle at rest, this reduces to Einstein's celebrated relation $E = mc^2$.

So far, we have just constructed a lagrangian $\bar{L}$ and relativistic Hamilton's principle that yield the correct relativistic equations of motion. We must at least show that a non-relativistic reduction gives the correct non-relativistic limit (NRL). The familiar Hamilton's principle is a statement

---

[8]With no forces, the particle velocity will be constant; however, see Probs. 6.2 and 8.9.

of stationary action

$$S = \int_{t_1}^{t_2} L\, dt \qquad ; \text{action}$$

$$\delta S = 0 \qquad\qquad \text{stationary action} \qquad (6.32)$$

With a change of variables, the action in Eq. (6.24) can be rewritten as

$$S = \int_{\tau_1}^{\tau_2} \bar{L}\, d\tau = \int_{t_1}^{t_2} \left(\bar{L}\, \frac{d\tau}{dt} \right) dt$$

$$= \int_{t_1}^{t_2} [\bar{L}\,(1 - \beta^2)^{1/2}]\, dt \qquad (6.33)$$

Here the relation between time and proper time in Eq. (6.20) has been used in the second line. This leads to an identification of the equivalent lagrangian L to be used in the Hamilton's principle of Eq. (6.32) as

$$L = \bar{L}\,(1 - \beta^2)^{1/2} = \bar{L} \left(1 - \frac{\vec{v}^2}{c^2} \right)^{1/2}$$

$$L = \bar{L} \left[1 - \frac{1}{c^2} \left(\frac{d\vec{x}}{dt} \right)^2 \right]^{1/2} \qquad (6.34)$$

This relation holds for all $\beta < 1$.

Now prepare to take the NRL as an expansion in $\beta^2 = \vec{v}^2/c^2$. From Eq. (6.23) and the above one has

$$\bar{L} = m g_{\mu\nu} u^\mu u^\nu = m g_{\mu\nu} \frac{dx^\mu}{d\tau} \frac{dx^\nu}{d\tau}$$

$$= m g_{\mu\nu} \frac{dx^\mu}{dt} \frac{dx^\nu}{dt} \left(\frac{dt}{d\tau} \right)^2 = \frac{m}{1 - \beta^2} (\vec{v}^2 - c^2)$$

$$= -mc^2 \frac{1 - \beta^2}{1 - \beta^2} = -mc^2 \qquad (6.35)$$

In changing variables from $(x^\mu, dx^\mu/d\tau\,; \tau)$ to $(\vec{x}, d\vec{x}/dt\,; t)$, the result for $\bar{L}$ becomes *independent of* $\vec{\beta}$. Equation (6.34) then gives

$$L = -mc^2 \left(1 - \frac{\vec{v}^2}{c^2} \right)^{1/2} \qquad (6.36)$$

This relation still holds for all $\beta < 1$. A subsequent expansion for small β^2

gives

$$L = -mc^2 + \frac{1}{2}m\vec{v}^2 + \cdots \qquad (6.37)$$

The first term is a constant and is irrelevant as far as Hamilton's principle is concerned. Thus the effective non-relativistic lagrangian is

$$L \doteq \frac{1}{2}m\vec{v}^2 = \frac{m}{2}\left(\frac{d\vec{x}}{dt}\right)^2 = T \qquad (6.38)$$

This is the familiar non-relativistic result for a free particle.

In summary, our relativistic Hamilton's principle in Eq. (6.24) is equivalent to

$$\delta \int_{\tau_1}^{\tau_2} \bar{L}\,d\tau = \delta \int_{t_1}^{t_2} L\,dt = 0 \qquad\qquad ; \text{ Hamilton's principle}$$

$$L = -mc^2\left[1 - \frac{1}{c^2}\left(\frac{d\vec{x}}{dt}\right)^2\right]^{1/2} \qquad (6.39)$$

In the non-relativistic limit this reduces to

$$L \doteq \frac{m}{2}\left(\frac{d\vec{x}}{dt}\right)^2 \qquad\qquad ; \text{ correct NRL} \qquad (6.40)$$

Here $\doteq$ neglects the constant term, and $L = T$ is the usual non-relativistic lagrangian for particle motion.

Let us examine Lagrange's equations for relativistic particle motion in a general four-dimensional *curvilinear coordinate system* with coordinates $(q^1, \cdots, q^4)$ and corresponding basis vectors $(\mathbf{e}_1, \cdots, \mathbf{e}_4)$. There is again a *local tangent space*; however, it is now not a euclidian space, but rather a *Minkowski space* with the metric in a cartesian basis of Eq. (6.5).[9] The line element and four velocity are now given by

$$d\mathbf{s} = \mathbf{e}_\mu \, dq^\mu$$

$$\mathbf{u} = \frac{d\mathbf{s}}{d\tau} = \mathbf{e}_\mu \, u^\mu = \mathbf{e}_\mu \frac{dq^\mu}{d\tau} \qquad (6.41)$$

In curvilinear coordinates the metric depends on position, and the invariant interval is expressed as

$$(d\mathbf{s})^2 = g_{\mu\nu}(q^1, \cdots, q^4)dq^\mu dq^\nu \qquad\qquad ; \text{ invariant interval} \quad (6.42)$$

---

[9] Here the tangent space is again identical to the whole space; later this will not be the case.

The relativistic lagrangian then takes the form

$$\bar{L} = m\left(\frac{d\mathbf{s}}{d\tau}\right)^2 = mg_{\mu\nu}(q^1,\cdots,q^4)\frac{dq^\mu}{d\tau}\frac{dq^\nu}{d\tau} \qquad (6.43)$$

The relativistic Hamilton's principle becomes

$$\delta\int_{\tau_1}^{\tau_2}\bar{L}\,d\tau = \delta\int_{\tau_1}^{\tau_2} m\left[g_{\mu\nu}(q^1,\cdots,q^4)\frac{dq^\mu}{d\tau}\frac{dq^\nu}{d\tau}\right]d\tau = 0 \qquad (6.44)$$

The problem is now reduced to ordinary mechanics and the Euler-Lagrange equations follow just as in Eq. (2.23)

$$\frac{du^\mu}{d\tau} + \Gamma^\mu_{\lambda\sigma}u^\lambda u^\sigma = 0 \qquad\qquad \text{; E-L equations}$$

or;
$$\frac{d^2q^\mu}{d\tau^2} + \Gamma^\mu_{\lambda\sigma}\frac{dq^\lambda}{d\tau}\frac{dq^\sigma}{d\tau} = 0 \qquad\qquad \text{curvilinear coordinates}$$

$$(d\mathbf{s})^2 = -c^2(d\tau)^2 \qquad (6.45)$$

The last relation follows since the E-L equations describe the particle motion along the actual path. The affine connection in Minkowski space appearing in these equations is given by

$$\Gamma^\mu_{\lambda\sigma}(q) = \frac{1}{2}g^{\mu\nu}(q)\left[\frac{\partial g_{\nu\lambda}(q)}{\partial q^\sigma} + \frac{\partial g_{\nu\sigma}(q)}{\partial q^\lambda} - \frac{\partial g_{\lambda\sigma}(q)}{\partial q^\nu}\right] \qquad (6.46)$$

The solution to Eqs. (6.45) gives the particle trajectory $q^\mu(\tau)$ expressed in the space of the generalized coordinates.

Geodesics are again obtained by minimizing the "length" of the curve C in Fig. 6.3. Here one is really minimizing the *proper time*.[10] If $\bar{\tau}$ represents the proper time along the varied path, and τ that along the actual path, then one wants to minimize

$$S_{21} \equiv \int_1^2 c\,d\bar{\tau} = \int_1^2 [(c\,d\bar{\tau})^2]^{1/2}$$

$$= \int_1^2 [-(d\mathbf{s})^2]^{1/2}$$

$$S_{21} = \int_{\tau_1}^{\tau_2}\left[-\left(\frac{d\mathbf{s}}{d\tau}\right)^2\right]^{1/2}d\tau \qquad (6.47)$$

---

[10]Or, more generally, making it stationary.

One just has to be careful with signs here because of the metric convention we have chosen [recall Eq. (6.14)]. Now substitute Eq. (6.42)

$$S_{21} = \int_{\tau_1}^{\tau_2} \left[-g_{\mu\nu} \frac{dq^\mu}{d\tau} \frac{dq^\nu}{d\tau} \right]^{1/2} d\tau \qquad (6.48)$$

Along the actual path C that solves the resulting E-L equations, the expression in square brackets satisfies Eq. (6.17) and thus is constant. Hence, exactly as with Eq. (2.45), one finds

$$\frac{d^2 q^\mu}{d\tau^2} + \Gamma^\mu_{\lambda\sigma} \frac{dq^\lambda}{d\tau} \frac{dq^\sigma}{d\tau} = 0 \qquad ; \text{ geodesics}$$
$$(ds)^2 = -c^2 (d\tau)^2 \qquad (6.49)$$

The resulting trajectories $q^\mu(\tau)$ are the geodesics in four-dimensional space-time, and these equations are again *identical* to the dynamical Eqs. (6.45).

So far we have gone through a lot of work to describe what is, after all, just straight-line motion; however, that work will soon pay off for us.

6.4 Lorentz Transformations

It was Einstein's genius to provide the *physical interpretation* of special relativity. He said that we live in a four-dimensional space-time continuum with cartesian coordinates $x^\mu = (x^1, x^2, x^3, ct)$ and the Lorentz metric of Eq. (6.5), and

Physics in different inertial frames is represented by coordinate transformations in this Minkowski space.

- One views the same *physics*, the same *events*, the same *line element ds*, the same *intervals* $(ds)^2$. They are just viewed in different coordinate systems;
- Different inertial frames are put into a one-to-one correspondence with transformed coordinate systems in this space. They are related by a *Lorentz transformation*;
- A Lorentz transformation is a rule that tells how the coordinates in one inertial frame $x^\mu = (x^1, x^2, x^3, ct)$ are related to those in a second inertial frame $\bar{x}^\mu = (\bar{x}^1, \bar{x}^2, \bar{x}^3, c\bar{t})$.

We note that now *time is different* in the two inertial frames. This is a major break from classical thinking and revolutionizes our understanding of how the Universe behaves!

We make use of all of our previous analysis of coordinate transformations. At a given point in space-time, in the tangent space, introduce a cartesian basis (Figs. 6.1-6.2). Call this frame f as indicated in Fig. 6.4. In the frame f' moving with velocity $\vec{v}$, we have a new cartesian basis at that point in the tangent space with coordinates $\bar{x}^\mu = (\bar{x}^1, \bar{x}^2, \bar{x}^3, c\bar{t})$ and corresponding basis vectors $\boldsymbol{\alpha}_\mu$ as indicated in Fig. 6.5.[11]

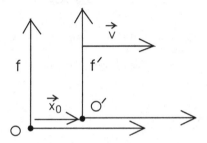

Fig. 6.4 Two different inertial frames f and f' moving with relative velocity $\vec{v}$. The origin of the second frame lies at a position $\vec{x}_0$ in the first frame.

In terms of the new coordinates, the line element, interval, and metric are given by

$$d\mathbf{s} = \boldsymbol{\alpha}_\mu \, d\bar{x}^\mu \qquad ; \text{ line element–invariant}$$
$$(d\mathbf{s})^2 = \bar{g}_{\mu\nu} d\bar{x}^\mu d\bar{x}^\nu \qquad \text{interval}$$
$$\bar{g}_{\mu\nu} = \boldsymbol{\alpha}_\mu \cdot \boldsymbol{\alpha}_\nu \qquad \text{metric} \qquad (6.50)$$

Since by our basic assumption all inertial frames are equivalent, one must be able to use a metric of the same form in the second inertial frame

$$\underline{\bar{g}} = \begin{bmatrix} 1 & 0 & 0 & 0 \\ 0 & 1 & 0 & 0 \\ 0 & 0 & 1 & 0 \\ 0 & 0 & 0 & -1 \end{bmatrix} \qquad ; \text{ metric in second frame} \qquad (6.51)$$

---

[11]It is easy to convince oneself, and it can be explicitly demonstrated, that if an observer in f sees the frame f' moving with velocity $\vec{v}$, then an observer in f' will see f moving with $-\vec{v}$ (see Prob. 6.4).

Hence we conclude that a *Lorentz transformation leaves the metric invariant*.

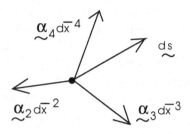

Fig. 6.5   New cartesian basis in second inertial frame f' moving with velocity $\vec{v}$ relative to the first frame f. The coordinates are now $\bar{x}^\mu = (\bar{x}^1, \bar{x}^2, \bar{x}^3, c\bar{t})$ and corresponding basis vectors $\boldsymbol{\alpha}_\mu$. The line element $d\mathbf{s}$ is also indicated. The spatial coordinate $\bar{x}^1$ is again suppressed.

We note the following:

- The space is *flat* (see Prob. 6.1), and here the tangent space is identical to the whole space;
- The cartesian basis vectors $\mathbf{e}_\mu$ are the same everywhere, they are *global* in the frame f;
- The cartesian basis vectors $\boldsymbol{\alpha}_\mu$ are the same everywhere, they are *global* in the frame f';
- The corresponding coordinate transformation, or Lorentz transformation, describes the physical transformation between the inertial frames (Einstein).

We emphasize that in a situation where the whole space is identical to the tangent space, we can talk *globally*.[12]

Let us both agree to start our clocks when the origins O and O' coincide (an event in space-time on which we can both clearly agree), as in Fig. 6.6.[13] This defines an *origin* for the four-vector $\mathbf{x}$.[14] Then the four-vector $\mathbf{x}$, a physical quantity, can be put into a one-to-one correspondence with a second point (second event) in space-time. The four-vector $\mathbf{x}$ can

---

[12]Later on, this will not be true. Note that global here means for all space-time.

[13]We use "events," upon which we can all agree, to locate points in space-time.

[14]And implies that the subsequent discussion is confined to *homogeneous* Lorentz transformations.

be decomposed in either frame

$$\mathbf{x} = \mathbf{e}_\mu \, x^\mu = x^i \, \mathbf{e}_i + ct \, \mathbf{e}_4$$
$$= \boldsymbol{\alpha}_\mu \, \bar{x}^\mu = \bar{x}^i \, \boldsymbol{\alpha}_i + c\bar{t} \, \boldsymbol{\alpha}_4 \tag{6.52}$$

Here repeated *Latin* indices are summed from 1 to 3. We further define these expressions as[15]

$$x^i \, \mathbf{e}_i + ct \, \mathbf{e}_4 \equiv \vec{x} + ct \, \mathbf{e}_4$$
$$\bar{x}^i \, \boldsymbol{\alpha}_i + c\bar{t} \, \boldsymbol{\alpha}_4 \equiv \vec{x}\,' + ct' \, \boldsymbol{\alpha}_4 \tag{6.53}$$

where $t' \equiv \bar{t}$.

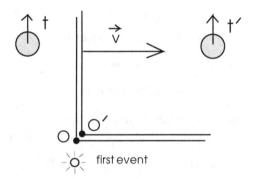

Fig. 6.6 We agree to start our clocks at the first event where the origins in Fig. 6.4 coincide.

- In frame f, we use the quantities $(\vec{x}, ct)$ to describe the four-vector $\mathbf{x}$ and corresponding space-time point;
- In frame f', we use the quantities $(\vec{x}\,', ct')$ to describe the four-vector $\mathbf{x}$ and corresponding space-time point;
- It is the *same physical four-vector* and *same space-time point*!

Now, after synchronizing our clocks when the origins coincide, consider the subsequent motion of the origin O' of the system f' as viewed first in f and then in f' (Fig. 6.4). A second event, again something on which we can all agree, happens at the space-time point locating the origin O' (Fig. 6.7).

---

[15]We will frequently use the shorthand notation $\mathbf{x} = (\vec{x}, ct)$ for the four-vector $\mathbf{x} = x^i \, \mathbf{e}_i + ct \, \mathbf{e}_4 = \vec{x} + ct \, \mathbf{e}_4$, and similarly for other four-vectors.

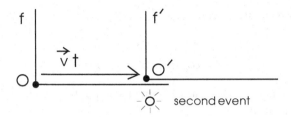

Fig. 6.7   A second event in space-time marking the location of the origin O'.

Let the four-vector $\mathbf{x}$ connect the two points in space-time marked by these two events. In frame f one would say

$$\mathbf{x} = \vec{x}_0 + ct\,\mathbf{e}_4$$
$$= \vec{v}\,t + ct\,\mathbf{e}_4 \qquad ;\ O'\ \text{in}\ f \qquad (6.54)$$

since the location of the origin O' in f is now at

$$\vec{x}_0 = \vec{v}\,t = (v^i\,\mathbf{e}_i)\,t \qquad (6.55)$$

In frame f', the second event occurs at the origin $\vec{x}' = \vec{0}$ and at a time t', and hence one says that the physical four-vector $\mathbf{x}$ is given by

$$\mathbf{x} = \vec{0} + ct'\,\boldsymbol{\alpha}_4 \qquad ;\ O'\ \text{in}\ f' \qquad (6.56)$$

These two expressions for $\mathbf{x}$ can now equated (it is the same four-vector) and solved for $\boldsymbol{\alpha}_4$. The result is

$$\boldsymbol{\alpha}_4 = \frac{t}{t'}(\vec{\beta} + \mathbf{e}_4) \qquad ;\ \vec{\beta} = \frac{\vec{v}}{c} \qquad (6.57)$$

This expression in Eq. (6.57) may now be dotted into itself and the following used[16]

$$\boldsymbol{\alpha}_4 \cdot \boldsymbol{\alpha}_4 = \bar{g}_{44} = -1$$
$$\mathbf{e}_4 \cdot \mathbf{e}_4 = g_{44} = -1$$
$$\vec{\beta} \cdot \mathbf{e}_4 = (\beta^i\,\mathbf{e}_i) \cdot \mathbf{e}_4 = \beta^i\,g_{i4} = 0 \qquad (6.58)$$

Hence

$$-1 = \left(\frac{t}{t'}\right)^2 (\beta^2 - 1) \qquad (6.59)$$

---

[16]Note that in the present approach, the complexity of Minkowski space is put into the fourth basis vector. For an alternative approach, see Prob. 6.12.

or

$$t = \frac{t'}{(1 - \beta^2)^{1/2}} \qquad ; \text{ time dilation} \qquad (6.60)$$

The relation between the basis vectors is then given by

$$\alpha_4 = \frac{\vec{\beta} + \mathbf{e}_4}{(1 - \beta^2)^{1/2}} \qquad ; \text{ relates basis vectors} \qquad (6.61)$$

Now note that the time t is well-defined everywhere in f, since it is an ordinary global laboratory system. The same holds for the time t' in f'. Equation (6.60) says, however, that the *clock in the frame f' at the origin O' runs slower than the clock in the frame f*. This is *time dilation*— a most non-intuitive result, but one that has been repeatedly verified over the years.[17]

Consider next an *arbitrary* four-vector $\mathbf{x}$ (Fig. 6.8).

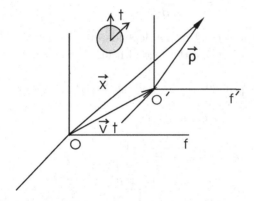

Fig. 6.8   Arbitrary four-vector $\mathbf{x}$ in space-time as viewed in two different inertial frames f and f' with origins that coincide when the clocks start. In f the four-vector is $\mathbf{x} = (\vec{x}, ct)$.

It can be expanded in either coordinate system (inertial frame) as

$$\mathbf{x} = \vec{x} + ct\,\mathbf{e}_4$$
$$= \vec{x}' + ct'\boldsymbol{\alpha}_4 \qquad (6.62)$$

---

[17]For example, by the extended lifetime of relativistic cosmic-ray muons, which allows them to reach the surface of the earth, or similarly, by their extended lifetime in storage rings, which permits a precise measurement of their magnetic moment.

Here the spatial three-vectors are defined in Eq. (6.53) as

$$\vec{x} = x^i \, \mathbf{e}_i \qquad\qquad ; \; \vec{x}' = \bar{x}^i \, \boldsymbol{\alpha}_i \qquad\qquad (6.63)$$

Dot $\boldsymbol{\alpha}_4$ into $\mathbf{x}$ in Eq. (6.62) and use, as with $\vec{\beta} \cdot \mathbf{e}_4$ in Eqs. (6.58),

$$\boldsymbol{\alpha}_4 \cdot \vec{x}' = \boldsymbol{\alpha}_4 \cdot (\bar{x}^i \, \boldsymbol{\alpha}_i) = 0 \qquad\qquad (6.64)$$

With the insertion of Eq. (6.61) for $\boldsymbol{\alpha}_4$, this leads to

$$\left[\frac{\vec{\beta} + \mathbf{e}_4}{(1 - \beta^2)^{1/2}} \right] \cdot (\vec{x} + ct \, \mathbf{e}_4) = -ct'$$

$$\frac{\vec{\beta} \cdot \vec{x} - ct}{(1 - \beta^2)^{1/2}} = -ct' \qquad\qquad (6.65)$$

This gives the general relation between the times t and t' in the two inertial frames

$$t' = \frac{t - \vec{\beta} \cdot \vec{x}/c}{(1 - \beta^2)^{1/2}} \qquad\qquad ; \; t' \equiv \bar{t} \qquad\qquad (6.66)$$

The time t' in the second frame depends on $(\vec{x}, ct)$ in the first frame. If one sits at the origin O' of the second frame where $\vec{x} = \vec{v}t$, this expression reduces to Eq. (6.60).

Next, solve Eq. (6.62) for $\vec{x}'$, and use Eq. (6.61) and the above

$$\vec{x}' = \bar{x}^i \, \boldsymbol{\alpha}_i = \vec{x} + ct \, \mathbf{e}_4 - ct' \, \boldsymbol{\alpha}_4$$

$$= \vec{x} + ct \, \mathbf{e}_4 - c \left[\frac{t - \vec{\beta} \cdot \vec{x}/c}{(1 - \beta^2)^{1/2}} \right] \left[\frac{\vec{\beta} + \mathbf{e}_4}{(1 - \beta^2)^{1/2}} \right]$$

$$= \vec{x} + ct \, \mathbf{e}_4 - \frac{\vec{v}t}{1 - \beta^2} + \frac{(\vec{v} \cdot \vec{x})\vec{v}/c^2}{1 - \beta^2} - \frac{ct \, \mathbf{e}_4}{1 - \beta^2} + \frac{(\vec{v} \cdot \vec{x}/c) \, \mathbf{e}_4}{1 - \beta^2}$$

$$\vec{x}' = \frac{\vec{x} - \vec{v}t - \beta^2 [\vec{x} - (\hat{\mathbf{v}} \cdot \vec{x})\hat{\mathbf{v}}]}{1 - \beta^2} + \frac{(\vec{v}/c) \cdot [\vec{x} - \vec{v}t]}{1 - \beta^2} \, \mathbf{e}_4 \qquad (6.67)$$

Here $\hat{\mathbf{v}} \equiv \vec{v}/\beta c$ is a unit vector in the direction of $\vec{v}$. Define the position of $\vec{x}$ relative to the origin O' in the frame f as (Fig. 6.8)

$$\vec{\rho} \equiv \vec{x} - \vec{v}t \qquad\qquad (6.68)$$

and use the following vector manipulations

$$(\vec{\rho} \times \vec{\beta}) \times \vec{\beta} = (\vec{x} \times \vec{\beta}) \times \vec{\beta} = -\vec{\beta} \times (\vec{x} \times \vec{\beta})$$

$$= -\vec{x} \, \beta^2 + \vec{\beta} \, (\vec{x} \cdot \vec{\beta}) \qquad\qquad (6.69)$$

Equation (6.67) then becomes

$$\vec{x}' = \bar{x}^i \, \boldsymbol{\alpha}_i = \frac{\vec{\rho} + (\vec{\rho} \times \vec{\beta}) \times \vec{\beta} + (\vec{\rho} \cdot \vec{\beta}) \, \mathbf{e}_4}{1 - \beta^2} \qquad (6.70)$$

Equations (6.70) and (6.66) are the *general Lorentz transformation* between two coordinate systems, the first with coordinates $x^\mu = (x^1, x^2, x^3, ct)$ and basis vectors $\mathbf{e}_\mu$, and the second with $\bar{x}^\mu = (\bar{x}^1, \bar{x}^2, \bar{x}^3, c\bar{t})$ and $\boldsymbol{\alpha}_\mu$, representing, according to Einstein, two inertial frames moving with a relative velocity $\vec{v} = \vec{\beta}c$, and with origins that coincide. The situation is illustrated in Fig. 6.9.

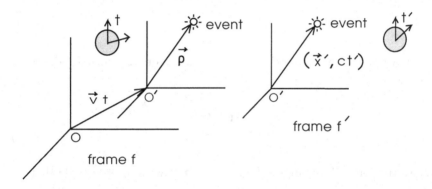

Fig. 6.9 A second event, after the first event in Fig. 6.6, as viewed in the two inertial frames f and f' moving with a relative velocity $\vec{v} = \vec{\beta}c$.

The second event occurs at a given space-time point specifying $\mathbf{x}$, which is then described as follows:

- In frame f:

$$\mathbf{x} = x^\mu \, \mathbf{e}_\mu \qquad ; \text{ with } \mathbf{x} = (\vec{v}\,t + \vec{\rho}, \; ct) \qquad (6.71)$$

- In frame f':

$$\mathbf{x} = \bar{x}^\mu \, \boldsymbol{\alpha}_\mu \qquad ; \text{ with } \mathbf{x} = (\vec{x}', \; ct') \qquad (6.72)$$

Here $t' \equiv \bar{t}$.

As an example, consider the case where the vector $\vec{\rho}$ in Fig. 6.9 is perpendicular to the velocity so that $\vec{\rho} \perp \vec{\beta}$ and $\vec{\rho} \cdot \vec{\beta} = 0$. In this case

$$\vec{x}' = \frac{\vec{\rho} + (\vec{\rho} \times \vec{\beta}) \times \vec{\beta}}{1 - \beta^2} = \vec{\rho}$$
$$t' = (1 - \beta^2)^{1/2} t \tag{6.73}$$

Transverse vectors are thus *unchanged* under a Lorentz transformation and for such vectors, time dilation is independent of $\vec{\rho}$.

We use the above results to further specify the relation between the basis vectors. We now choose to put $\mathbf{e}_3$ along the direction of the velocity so that

$$\vec{\beta} = \beta \, \mathbf{e}_3 \tag{6.74}$$

We have just shown that transverse spatial vectors are unchanged, so one can take

$$\boldsymbol{\alpha}_1 = \mathbf{e}_1$$
$$\boldsymbol{\alpha}_2 = \mathbf{e}_2 \tag{6.75}$$

The situation is illustrated in Fig. 6.10. It follows from Eq. (6.61) that

$$\boldsymbol{\alpha}_4 = \frac{\beta \, \mathbf{e}_3 + \mathbf{e}_4}{(1 - \beta^2)^{1/2}} \tag{6.76}$$

In order to satisfy $\boldsymbol{\alpha}_3 \cdot \boldsymbol{\alpha}_4 = 0$ and $\boldsymbol{\alpha}_3 \cdot \boldsymbol{\alpha}_3 = 1$, we take

$$\boldsymbol{\alpha}_3 = \frac{\mathbf{e}_3 + \beta \, \mathbf{e}_4}{(1 - \beta^2)^{1/2}} \tag{6.77}$$

The metric in the first frame is

$$g_{\mu\nu} = \mathbf{e}_\mu \cdot \mathbf{e}_\nu$$
$$\underline{g} = \begin{bmatrix} 1 & 0 & 0 & 0 \\ 0 & 1 & 0 & 0 \\ 0 & 0 & 1 & 0 \\ 0 & 0 & 0 & -1 \end{bmatrix} \quad ; \text{ metric in first frame} \tag{6.78}$$

It is then readily verified that the metric in the new basis is

$$\bar{g}_{\mu\nu} = \boldsymbol{\alpha}_\mu \cdot \boldsymbol{\alpha}_\nu$$

$$\underline{\bar{g}} = \begin{bmatrix} 1 & 0 & 0 & 0 \\ 0 & 1 & 0 & 0 \\ 0 & 0 & 1 & 0 \\ 0 & 0 & 0 & -1 \end{bmatrix} \qquad ; \text{ metric in second frame} \qquad (6.79)$$

and, indeed, the Lorentz metric is preserved under this coordinate transformation. We point out once again that in this case the relation between the basis vectors is *global*. Note again also that both metrics are diagonal and

$$g^{\mu\nu} = g_{\mu\nu} \qquad\qquad ; \ \bar{g}^{\mu\nu} = \bar{g}_{\mu\nu} \qquad\qquad (6.80)$$

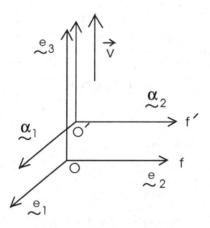

Fig. 6.10 Spatial configuration of Eqs. (6.74) and (6.75).

Now from our general theory of coordinate transformations, the relation between the basis vectors is given by [see Eq. (3.39)]

$$\boldsymbol{\alpha}_\nu = a^\mu{}_\nu \, \mathbf{e}_\mu \qquad\qquad (6.81)$$

Hence one can just read off the Lorentz transformation *matrix* $\underline{a}$ from the relation between the basis vectors

$$[\underline{a}]_{\mu\nu} \equiv a^\mu{}_\nu$$

$$\underline{a} = \frac{1}{(1-\beta^2)^{1/2}} \begin{bmatrix} (1-\beta^2)^{1/2} & 0 & 0 & 0 \\ 0 & (1-\beta^2)^{1/2} & 0 & 0 \\ 0 & 0 & 1 & \beta \\ 0 & 0 & \beta & 1 \end{bmatrix} \; ; \; \vec{\beta} = \beta\, \mathbf{e}_3 \quad (6.82)$$

From the general analysis we also know that

$$a^\nu{}_\mu = \bar{a}_\mu{}^\nu$$
$$\bar{a}^\mu{}_\nu = \bar{g}^{\mu\lambda}\, g_{\nu\sigma}\, \bar{a}_\lambda{}^\sigma \quad (6.83)$$

Thus with the matrix notation

$$[\bar{\underline{a}}]_{\mu\nu} \equiv \bar{a}^\mu{}_\nu \quad (6.84)$$

one has[18]

$$\bar{\underline{a}} = \frac{1}{(1-\beta^2)^{1/2}} \begin{bmatrix} (1-\beta^2)^{1/2} & 0 & 0 & 0 \\ 0 & (1-\beta^2)^{1/2} & 0 & 0 \\ 0 & 0 & 1 & -\beta \\ 0 & 0 & -\beta & 1 \end{bmatrix} \; ; \; \vec{\beta} = \beta\, \mathbf{e}_3 \quad (6.85)$$

We also know from our general analysis that

$$a^\mu{}_\lambda \bar{a}^\lambda{}_\nu = \delta^\mu{}_\nu \quad (6.86)$$

In matrix language this reads

$$\underline{a}\,\bar{\underline{a}} = \underline{1}$$
$$\text{or;} \qquad \bar{\underline{a}} = \underline{a}^{-1} \quad (6.87)$$

The first relation is readily verified by direct matrix multiplication of the above.

Consider the four-velocity [see Eqs. (6.21)] of the origin O' as observed in f. In the coordinate system of Fig. 6.10 it is given by

$$\frac{1}{c}u^\mu = \frac{1}{(1-\beta^2)^{1/2}}[0,0,\beta,1] \quad (6.88)$$

---

[18]Note that the effect of raising and lowering the indices in Eq. (6.83) is to simply multiply both the fourth row and fourth column of $\underline{a}$ by (-1).

We note that the fourth column and fourth row of the Lorentz transformation matrix $\underline{a}$ in Eq. (6.82) are given by

$$[\underline{a}]_{\mu 4} = a^\mu{}_4 = \frac{1}{c} u^\mu = [\underline{a}]_{4\mu} \qquad ; \mu = 1, \cdots, 4 \qquad (6.89)$$

More generally, if $\vec{\beta}$ has an arbitrary direction in f, then a simple rotation gives for the four-velocity

$$\frac{1}{c} u^\mu = \frac{1}{(1 - \beta^2)^{1/2}} [\vec{\beta}, 1] \qquad (6.90)$$

These relations will be of use to us later.

6.5 General Tensor Transformation Law

From our previous analysis of coordinate transformations, the relation between a general tensor T in the frame f, and $\bar{T}$ in the new frame f', is now

$$T^{\mu\nu\rho\cdots} = a^\mu{}_{\mu'} a^\nu{}_{\nu'} a^\rho{}_{\rho'} \cdots \bar{T}^{\mu'\nu'\rho'\cdots}$$
$$\bar{T}^{\mu\nu\rho\cdots} = \bar{a}^\mu{}_{\mu'} \bar{a}^\nu{}_{\nu'} \bar{a}^\rho{}_{\rho'} \cdots T^{\mu'\nu'\rho'\cdots} \qquad (6.91)$$

As an example, consider a four-vector $\mathbf{x}$ that locates a second event along the z-axis in Fig. 6.10 occurring at (z, t) in f, and at $(\bar{z}, \bar{t})$ in f'

$$x^\mu = [0, 0, z, ct]$$
$$\bar{x}^\mu = \bar{a}^\mu{}_\nu x^\nu = [0, 0, \bar{z}, c\bar{t}] \qquad (6.92)$$

Given Eqs. (6.84-6.85), $\bar{x}^\mu$ is simply obtained by matrix multiplication

$$\begin{bmatrix} 0 \\ 0 \\ \bar{z} \\ c\bar{t} \end{bmatrix} = \frac{1}{(1 - \beta^2)^{1/2}} \begin{bmatrix} (1 - \beta^2)^{1/2} & 0 & 0 & 0 \\ 0 & (1 - \beta^2)^{1/2} & 0 & 0 \\ 0 & 0 & 1 & -\beta \\ 0 & 0 & -\beta & 1 \end{bmatrix} \begin{bmatrix} 0 \\ 0 \\ z \\ ct \end{bmatrix}$$

$$\bar{z} = \frac{z - vt}{(1 - \beta^2)^{1/2}}$$

$$\bar{t} = \frac{t - vz/c^2}{(1 - \beta^2)^{1/2}} \qquad (6.93)$$

Suppose that the event occurs at the position of the origin O'. Then

$$z = vt \qquad\qquad ; \text{ position of } O'$$

$$\Rightarrow \qquad \bar{z} = 0$$

$$\bar{t} = t\,(1 - \beta^2)^{1/2} \qquad \text{time dilation} \qquad (6.94)$$

We recover our previous result for time dilation.

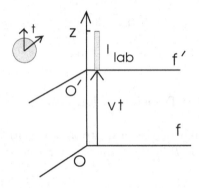

Fig. 6.11   A bar with length l at rest in the frame f' and oriented along the z-axis. Here (z, t) measures the position of the leading end of the bar at time t in the frame f (the "lab") and (vt, t) the position of the trailing end of the bar at that same time.

Suppose one has a bar of length l at rest in the frame f' and oriented along the z-axis starting at O'. Let a second event occur when the leading end of the bar reaches some point z in the frame f (the "lab"). This will occur at some time t in f. The position of the trailing end of the bar in f at the time t will simply be vt. The situation is illustrated in Fig. 6.11. In this case

$$\bar{z} = l \qquad\qquad ; \text{ length in } f'$$

$$l_{\text{lab}} = z - vt$$

$$= l\,(1 - \beta^2)^{1/2} \qquad (6.95)$$

The bar appears *shorter* in the frame f. There is a *Lorentz contraction* of its length in the direction of motion, with

$$l_{\text{lab}} = l\,(1 - \beta^2)^{1/2} \qquad ; \text{Lorentz contraction} \qquad (6.96)$$

This effect also been repeatedly verified over the years.[19]

Both length and time change from one inertial frame to another in SR, but the coordinate changes in the Lorentz transformation are just such as to keep the *speed of light constant* in the two inertial frames. To see this, suppose one has two events in space-time that can be connected with a light signal. For example, light is emitted by one oscillator at one point and absorbed by a second oscillator at another point.[20] Start in the inertial frame f. An observer there says the first event occurred at $\mathbf{x}_1 = (\vec{x}_1, ct_1)$ and the second at $\mathbf{x}_2 = (\vec{x}_2, ct_2)$. The observer then measures a velocity c for the light signal (distance/time). This implies that the global *interval* connecting these two events vanishes in f

$$
\begin{aligned}
\mathbf{s}_{12}^2 &= (\mathbf{x}_1 - \mathbf{x}_2)^2 \\
&= (\vec{x}_1 - \vec{x}_2)^2 - c^2(t_1 - t_2)^2 = 0 \quad ; \text{ light-like interval} \quad (6.97)
\end{aligned}
$$

The interval is a *physical quantity* that is preserved under a coordinate transformation. Now make a Lorentz transformation to a second inertial frame f'. An observer in this frame says the first event occurs at $\mathbf{x}_1 = (\vec{x}_1', ct_1')$ and the second at $\mathbf{x}_2 = (\vec{x}_2', ct_2')$, but since the interval again vanishes

$$
\begin{aligned}
\mathbf{s}_{12}^2 &= (\mathbf{x}_1 - \mathbf{x}_2)^2 \\
&= (\vec{x}_1' - \vec{x}_2')^2 - c^2(t_1' - t_2')^2 = 0 \quad (6.98)
\end{aligned}
$$

the second observer measures the *same velocity for the light signal* (same distance/time). Note that the minus sign in the fourth component of the Lorentz metric in Eq. (6.5) is what leads to this result.

6.6 Relativistic Hydrodynamics

We start with a global analysis, and will later make it local.

Consider an isotropic fluid with no shear forces. Introduce a set of cartesian spatial basis vectors $(\mathbf{e}_1, \mathbf{e}_2, \mathbf{e}_3)$ with origin O in the laboratory frame f. Assume, to start with, a velocity field $\vec{v} = v\,\mathbf{e}_3$ where $\vec{v} = $ constant. Now go to the rest frame of the fluid f', and introduce a corresponding set

---

[19]For example, through the length of a linear accelerator as seen by highly relativistic electrons, which plays an important role in our ability to steer them.

[20]Compare Fig. 6.9 where the first event occurs at the origin and the second event, illustrated there, is one connected to the origin by a light signal.

of spatial basis vectors $(\alpha_1, \alpha_2, \alpha_3)$ with an origin O' that coincides with O when the clocks start.

In special relativity there is a second-rank *energy-momentum tensor* for the fluid[21]

$$\underline{\underline{T}} = T^{\mu\nu}\,\mathbf{e}_\mu\mathbf{e}_\nu = \bar{T}^{\mu\nu}\,\alpha_\mu\alpha_\nu \quad ; \text{ energy-momentum tensor} \quad (6.99)$$

The energy-momentum tensor $\bar{T}_{\mu\nu}$ in the rest frame of the fluid f' is *defined* as

$$\bar{T}^{ij} = P\,\delta^{ij} \qquad\qquad ; \text{ pressure}$$
$$\bar{T}^{44} = \rho c^2 \qquad\qquad \text{proper energy density} \qquad (6.100)$$

where, once again, the Latin indices refer to the spatial components.[22] Here P is the pressure, ρ is the proper mass density, and ρc^2 is the proper energy density.

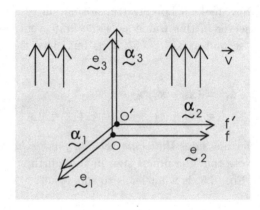

Fig. 6.12　Isotropic fluid with no shear forces moving with a uniform velocity $\vec{v} = v\,\mathbf{e}_3$. Here $(\mathbf{e}_1, \mathbf{e}_2, \mathbf{e}_3)$ are a set of cartesian spatial basis vectors in the laboratory frame f, and $(\alpha_1, \alpha_2, \alpha_3)$ are a set of cartesian spatial basis vectors in the rest frame of the fluid f'. The origins O and O' coincide when the clocks start. The third and fourth components of the basis vectors are related by Eqs. (6.76) and (6.77).

The tensor transformation law allows us to re-express this tensor in the

---

[21] See [Fetter and Walecka (2003)] for a discussion of non-relativistic fluid mechanics. In the present work, we use interchangeably the terms *stress tensor* and *energy-momentum tensor* for the fluid.

[22] Since the spatial part of the Lorentz metric in a cartesian basis is just the identity matrix, it does not matter whether the *spatial indices* are up or down in this case.

laboratory frame f

$$T^{\mu\nu} = a^{\mu}{}_{\mu'} a^{\nu}{}_{\nu'} \bar{T}^{\mu'\nu'} \tag{6.101}$$

Upon substitution of Eq. (6.100), this become

$$T^{\mu\nu} = P a^{\mu}{}_i a^{\nu}{}_i + \rho c^2 a^{\mu}{}_4 a^{\nu}{}_4 \tag{6.102}$$

Equation (6.89) can now be used to rewrite the second contribution in terms of the four-velocity of the fluid

$$a^{\mu}{}_4 a^{\nu}{}_4 = \frac{1}{c^2} u^{\mu} u^{\nu}$$

$$\frac{1}{c} u^{\mu} = \frac{1}{(1-\beta^2)^{1/2}} [0, 0, \beta, 1] \qquad ; \text{ four-velocity of fluid} \tag{6.103}$$

If the fluid is moving in an arbitrary direction, the four-velocity is then given by Eq. (6.90)

$$\frac{1}{c} u^{\mu} = \frac{1}{(1-\beta^2)^{1/2}} [\vec{\beta}, 1] \qquad ; \text{ four-velocity of fluid} \tag{6.104}$$

where $\vec{\beta} = \vec{v}/c$.

For the first term in Eq. (6.102), we claim that for the present configuration

$$a^{\mu}{}_i a^{\nu}{}_i = g^{\mu\nu} + \frac{1}{c^2} u^{\mu} u^{\nu} \tag{6.105}$$

Our rather inelegant proof is simply a direct evaluation using Eq. (6.82)[23]

$$a^1{}_i a^1{}_i = a^2{}_i a^2{}_i = 1$$
$$a^1{}_i a^2{}_i = a^1{}_i a^3{}_i = a^1{}_i a^4{}_i = a^2{}_i a^1{}_i = a^2{}_i a^3{}_i = 0 \qquad ; \text{ etc.}$$
$$a^4{}_i a^4{}_i = \frac{\beta^2}{1-\beta^2} = -1 + \frac{1}{1-\beta^2}$$
$$a^3{}_i a^3{}_i = \frac{1}{1-\beta^2} = 1 + \frac{\beta^2}{1-\beta^2}$$
$$a^3{}_i a^4{}_i = a^4{}_i a^3{}_i = \frac{\beta}{1-\beta^2} \tag{6.106}$$

---

[23]See, however, Prob. 6.16.

This is now to be compared with

$$g^{\mu\nu} + \frac{1}{c^2}u^\mu u^\nu = \begin{bmatrix} 1 & 0 & 0 & 0 \\ 0 & 1 & 0 & 0 \\ 0 & 0 & 1 + \frac{\beta^2}{1-\beta^2} & \frac{\beta}{1-\beta^2} \\ 0 & 0 & \frac{\beta}{1-\beta^2} & -1 + \frac{1}{1-\beta^2} \end{bmatrix} \quad (6.107)$$

This is identical to Eqs. (6.106), and the result is established. The condition $\vec{v} = v\,\mathbf{e}_3$ can now be relaxed by using Eq. (6.104).

In summary, the energy-momentum tensor for an isotropic fluid with no shear forces moving with a four velocity u^μ is given by

$$T^{\mu\nu} = Pg^{\mu\nu} + \left(\rho + \frac{P}{c^2}\right)u^\mu u^\nu \quad ; \text{ energy-momentum tensor} \quad (6.108)$$

We know that the four-velocity u^μ is a four-vector from the way it was defined. The metric is a second-rank tensor. The pressure and proper energy density are *defined in the rest frame of the fluid* and *do not transform*. They are Lorentz scalars. Hence $T^{\mu\nu}$ is indeed a second-rank Lorentz tensor, as advertised. Note that it is obviously symmetric

$$T^{\mu\nu} = T^{\nu\mu} \qquad ; \text{ symmetric} \qquad (6.109)$$

A classical fluid is made up of particles, which are neither created nor destroyed. The *particle current* forms a four-vector in special relativity. Still working on a global basis, one can define the particle density and current in the rest frame of the fluid as

$$\bar{N}^4 = n \qquad\qquad ; \text{ particle density}$$
$$\bar{N}^i = 0 \qquad\qquad\qquad\qquad\qquad\qquad (6.110)$$

In the laboratory frame, the transformed current is then given by

$$N^\mu = a^\mu{}_\nu \bar{N}^\nu = na^\mu{}_4 \qquad (6.111)$$

Thus

$$N^\mu = \frac{n}{c}u^\mu \qquad ; \text{ current} \qquad (6.112)$$

Here u^μ is the four-velocity of the fluid, and since the particle density n is again *defined in the rest frame of the fluid* and *does not transform*, this particle current is indeed a four-vector, as claimed.

The above relations have all been derived as *global* statements, where one assumes a common rest frame for the fluid and then Lorentz transforms with a given $\vec{v} = c\vec{\beta}$ to the laboratory frame. In fact they are *local* in the sense that if things *vary*, so that one has a fluid density $n(x)$ and velocity field $\vec{v}(x)$ where $x \equiv (\vec{x}, t)$, then one can just assign a current $n(x)u^{\mu}(x)/c$ and energy-momentum tensor $T^{\mu\nu}(x)$ at each point.

There will be a *local rest frame* for the fluid in which the fluid velocity, and hence the current, vanishes at the origin (Fig. 6.13). We start in that inertial frame and then carry out a coordinate transformation (Lorentz transformation) back to the laboratory frame. We proceed to establish a set of *conservation laws* for these local quantities.

Consider first the current of Eq. (6.112), which at the point $x = (\vec{x}, t)$ in the laboratory frame now has the form

$$N^{\mu} = \frac{n(x)}{c} u^{\mu}(x) \tag{6.113}$$

Current conservation here reflects the conservation of the number of the particles composing the fluid.[24] In the local rest frame the particle current $\bar{S}^{i}$ vanishes at the origin, but since it now varies from point to point, its spatial divergence $\partial \bar{S}^{i}/\partial \bar{x}^{i}$ may not vanish there, and hence $\partial n/\partial \bar{t}$ may be non-zero.

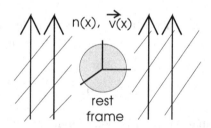

Fig. 6.13 Transition from global to local analysis. The fluid density and velocity field now vary with $x = (\vec{x}, t)$. There is a local frame in which the fluid is at rest at the origin.

Close to the origin in the rest frame all velocities are small, and thus we can use the familiar non-relativistic analysis.[25] The continuity equation in

---

[24]This is exactly analogous to electromagnetic current conservation in E&M, where the underlying particles are electrons.

[25]It is an essential assumption here that things vary smoothly from point to point in the fluid. Note that local here means local in space-time.

the local rest frame then states that

$$\frac{\partial n}{\partial \bar{t}} + \frac{\partial}{\partial \bar{x}^i} \bar{S}^i = 0 \qquad ; \text{continuity equation} \qquad (6.114)$$

With the introduction of the current four-vector

$$\bar{N}^\mu \equiv \left(\frac{1}{c}\bar{S}^i, \, n\right) \qquad (6.115)$$

Eq. (6.114) can be written as

$$\frac{\partial}{\partial \bar{x}^\mu} \bar{N}^\mu = 0 \qquad (6.116)$$

The bars again indicate quantities in the local rest frame. This is the statement of current conservation at the origin, which can also be written with the gradient of Eq. (3.61) as

$$\bar{\nabla}_\mu \bar{N}^\mu = 0 \qquad (6.117)$$

Now the *affine connection* vanishes in a space where the metric in cartesian coordinates is everywhere given by Eq. (6.79)

$$\bar{\Gamma}^\lambda_{\mu\nu} = 0 \qquad (6.118)$$

Hence the *ordinary* derivative is also the *covariant* derivative in this case. Thus the l.h.s. of Eq. (6.116) can just as well be written as a covariant divergence

$$\bar{N}^\mu_{\,;\mu} = 0 \qquad (6.119)$$

Now make the coordinate transformation (Lorentz transformation) from the local rest frame back to the laboratory frame. The covariant divergence transforms as a scalar under a coordinate transformation, and thus one immediately has the corresponding statement of current conservation in the laboratory frame as[26]

$$N^\mu_{\,;\mu} = \bar{N}^\mu_{\,;\mu} = 0 \qquad (6.120)$$

---

[26]There is a single Lorentz transformation to and from the rest frame of the fluid determined by the fluid velocity at the point x; there are no additional derivatives in Eq. (6.120). A simple translation back to a common origin in the laboratory frame is implied at the end.

Again, the affine connection vanishes with the metric of Eq. (6.78), and the covariant divergence in the lab frame reduces to the ordinary divergence[27]

$$N^{\mu}{}_{;\mu} = \frac{\partial}{\partial x^{\mu}} N^{\mu} = 0 \tag{6.121}$$

With the substitution of Eq. (6.113), this relation becomes

$$\frac{\partial}{\partial x^{\mu}} \left[\frac{n(x)}{c} u^{\mu}(x) \right] = 0 \qquad ; \text{ current conservation} \tag{6.122}$$

This is the statement of conservation of particles in the laboratory frame in special relativity.[28]

In detail, with the substitution of Eq. (6.90), this relation states

$$\vec{\nabla} \cdot \left[\frac{n\,\vec{\beta}}{(1 - \beta^2)^{1/2}} \right] + \frac{1}{c} \frac{\partial}{\partial t} \left[\frac{n}{(1 - \beta^2)^{1/2}} \right] = 0$$
$$; \text{ current conservation} \tag{6.123}$$

Here $n(\vec{x}, t)$ is the local particle density, defined in the local rest frame, and $\vec{v}(\vec{x}, t) = c\,\vec{\beta}(\vec{x}, t)$ is the local laboratory fluid velocity. The non-relativistic limit (NRL) of this expression simplifies considerably and reduces to the familiar result

$$\frac{\partial n}{\partial t} + \vec{\nabla} \cdot (n\,\vec{v}) = 0 \qquad ; \text{ current conservation–NRL} \tag{6.124}$$

Consider next the conservation of the *energy-momentum tensor*. Go to the rest frame of the fluid, and in the vicinity of the origin in the rest frame one can again use the non-relativistic analysis. With $u^{\mu} \to (\vec{v}, c)$ and $P/\rho c^2 \to 0$, Eq. (6.108) reduces to

$$\bar{T}^{ij} = P\delta^{ij} + \rho\,\bar{v}^i\bar{v}^j$$
$$\bar{T}^{4i} = \rho c\,\bar{v}^i \qquad\qquad ; \bar{T}^{44} = \rho c^2 \tag{6.125}$$

---

[27]While proceeding through the covariant divergence may seem pedantic here, soon it will be essential.

[28]For an alternate derivation of current conservation in the laboratory frame, see Prob. 6.9. We leave it as an exercise for the dedicated reader to provide the equivalent alternate derivation of the conservation of the stress tensor in Eq. (6.130).

and the non-relativistic form of the conservation laws is[29]

$$\frac{\partial}{\partial \bar{x}^j}\bar{T}^{ij} + \frac{\partial}{\partial c\bar{t}}\,\bar{T}^{i4} = 0 \quad ; \text{ Newton's law}$$

$$\frac{\partial}{\partial \bar{x}^j}\bar{T}^{4j} + \frac{\partial}{\partial c\bar{t}}\,\bar{T}^{44} = 0 \quad \text{ energy conservation} \qquad (6.126)$$

The first relation is just Newton's law for the fluid, which leads to the basic equation of hydrodynamics (as verified through the NRL below). The second relation is the continuity equation for the energy density ρc^2, which expresses the conservation of energy (also verified below).[30] These relations can be combined to express energy-momentum conservation in the rest frame as the tensor relation

$$\frac{\partial}{\partial \bar{x}^\nu}\bar{T}^{\mu\nu} = \bar{\nabla}_\nu \bar{T}^{\mu\nu} = 0 \qquad ; \mu = 1, \cdots, 4 \qquad (6.127)$$

The affine connection vanishes, and this is the same as a statement on the covariant divergence

$$\bar{T}^{\mu\nu}{}_{;\nu} = 0 \qquad\qquad\qquad (6.128)$$

Now, as before, carry out a coordinate transformation (Lorentz transformation) back to the laboratory frame. The contracted indices do not transform, and thus

$$T^{\mu\nu}{}_{;\nu} = a^\mu{}_{\mu'}\bar{T}^{\mu'\nu}{}_{;\nu} = 0 \qquad\qquad (6.129)$$

The covariant divergence is again the same as the ordinary divergence, and the statement of energy-momentum conservation in the laboratory frame becomes

$$\frac{\partial}{\partial x^\nu}T^{\mu\nu} = 0 \qquad ; \text{ energy-momentum conservation}$$

$$\mu = 1, \cdots 4 \qquad\qquad (6.130)$$

The local stress tensor is given by

$$T^{\mu\nu}(x) = P(x)g^{\mu\nu} + \left[\rho(x) + \frac{P(x)}{c^2}\right]u^\mu(x)u^\nu(x) \qquad (6.131)$$

---

[29] See [Fetter and Walecka (2003)] pp. 297, 300. Note that we are here using the terms *stress tensor* and *energy-momentum tensor* interchangeably.

[30] Just as the continuity equation for the particle density expresses conservation of the number of particles.

and the local four-velocity is

$$\frac{1}{c}u^\mu(x) = \frac{1}{[1 - \beta^2(x)]^{1/2}}[\,\vec{\beta}(x),\,1] \qquad ; \vec{v} = \vec{\beta}c \qquad (6.132)$$

We proceed to write Eq. (6.130) out in detail.

(1) For $\mu = i$, a spatial index, one has

$$\frac{\partial}{\partial x^i}P + \frac{\partial}{\partial x^j}\left[\left(\rho + \frac{P}{c^2}\right)u^i u^j\right] + \frac{\partial}{\partial ct}\left[\left(\rho + \frac{P}{c^2}\right)\frac{u^i c}{(1 - \beta^2)^{1/2}}\right] = 0$$

$$\Rightarrow \quad \vec{\nabla}P + \frac{\partial}{\partial x^j}\left[\left(\rho + \frac{P}{c^2}\right)\frac{\vec{v}\,v^j}{1 - \beta^2}\right] + \frac{\partial}{\partial ct}\left[\left(\rho + \frac{P}{c^2}\right)\frac{\vec{v}\,c}{1 - \beta^2}\right] = 0$$

$$(6.133)$$

(2) For $\mu = 4$, one finds

$$\frac{\partial}{\partial ct}(-P) + \frac{\partial}{\partial x^j}\left[\left(\rho + \frac{P}{c^2}\right)u^4 u^j\right] + \frac{\partial}{\partial ct}\left[\left(\rho + \frac{P}{c^2}\right)\frac{c^2}{1 - \beta^2}\right] = 0$$

$$\Rightarrow \quad \frac{\partial}{\partial ct}(-P) + \frac{\partial}{\partial x^j}\left[\left(\rho + \frac{P}{c^2}\right)\frac{v^j c}{1 - \beta^2}\right] + \frac{\partial}{\partial ct}\left[\left(\rho + \frac{P}{c^2}\right)\frac{c^2}{1 - \beta^2}\right] = 0$$

$$(6.134)$$

Now combine these two relations by taking $(1) - (\vec{v}/c)(2)$:

$$\vec{\nabla}P + \left(\rho + \frac{P}{c^2}\right)\frac{(\vec{v}\cdot\vec{\nabla})\,\vec{v}}{1 - \beta^2} + \left(\rho + \frac{P}{c^2}\right)\frac{1}{1 - \beta^2}\frac{\partial\vec{v}}{\partial t} + \frac{\vec{v}}{c}\frac{\partial P}{\partial ct} = 0 \quad (6.135)$$

This is the *relativistic Euler equation* for fluid flow

$$\frac{\partial\vec{v}}{\partial t} + (\vec{v}\cdot\vec{\nabla})\,\vec{v} = -\frac{(1 - \beta^2)}{(\rho + P/c^2)}\left[\vec{\nabla}P + \vec{v}\frac{\partial}{\partial t}\left(\frac{P}{c^2}\right)\right]$$

$$; \text{ relativistic Euler eqn} \quad (6.136)$$

The non-relativistic limit (NRL) yields the familiar *basic equation of hydrodynamics* for the velocity field $\vec{v}(\vec{x}, t)$

$$\frac{d\vec{v}}{dt} = \frac{\partial\vec{v}}{\partial t} + (\vec{v}\cdot\vec{\nabla})\,\vec{v} = -\frac{1}{\rho}\vec{\nabla}P + \vec{f}_{app} \quad ; \text{ hydrodynamics–NRL} \quad (6.137)$$

Here we have included the possibility of an additional (applied force/unit mass) $\vec{f}_{app}$.

The expression in (2) above [Eq. (6.134)] is the relativistic generalization of the continuity equation for the energy density. Multiplication by c and a subsequent expansion in $1/c^2$ provides the non-relativistic limit for *energy conservation* as (see Prob. 6.10)

$$\frac{\partial}{\partial t}\left(\rho\epsilon + \frac{1}{2}\rho v^2\right) + \vec{\nabla}\cdot\left[\vec{v}\left(\rho\epsilon + \frac{1}{2}\rho v^2 + P\right)\right] = \rho\,\vec{v}\cdot\vec{f}_{\text{app}}$$

$$;\text{ energy conservation–NRL} \quad (6.138)$$

Here $\rho\epsilon$ is the internal energy density in the laboratory system, which in leading order is just $\rho\epsilon = \rho c^2$ [see the second of Eqs. (6.126)]. Here we have again included the possibility of an additional force $\vec{f}_{\text{app}}$ doing work on the system. Equation (6.138) allows us to identify both the energy density and energy flux for the moving fluid through this order.

We remind the reader that we now have a collection of *proper* quantities that are defined in the *rest frame* and are scalars under Lorentz transformations: time τ, particle density n, mass density ρ, energy density ρc^2, and pressure P.

With this introduction to his theory of *special relativity*, we now move on to the central topic of this book, Einstein's theory of *general relativity*.

6.7 Transition to General Relativity

We know from freshman physics that if Newton's law of gravitation is combined with his second law of motion, then a non-relativistic point particle of mass m at a position $\vec{r}$ moving about a spherically symmetric gravitational force center of mass M satisfies the following equation of motion

$$\frac{d^2\vec{r}}{dt^2} = -MG\frac{\vec{r}}{r^3} \quad (6.139)$$

Here G is Newton's gravitational constant (see later). The mass m of the moving particle has disappeared entirely from this relation. All particles, regardless of their mass, follow the same trajectory.[31] Now this is actually a truly amazing result, one that ordinarily just passes by and is readily accepted at the freshman level.[32] The mass m_{inertial} appearing in Newton's second law has to do with acceleration relative to the primary inertial coordinate system, and the mass m_{gravity} appearing in his universal law of gravitation has to do with the strength of the force between two mutually

---

[31]This was observed long ago by Galileo in his classic experiment where various objects were dropped from the leaning tower of Pisa and observed to fall with the same acceleration.

[32]It was Einstein, of course, who recognized and explored its profound implications.

attracting gravitational bodies. Why should these quantities have anything at all to do with each other??

We have seen, however, a situation in which just this phenomenon occurs. In our introductory example[33] of a particle moving without friction on a surface of arbitrary shape, all particles follow the same trajectory. It is a trajectory determined purely by geometry, as particles follow the geodesics on the surface. The surface, in turn, is characterized by its metric, and the associated affine connection and Riemann curvature tensor. Is it possible that the same situation could hold for particles moving about a gravitating body? Could four-dimensional space-time be such that particles simply follow geodesics in that space, with a metric determined by the presence of the mass M in just such a way as to produce the correct orbits? We have already seen that, in the absence of a distorting mass M, we live in a space-time continuum with a metric, the Lorentz metric, that is more than a little surprising. We do not live in a euclidian space, but rather a Minkowski space. It does not require too much stretch of the imagination to have the space further "curved" by the presence of mass and energy. Of course, one has to incorporate the successful theory of special relativity, and, at least locally, there must be a flat Minkowski tangent space in which the principles of special relativity hold.[34]

If it were possible to formulate such a theory, one would have a unified understanding of both Newton's second law and his universal law of gravitation, two of the fundamental pillars of modern science. This is exactly the problem solved by Einstein's theory of general relativity, and we now turn to the basic principles of that theory.

---

[33] Admittedly, selected with foresight.

[34] It is worth emphasizing that our physical perceptions, which provide most of us with our physical intuition, sense only non-relativistic motion in local regions of space and for short periods of time over which the speed of light is essentially infinite.

Chapter 7

General Relativity

We proceed to a discussion of Einstein's theory of general relativity [Einstein (1916)].

7.1 Einstein's Theory

We present the basic concepts in the theory as a set of three principles:

I. *We live in a four-dimensional riemannian space.*

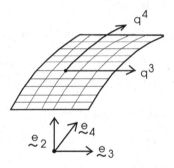

Fig. 7.1 Sketch of the four-dimensional riemannian space in which we live. The surface suppresses the two dimensions (q^1, q^2), and the set of basis vectors suppresses $\mathbf{e}_1$.

- There is a set of generalized coordinates $(q^1, \cdots, q^4)$ and corresponding set of basis vectors $(\mathbf{e}_1, \cdots, \mathbf{e}_4)$ in this space (Fig. 7.1). The *line element* is

$$ d\mathbf{s} = \mathbf{e}_\mu \, dq^\mu \qquad ; \text{ line element} \qquad (7.1) $$

The *physical interval* is given in terms of the *metric* by

$$(d\mathbf{s})^2 = g_{\mu\nu}dq^\mu dq^\nu \qquad ; \text{physical interval}$$

$$g_{\mu\nu} = \mathbf{e}_\mu \cdot \mathbf{e}_\nu \qquad\qquad \text{metric} \tag{7.2}$$

- Free, flat space is a Minkowski space with the Lorentz metric in a cartesian basis of

$$\underline{g} = \begin{bmatrix} 1 & 0 & 0 & 0 \\ 0 & 1 & 0 & 0 \\ 0 & 0 & 1 & 0 \\ 0 & 0 & 0 & -1 \end{bmatrix} \qquad ; \text{flat space} \tag{7.3}$$

- At each point in the *curved* space there is a *flat tangent space* with the metric of Eq. (7.3).

II. *The structure of this space is given by the Einstein field equations.*

$$\underline{\underline{G}} = \kappa \underline{\underline{T}} \qquad\qquad ; \text{Einstein field equations}$$

or ; $\qquad G^{\mu\nu} = \kappa T^{\mu\nu} \qquad\qquad \mu, \nu = 1, \cdots, 4 \tag{7.4}$

- Here $\underline{\underline{G}}$ is the *Einstein tensor*

$$\underline{\underline{G}} = G^{\mu\nu}\mathbf{e}_\mu\,\mathbf{e}_\nu \qquad ; \text{Einstein tensor} \tag{7.5}$$

It is derived from the *Riemann curvature tensor* in the space. The Einstein tensor is a second-order, nonlinear, differential form in the metric.

- The source $\underline{\underline{T}}$ is the *energy-momentum tensor*

$$\underline{\underline{T}} = T^{\mu\nu}\mathbf{e}_\mu\,\mathbf{e}_\nu \qquad ; \text{energy-momentum tensor} \tag{7.6}$$

for the system under consideration. The strength κ is a constant to be determined.

- Equation (7.4) is a tensor relation between physical quantities. It is a *local* relation holding at each point in space-time. It provides a differential equation for the metric, which is then determined by the energy-momentum present. The solution to these equations determines the metric and associated *curvature* of the riemannian space.

III. *Particles move along geodesics in this space.*

- Along the *actual* path of a particle, the physical interval is related to the proper time by (Fig. 7.2)

$$(d\mathbf{s})^2 = -c^2(d\tau)^2 \qquad ; \text{ actual path} \qquad (7.7)$$

- The geodesics are then determined by the equations

$$\frac{d^2q^\mu}{d\tau^2} + \Gamma^\mu_{\lambda\sigma} \frac{dq^\lambda}{d\tau} \frac{dq^\sigma}{d\tau} = 0 \qquad ; \text{ geodesics}$$

$$\mu = 1, \cdots, 4 \qquad (7.8)$$

Here $\Gamma^\mu_{\lambda\sigma}$ is the *affine connection* in the space. The solution to Eqs. (7.8) for a given set of initial conditions determines the geodesic $q^\mu(\tau)$ in the space of the generalized coordinates.

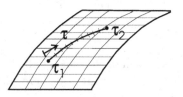

Fig. 7.2 Particle trajectory (geodesic) in the four-dimensional riemannian space in which we live. Here τ is the proper time along the actual path.

Let us enlarge a little on these principles. Recall from our general tensor analysis that the *Einstein tensor $G^{\mu\nu}$* is obtained from the *Ricci tensor $R^{\mu\nu}$*, *scalar curvature R*, and *Riemann curvature tensor $R^\lambda_{\mu\rho\nu}$* through the following relations

$$G^{\mu\nu} = R^{\mu\nu} - \frac{1}{2}R\,g^{\mu\nu} \qquad ; R \equiv R^\mu_\mu$$

$$R_{\mu\nu} = R^\lambda_{\mu\lambda\nu} \qquad (7.9)$$

The *Riemann curvature tensor* is related to the *affine connection* by

$$R^\lambda_{\mu\rho\nu} = \frac{\partial}{\partial q^\rho}\Gamma^\lambda_{\mu\nu} + \Gamma^\lambda_{\rho\sigma}\Gamma^\sigma_{\mu\nu} - (\rho \rightleftharpoons \nu) \qquad (7.10)$$

The affine connection is, in turn, obtained from the metric according to

$$\Gamma^\lambda_{\mu\nu} = \frac{1}{2}g^{\lambda\sigma}\left[\frac{\partial g_{\sigma\nu}}{\partial q^\mu} + \frac{\partial g_{\sigma\mu}}{\partial q^\nu} - \frac{\partial g_{\mu\nu}}{\partial q^\sigma}\right] \qquad (7.11)$$

All the quantities appearing in the basic principles (I-III) are now well defined.

With the substitution of Eq. (7.9) and the lowering of an index, the Einstein field equations become

$$R^\mu{}_\nu - \frac{1}{2} R \, g^\mu{}_\nu = \kappa T^\mu{}_\nu \tag{7.12}$$

Now set $\mu = \nu$ and sum from 1 to 4 (recall that $g^\mu{}_\nu = \delta^\mu{}_\nu$)

$$R - \frac{4}{2} R = \kappa T \qquad ; \quad T \equiv T^\mu{}_\mu$$
$$-R = \kappa T \tag{7.13}$$

Hence, upon substituting this expression for R on the l.h.s. of Eq. (7.12) and then taking the term to the r.h.s. and raising an index, one has

$$R^{\mu\nu} = \kappa \left(T^{\mu\nu} - \frac{1}{2} T \, g^{\mu\nu} \right)$$
$$T = T^\lambda{}_\lambda \tag{7.14}$$

This provides an equivalent form of the Einstein field equations. The Ricci tensor, and associated scalar curvature, are here directly related to the energy-momentum tensor.

Some insight into the Einstein field equations can be obtained in the following way. If mass and energy are to determine the structure of space-time, the most appropriate source is the energy-momentum tensor. This immediately leads one to a second-rank tensor relation between physical quantities. The structure of space-time should then be described with a rank-two tensor differential form in the metric. Since the energy-momentum tensor is symmetric and has vanishing covariant divergence, this second tensor must share those properties. Thus one is led to the Einstein tensor to provide the appropriate differential form in the metric.

We proceed to discuss implications of the three basic principles given above.

7.2   Newtonian Limit

We must first try to make a connection with what we know about gravity and Newton's laws. Consider the configuration in Fig. 7.3, and work with the situation where the space is *almost flat*, that is, where one is far enough away from the source.

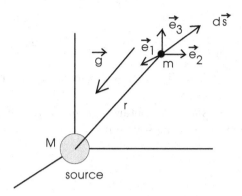

Fig. 7.3 Configuration for newtonian limit. There is a spherically symmetric gravitational source of mass M at the origin, which gives rise to a gravitational potential $\phi = -MG/r$ and gravitational field $\vec{g} = -\vec{\nabla}\phi = -g\,\hat{r}$. A point mass m moves in this field. Far away from the source the space is *almost flat*.

Introduce a cartesian basis so that the coordinates are $(q^1, q^2, q^3, q^4) = (x^1, x^2, x^3, ct)$. We put the complexity of the Minkowski space into the basis vectors so the metric is given by the second of Eqs. (7.2), and in flat space the metric is that of Eq. (7.3). Call this Lorentz metric $g^0_{\mu\nu}$. If the space is almost flat, there will be only a small modification of this metric, and we can write the first-order change as

$$g_{\mu\nu} = g^0_{\mu\nu} + \gamma_{\mu\nu}(\vec{x})$$
$$|\gamma_{\mu\nu}(\vec{x})| \ll 1 \tag{7.15}$$

We specifically consider a *static* situation where the modification $\gamma_{\mu\nu}(\vec{x})$ depends only on position.

The interval is given by the first of Eqs. (7.2), and hence one has

$$(ds)^2 = g_{\mu\nu}dx^\mu dx^\nu$$
$$= \left[g^0_{\mu\nu} + \gamma_{\mu\nu}(\vec{x})\right] dx^\mu dx^\nu \tag{7.16}$$

Division by $[d(ct)]^2$ and the use of Eq. (7.7) leads to following relation between time and proper time

$$\left(\frac{ds}{dct}\right)^2 = -\left(\frac{d\tau}{dt}\right)^2 = \left[g^0_{\mu\nu} + \gamma_{\mu\nu}(\vec{x})\right] \frac{dx^\mu}{dct}\frac{dx^\nu}{dct} \tag{7.17}$$

Now define the *newtonian limit* by letting the velocity of light become infinite $c \to \infty$, which implies $\vec{\beta} = \vec{v}/c \to 0$. In this case, only the coordinate $x^4 = ct$ contributes in Eq. (7.17), and hence that expression reduces to

$$\frac{d\tau}{dt} \approx [1 - \gamma_{44}(\vec{x})]^{1/2} \approx 1 - \frac{1}{2}\gamma_{44}(\vec{x}) \qquad (7.18)$$

We will find that $\gamma_{44}(\vec{x}) = O(1/c^2)$, and hence time and proper time are indistinguishable in the newtonian limit.[1]

Let us now invoke some elementary *newtonian physics*. Let ϕ be the gravitational potential from the source in Fig. 7.3. Then the gravitational field is

$$\vec{g} = -\vec{\nabla}\phi \qquad (7.19)$$

For a point mass at the origin (or outside of a spherically symmetric mass), the potential takes the form

$$\phi = -\frac{MG}{r} \qquad (7.20)$$

Here G is Newton's constant and is given by[2]

$$G = 6.673 \times 10^{-11} \ \text{m}^3/\text{kg-sec}^2$$
$$= 5.905 \times 10^{-39} \ \hbar c/\text{m}_\text{p}^2 \qquad (7.21)$$

Newton's second law plus his law of gravitation give for the particle motion

$$m_i \frac{d^2\vec{x}}{dt^2} = -m_g \vec{\nabla}\phi \qquad (7.22)$$

Here m_i is the inertial mass, the one entering into the second law, and m_g is the gravitational mass, the one entering into his law of gravitation. We now assume these masses are *identical*, and cancel them from this expression. This is known as the *equivalence principle*, and we shall have much more to say about this relation shortly

$$m_i = m_g \qquad \qquad \text{; equivalence principle} \qquad (7.23)$$

Equation (7.22) then reduces to

$$\frac{d^2\vec{x}}{dt^2} = -\vec{\nabla}\phi \qquad (7.24)$$

---

[1] That $\gamma_{44}(\vec{x}) = O(1/c^2)$ is verified in Eq. (7.31).

[2] In the second line $\hbar = h/2\pi$ where h is Planck's constant, and m_p is the proton mass. Although there is no quantum mechanics in anything we have done, this turns out to be an illuminating way to re-express G. For example, it clearly manifests the weakness of gravity in the atomic domain.

The orbit becomes only a function of the strength of the source and *geometry*.

Now go to the newtonian limit where $c \to \infty$

$$c \to \infty \qquad\qquad ; \text{ newtonian limit}$$

$$\Rightarrow \qquad \vec{\beta} \to 0 \;, \qquad \frac{2\phi}{c^2} \to 0 \qquad\qquad (7.25)$$

The last expression is a *dimensionless ratio* characterizing the strength of the gravitational field.

Return now to principle III of general relativity where the particle follows a *geodesic* in the space. We have argued above that time and proper time are indistinguishable in the newtonian limit, and in this limit, with the cartesian coordinates (x^1, x^2, x^3, ct), the geodesic equations take the form

$$\frac{d^2 x^\mu}{dt^2} + \Gamma^\mu_{\lambda\sigma} \frac{dx^\lambda}{dt} \frac{dx^\sigma}{dt} = 0 \qquad ; \text{ geodesic} \qquad (7.26)$$

Again, in the newtonian limit of Eq. (7.25), *only the coordinate* $x^4 = ct$ *contributes to the second term*, and this relation becomes[3]

$$\frac{d^2 x^\mu}{dt^2} + \Gamma^\mu_{44} c^2 \approx 0 \qquad\qquad (7.27)$$

We thus require only Γ^μ_{44} from the affine connection of Eq. (7.11). To first order in $\gamma_{\mu\nu}(\vec{x})$, this is given by

$$\Gamma^\mu_{44} \approx \frac{1}{2} \left[g^0\right]^{\mu\sigma} \left[\frac{\partial \gamma_{\sigma 4}(\vec{x})}{\partial(ct)} + \frac{\partial \gamma_{\sigma 4}(\vec{x})}{\partial(ct)} - \frac{\partial}{\partial x^\sigma} \gamma_{44}(\vec{x}) \right]$$

$$= -\frac{1}{2} \left[g^0\right]^{\mu\sigma} \left[\frac{\partial}{\partial x^\sigma} \gamma_{44}(\vec{x}) \right] \qquad\qquad (7.28)$$

Here we have used the fact that $\gamma_{\mu\nu}(\vec{x})$ is independent of time in arriving at the second line. This gives for the spatial indices $\mu = i = 1, 2, 3$, and for the time component $\mu = 4$,

$$\Gamma^i_{44} = -\frac{1}{2} \frac{\partial}{\partial x^i} \gamma_{44}(\vec{x}) \qquad\qquad ; \Gamma^4_{44} = 0 \qquad\qquad (7.29)$$

Hence the geodesic Eq. (7.27) reduces in the newtonian limit to

$$\frac{d^2 \vec{x}}{dt^2} = \vec{\nabla} \left[\frac{c^2}{2} \gamma_{44}(\vec{x}) \right] \qquad\qquad (7.30)$$

---

[3]Note that $\Gamma^\mu_{\lambda\sigma}$, since it involves derivatives of the metric, is of $O(\gamma)$.

To make the connection with newtonian physics, we *define*

$$\gamma_{44}(\vec{x}) \equiv -\frac{2\phi}{c^2} = \frac{2MG}{c^2 r} \qquad (7.31)$$

This is indeed a small quantity by our assumption in Eq. (7.25).[4] Equation (7.30) is now identical to Eq. (7.24).

We have thus accomplished part of the goal we set out for ourselves in section 6.7.

We have obtained Newton's second law and his law of gravitation from the basic principles of general relativity and a small, static modification of the metric as given in Eq. (7.31).

It is interesting to look back and see just how this comes about:

- We assume a small, static, additional term in the metric $\gamma_{\mu\nu}(\vec{x}) = O(1/c^2)$ in the newtonian limit;
- Since the fourth component of coordinates is $x^4 = ct$, and since we are taking the limit $c \to \infty$, it is only $c^2 \Gamma^{\mu}_{44}$ that enters into the geodesic equation;
- Γ^4_{44} vanishes, consistently implying $d^2(x^4)/dt^2 = 0$;
- The affine connection Γ^i_{44} then relates the acceleration $d^2\vec{x}/dt^2$ to the spatial gradient of the modification of the metric $\vec{\nabla}[c^2\gamma_{44}(\vec{x})/2] = -\vec{\nabla}\phi$, through which we establish the connection to Newton's laws;
- It is the modification of the metric $g^0_{44} \to g^0_{44} + \gamma_{44}(\vec{x})$ that provides the appropriate newtonian limit (Fig. 7.4). The corresponding modification of the *interval* is (recall $g^0_{44} = -1$)

$$(ds)^2 \approx (d\vec{x})^2 - \left(1 + \frac{2\phi}{c^2}\right)(cdt)^2 \quad ; \text{ so far} \qquad (7.32)$$

If the time component of the interval takes this form as $c \to \infty$, then one recovers both Newton's second law, and his law of gravitation.

- Note that we have taken a non-trivial journey through the time component of the metric and the affine connection in the riemannian space to arrive at this result!

---

[4]Note, however, that the limit $2\phi/c^2 \to 0$ is a rather subtle one. For large, but finite c, it is always satisfied as $r \to \infty$, that is, at large distances where the gravitational field is weak. On the other hand, it is always *violated* with a point source (or outside of a spherically symmetric source) as $r \to 0$. This means that the newtonian limit does *not* hold with a point source if one gets close enough to it. This will come back to us in many ways in GR.

• We then have just *classical, newtonian physics.*[5]

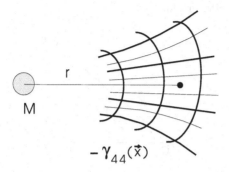

Fig. 7.4   The gravitational field outside a spherically symmetric mass M in the newtonian limit arises from the gradient of the time component of the metric $\vec{\nabla}[c^2\gamma_{44}(\vec{x})/2]$ where $\gamma_{44}(\vec{x}) = -2\phi/c^2 = 2MG/c^2r$.

It remains, of course, to demonstrate that we indeed obtain the limiting form of the metric in Eq. (7.32) from the Einstein field equations.

7.3   The Equivalence Principle

We recall the statement of the *equivalence principle*

$$m_i = m_g \qquad ; \text{equivalence principle} \qquad (7.33)$$

• m_i is the inertial mass, the one entering into Newton's second law describing acceleration with respect to the fixed stars;
• m_g is the gravitational mass, the one describing the coupling strength of the gravitational force;
• The equivalence principle implies that the mass drops out of the equation of motion for a particle in a gravitational field and reduces the problem to one of pure *geometry* (Einstein).

---

[5]Note carefully where the c^2 appears in Eq. (7.32). The term $(2\phi/c^2)(cdt)^2$ is finite as $c^2 \to \infty$. While the distinction between time and proper time vanishes in the newtonian limit [Eq. (7.18)], there is a finite modification of the interval, and corresponding finite contribution to the geodesic, resulting in Newton's laws. Note also that any $O(1/c^2)$ modification of the spatial part of the metric (see later) does not affect this argument.

Let us return to ordinary classical mechanics, and consider the motion of a particle in an *accelerated reference frame*.[6] Consider the configuration in Fig. 7.5 where the frame with origin O' is accelerating with respect to the inertial frame with origin O.

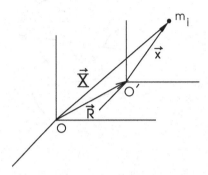

Fig. 7.5   Motion in an accelerated coordinate system in classical mechanics. The frame with origin O' is accelerating with respect to the inertial frame with origin O.

Let

$$\vec{X} = \vec{R} + \vec{x} \tag{7.34}$$

Two differentiations with respect to time give (there is only one time in classical mechanics)

$$\frac{d^2\vec{X}}{dt^2} = \frac{d^2\vec{R}}{dt^2} + \frac{d^2\vec{x}}{dt^2} \tag{7.35}$$

Now identify the acceleration of O' as

$$\frac{d^2\vec{R}}{dt^2} = \vec{a} \tag{7.36}$$

Newton's second law for a particle of mass m_i acted on by a force $\vec{\mathcal{F}}_{app}$ in the inertial frame with origin O states

$$m_i \frac{d^2\vec{X}}{dt^2} = \vec{\mathcal{F}}_{app} \tag{7.37}$$

With the substitution of the above, this becomes

$$m_i \frac{d^2\vec{x}}{dt^2} = \vec{\mathcal{F}}_{app} - m_i \vec{a} \tag{7.38}$$

---

[6]See [Fetter and Walecka (2003)] chap. 2.

There is an additional *inertial force* in the accelerated coordinate system given by $-m_i\vec{a}$.

Now suppose there is an additional *local* gravitational field $\vec{g}$ present, to which the particle couples with strength m_g, so that the applied force becomes

$$\vec{\mathcal{F}}_{\text{app}} = m_g\,\vec{g} + \delta\vec{\mathcal{F}}_{\text{app}} \tag{7.39}$$

Then

$$m_i\frac{d^2\vec{x}}{dt^2} = m_g\,\vec{g} - m_i\vec{a} + \delta\vec{\mathcal{F}}_{\text{app}}$$
$$m_i\frac{d^2\vec{x}}{dt^2} = m_i(\vec{g} - \vec{a}) + \delta\vec{\mathcal{F}}_{\text{app}} \tag{7.40}$$

Here the equivalence principle has been invoked in the second line. Now, if

$$\vec{a} = \vec{g} \tag{7.41}$$

then

- The effect of the gravitational field is *exactly canceled* in Eq. (7.40);
- The effect of the inertial force is *exactly canceled* in Eq. (7.40).

The equation of motion in the *accelerated frame* now becomes

$$m_i\frac{d^2\vec{x}}{dt^2} = \delta\vec{\mathcal{F}}_{\text{app}} \tag{7.42}$$

which is identical to Newton's law in an *inertial frame*!

How does one achieve the situation in Eq. (7.41) where the acceleration of the reference frame (or "laboratory") is identical to the acceleration of gravity? Just let the frame *fall freely* in the gravitational field!

7.4 Local Freely Falling Frame (LF^3)

Based upon the above arguments, we make the central observation in general relativity that

> In the flat tangent space there is a "local freely falling frame (LF^3)" in which the effect of a local gravitational field is exactly canceled by the inertial force, and one has just a free, flat space plus the laws of special relativity.

- The LF^3 will provide the *key* for our physical interpretation of general relativity;
- The interval and metric in cartesian coordinates in the LF^3 are just those of flat Minkowski space

$$(ds)^2 = (d\vec{x})^2 - (cdt)^2 \qquad ; \text{ in } LF^3$$

$$\underline{g} = \begin{bmatrix} 1 & 0 & 0 & 0 \\ 0 & 1 & 0 & 0 \\ 0 & 0 & 1 & 0 \\ 0 & 0 & 0 & -1 \end{bmatrix} \qquad (7.43)$$

- In the LF^3 one has just free, flat space and the laws of special relativity; there is *no* gravitational field;
- The LF^3 is just held and let go at a given point in space-time; there is no initial *velocity*, but there is an initial *acceleration* of the frame;
- It is essential to realize that this is a *local* statement. It only holds over a small (strictly infinitesimal) region in space-time. In general the gravitational field depends on $x = (\vec{x}, t)$, and one can only cancel out its effects *locally*.

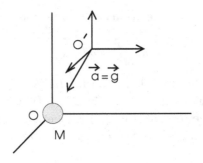

Fig. 7.6   The local freely falling frame (LF^3) in the gravitational field of a point source of mass M.

We can summarize the situation by presenting two *basic concepts* of special and general relativity (see Fig. 7.7):

(1) *No experiment we can do inside a closed laboratory will tell us whether or not we are moving with a constant velocity $\vec{v}$ with respect to the fixed stars;*

(2) *No experiment we can do inside a small closed laboratory will distinguish between*
    (a) *an acceleration $\vec{a}$ with respect to the fixed stars;*
    (b) *a local gravitational field $\vec{g} = -\vec{a}$.*

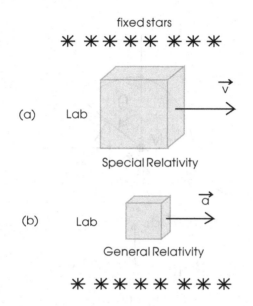

Fig. 7.7   (a) A global closed lab moving with constant velocity $\vec{v}$ with respect to the fixed stars in special relativity; (b) A local closed lab moving with acceleration $\vec{a}$ with respect to the fixed stars in general relativity.

7.5   Spherically Symmetric Solution to Field Equations

As the next step in making a connection to what we already know, let us seek a solution to the Einstein field equations outside of a spherically symmetric mass distribution. Let us see if that solution indeed reproduces the form of the metric assumed in Eq. (7.32). It is evident from the form of Eqs. (7.14) that in the source-free region we seek a solution to[7]

$$R^{\mu\nu} = 0 \qquad ; \text{ source-free region} \qquad (7.44)$$

---

[7]We already had a hint that we will be able to find a non-trivial solution to the source-free field equations $G_{\mu\nu} = 0$ from the fact that the Einstein tensor vanishes on the surface of a sphere, even though it is a curved space.

We shall again consider a static situation so that

$$g_{\mu\nu} = g_{\mu\nu}(\vec{x}) \qquad ; \text{ static solution} \tag{7.45}$$

Introduce spherical coordinates (Fig. 7.8) so that the coordinates are now $(q^1, \cdots, q^4) = (r, \theta, \phi, ct)$.

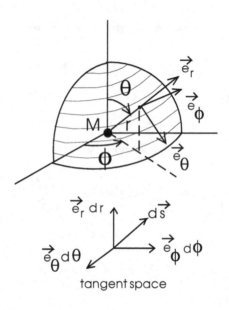

tangent space

Fig. 7.8 Spatial coordinate system for spherically symmetric solution to Einstein field equations.

The line element is then[8]

$$d\mathbf{s} = \mathbf{e}_r \, dr + \mathbf{e}_\theta \, d\theta + \mathbf{e}_\phi \, d\phi + \mathbf{e}_4 \, d(ct) \tag{7.46}$$

The basis vectors here are orthogonal.

From the spherical symmetry of the problem, all angles (θ, ϕ) are equivalent, and one would expect the angular part of the metric simply to be that of spherical coordinates. On the other hand, we expect the source to modify the radial part of the metric, and we know from our discussion of the newtonian limit that there should also be a modification of the time component. Therefore, let us seek a metric that yields an interval of the

---

[8]Here we change to a slightly more illuminating notation. Instead of using $\mu = (1, \cdots, 4)$ for subscripts and superscripts, we use $\mu = (r, \theta, \phi, 4)$.

form

$$(ds)^2 = A(r)(dr)^2 + r^2(d\theta)^2 + r^2 \sin^2 \theta (d\phi)^2 - B(r)(cdt)^2 \quad (7.47)$$

Here $[A(r), B(r)]$ are functions to be determined. Note that if $dr = dt = 0$, this is just the usual interval on the surface of a sphere. Consequently, the equatorial circumference of a sphere with radial coordinate r is, as usual, $2\pi r$, and the area of that sphere is $4\pi r^2$.

The metric is readily identified from the interval in Eq. (7.47). The metric matrix is diagonal, with diagonal elements

$$g_{rr} = A(r)$$
$$g_{\theta\theta} = r^2$$
$$g_{\phi\phi} = r^2 \sin^2 \theta$$
$$g_{44} = -B(r) \quad (7.48)$$

The inverse of the metric matrix is therefore also diagonal, and the diagonal elements of the inverse are

$$g^{rr} = \frac{1}{A(r)}$$
$$g^{\theta\theta} = \frac{1}{r^2}$$
$$g^{\phi\phi} = \frac{1}{r^2 \sin^2 \theta}$$
$$g^{44} = -\frac{1}{B(r)} \quad (7.49)$$

The functional dependence of the metric is $[g_{rr}(r), g_{\theta\theta}(r), g_{\phi\phi}(r, \theta), g_{44}(r)]$; there is no (ϕ, ct) dependence in the metric.

The next step is to compute the affine connection

$$\Gamma^{\lambda}_{\mu\nu} = \frac{1}{2} g^{\lambda\sigma} \left[\frac{\partial g_{\sigma\nu}}{\partial q^\mu} + \frac{\partial g_{\sigma\mu}}{\partial q^\nu} - \frac{\partial g_{\mu\nu}}{\partial q^\sigma} \right] \quad (7.50)$$

There are $4 \times 4 \times 4 = 64$ elements in the affine connection. Given the fact that the metric and its inverse are diagonal, it follows that the affine connection vanishes if all the indices are different

$$\Gamma^{\lambda}_{\mu\nu} = 0 \qquad ; \lambda \neq \mu \neq \nu$$

$$24 \text{ terms} \quad (7.51)$$

There are $4 \times 3 \times 2 = 24$ such terms. There are therefore 40 terms which have two or more identical indices, 10 for each value of λ. Of these 10, there will be 6 which have lower indices that differ from one another; however, we know that the affine connection is symmetric in the lower two indices, so we need calculate only half of these. Thus there are a total of $4 \times 7 = 28$ actual calculations that need to be carried out. These calculations are essential for an understanding of the theory and for subsequent developments. Thus we carry them out here in detail. We group together the 10 terms for each value of λ.[9] The results for the affine connection are summarized in Table 7.1.

1) $\underline{\lambda = r}$ (10 terms).

$$\Gamma^r_{rr} = \frac{1}{2} g^{rr} \left[\frac{\partial g_{rr}}{\partial r} + \frac{\partial g_{rr}}{\partial r} - \frac{\partial g_{rr}}{\partial r} \right] = \frac{1}{2A(r)} \left[\frac{dA(r)}{dr} \right]$$

$$\Gamma^r_{r\phi} = \frac{1}{2} g^{rr} \left[\frac{\partial g_{r\phi}}{\partial r} + \frac{\partial g_{rr}}{\partial \phi} - \frac{\partial g_{r\phi}}{\partial r} \right] = 0 = \Gamma^r_{r\theta} = \Gamma^r_{r4}$$

$$\Gamma^r_{\phi\phi} = \frac{1}{2} g^{rr} \left[\frac{\partial g_{r\phi}}{\partial \phi} + \frac{\partial g_{r\phi}}{\partial \phi} - \frac{\partial g_{\phi\phi}}{\partial r} \right] = -\frac{1}{2A(r)} (2r \sin^2 \theta)$$

$$\Gamma^r_{\theta\theta} = -\frac{1}{2A(r)} (2r)$$

$$\Gamma^r_{44} = -\frac{1}{2A(r)} \left[-\frac{dB(r)}{dr} \right] \tag{7.52}$$

2) $\underline{\lambda = \theta}$ (10 terms).

$$\Gamma^\theta_{\theta\theta} = \frac{1}{2} g^{\theta\theta} \left[\frac{\partial g_{\theta\theta}}{\partial \theta} + \frac{\partial g_{\theta\theta}}{\partial \theta} - \frac{\partial g_{\theta\theta}}{\partial \theta} \right] = 0$$

$$\Gamma^\theta_{rr} = \Gamma^\theta_{44} = 0$$

$$\Gamma^\theta_{\phi\phi} = \frac{1}{2} g^{\theta\theta} \left[\frac{\partial g_{\theta\phi}}{\partial \phi} + \frac{\partial g_{\theta\phi}}{\partial \phi} - \frac{\partial g_{\phi\phi}}{\partial \theta} \right] = \frac{1}{2r^2} (-2r^2 \sin \theta \cos \theta)$$

$$\Gamma^\theta_{\phi\theta} = 0$$

$$\Gamma^\theta_{r\theta} = \frac{1}{2} g^{\theta\theta} \left[\frac{\partial g_{\theta\theta}}{\partial r} + \frac{\partial g_{\theta r}}{\partial \theta} - \frac{\partial g_{r\theta}}{\partial \theta} \right] = \frac{1}{2r^2} (2r)$$

$$\Gamma^\theta_{4\theta} = 0 \tag{7.53}$$

---

[9] When it is felt that results follow immediately from the previous expression for the affine connection in terms of the metric, that expression may be suppressed.

3) $\underline{\lambda = \phi}$ (10 terms).

$$\Gamma^{\phi}_{\phi\phi} = 0$$

$$\Gamma^{\phi}_{rr} = \Gamma^{\phi}_{44} = \Gamma^{\phi}_{\theta\theta} = 0$$

$$\Gamma^{\phi}_{r\phi} = \frac{1}{2}g^{\phi\phi}\left[\frac{\partial g_{\phi\phi}}{\partial r} + \frac{\partial g_{\phi r}}{\partial \phi} - \frac{\partial g_{r\phi}}{\partial \phi}\right] = \frac{1}{2r^2\sin^2\theta}(2r\sin^2\theta)$$

$$\Gamma^{\phi}_{\theta\phi} = \frac{1}{2r^2\sin^2\theta}(2r^2\sin\theta\cos\theta)$$

$$\Gamma^{\phi}_{4\phi} = 0 \tag{7.54}$$

4) $\underline{\lambda = 4}$ (10 terms).

$$\Gamma^{4}_{44} = \Gamma^{4}_{rr} = \Gamma^{4}_{\theta\theta} = \Gamma^{4}_{\phi\phi} = \Gamma^{4}_{4\theta} = \Gamma^{4}_{4\phi} = 0$$

$$\Gamma^{4}_{r4} = \frac{1}{2}g^{44}\left[\frac{\partial g_{44}}{\partial r}\right] = -\frac{1}{2B(r)}\left[-\frac{dB(r)}{dr}\right] \tag{7.55}$$

Table 7.1   SUMMARY —Affine connection for spherically symmetric source with coordinates $(q^1, \cdots, q^4) = (r, \theta, \phi, ct)$ and the metric of Eqs. (7.48). It is symmetric in its lower two indices so that $\Gamma^{\lambda}_{\mu\nu} = \Gamma^{\lambda}_{\nu\mu}$. Here $A' = dA(r)/dr$ and $B' = dB(r)/dr$.

$\Gamma^{r}_{rr} = \dfrac{A'}{2A}$	$\Gamma^{r}_{\phi\phi} = -\dfrac{r\sin^2\theta}{A}$	$\Gamma^{r}_{\theta\theta} = -\dfrac{r}{A}$
$\Gamma^{r}_{44} = \dfrac{B'}{2A}$	$\Gamma^{\theta}_{\phi\phi} = -\sin\theta\cos\theta$	$\Gamma^{\theta}_{r\theta} = \dfrac{1}{r}$
$\Gamma^{\phi}_{r\phi} = \dfrac{1}{r}$	$\Gamma^{\phi}_{\theta\phi} = \dfrac{\cos\theta}{\sin\theta}$	$\Gamma^{4}_{r4} = \dfrac{B'}{2B}$
All others vanish		

The next steps are to compute the Riemann curvature tensor $R^{\lambda}{}_{\mu\rho\nu}$, and then the Ricci tensor $R_{\mu\nu} = R^{\lambda}{}_{\mu\lambda\nu}$. We here go right to the Ricci tensor[10]

$$R_{\mu\nu} = R^{\lambda}{}_{\mu\lambda\nu}$$

$$R_{\mu\nu} = \frac{\partial}{\partial q^{\lambda}}\Gamma^{\lambda}_{\mu\nu} + \Gamma^{\lambda}_{\lambda\sigma}\Gamma^{\sigma}_{\mu\nu} - \left[\frac{\partial}{\partial q^{\nu}}\Gamma^{\lambda}_{\mu\lambda} + \Gamma^{\lambda}_{\nu\sigma}\Gamma^{\sigma}_{\mu\lambda}\right] \tag{7.56}$$

Note that the curvature of the space is given by the Riemann tensor. The Einstein field equations are local relations and state that the Ricci tensor

---

[10] Recall that the Riemann tensor is antisymmetric in the last two indices.

vanishes wherever the energy-momentum tensor vanishes. This does not imply, however, that the space is not curved!

We know that the Ricci tensor is symmetric in its indices

$$R_{\mu\nu} = R_{\nu\mu} \tag{7.57}$$

Thus it has $(4 \times 5)/2 = 10$ independent components. Again, it is of sufficient importance that we here carry out the calculation of the Ricci tensor in detail. We present a consistent procedure for the evaluation. Consider the element $R_{r\theta}$

$$R_{r\theta} = \frac{\partial}{\partial q^\lambda} \Gamma^\lambda_{\underline{r\theta}} - \frac{\partial}{\partial \theta} \Gamma^\lambda_{\underline{r\lambda}} + \Gamma^\lambda_{\lambda\sigma} \Gamma^\sigma_{\underline{r\theta}} - \Gamma^\lambda_{\underline{\theta}\sigma} \Gamma^\sigma_{r\lambda} \tag{7.58}$$

First, search the rows of Table 7.1 for non-zero elements with the underlined indices, carry out the derivatives where appropriate, and then search the table on the remaining indices. This gives, in this case,

1) $\underline{R_{r\theta}}$:

$$R_{r\theta} = 0 - 0 + \frac{1}{r}\Gamma^\lambda_{\lambda\theta} - \left[\left(\frac{-r}{A} \right) \Gamma^\theta_{rr} + \left(\frac{1}{r} \right) \Gamma^r_{r\theta} + \cot\theta\, \Gamma^\phi_{r\phi} \right]$$

$$= \frac{1}{r}\cot\theta - \frac{1}{r}\cot\theta = 0 \tag{7.59}$$

This illustrates the method. The other elements are evaluated in the same manner.

2) $\underline{R_{r\phi}}$:

$$R_{r\phi} = 0 - 0 + \frac{1}{r}\Gamma^\lambda_{\lambda\phi} \tag{7.60}$$

$$- \left[\left(\frac{-r\sin^2\theta}{A} \right) \Gamma^\phi_{rr} - \sin\theta\cos\theta\, \Gamma^\phi_{r\theta} + \left(\frac{1}{r} \right) \Gamma^r_{r\phi} + \cot\theta\, \Gamma^\theta_{r\phi} \right] = 0$$

3) $\underline{R_{r4}}$:

$$R_{r4} = 0 - 0 + \frac{B'}{2B}\Gamma^\lambda_{\lambda 4} - \left[\left(\frac{B'}{2A} \right) \Gamma^4_{rr} + \left(\frac{B'}{2B} \right) \Gamma^r_{r4} \right] = 0 \tag{7.61}$$

4)$R_{\theta\phi}$:

$$R_{\theta\phi} = \frac{\partial}{\partial q^\lambda}\Gamma^\lambda_{\underline{\theta\phi}} - \frac{\partial}{\partial\phi}\Gamma^\lambda_{\underline{\theta\lambda}} + \Gamma^\lambda_{\lambda\sigma}\Gamma^\sigma_{\underline{\theta\phi}} - \Gamma^\lambda_{\underline{\phi\sigma}}\Gamma^\sigma_{\underline{\theta\lambda}}$$

$$= 0 - 0 + \cot\theta\,\Gamma^\lambda_{\lambda\phi} \qquad (7.62)$$

$$- \left[\left(\frac{-r\sin^2\theta}{A}\right)\Gamma^\phi_{\theta r} - \sin\theta\cos\theta\,\Gamma^\phi_{\theta\theta} + \left(\frac{1}{r}\right)\Gamma^r_{\phi\theta} + \cot\theta\,\Gamma^\theta_{\phi\theta}\right] = 0$$

5) $R_{\theta 4}$:

$$R_{\theta 4} = 0 - 0 + 0 - \left[\left(\frac{B'}{2A}\right)\Gamma^4_{r\theta} + \left(\frac{B'}{2B}\right)\Gamma^r_{4\theta}\right] = 0 \qquad (7.63)$$

6) $R_{\phi 4}$:

$$R_{\phi 4} = 0 - 0 + 0 - \left[\left(\frac{B'}{2A}\right)\Gamma^4_{r\phi} + \left(\frac{B'}{2B}\right)\Gamma^r_{4\phi}\right] = 0 \qquad (7.64)$$

In summary to this point, the Ricci tensor *vanishes identically* outside a spherically symmetric source if its two indices differ

$$R_{\mu\nu} = 0 \qquad ; \mu \neq \nu \qquad (7.65)$$

The non-trivial results then come from the four diagonal elements, and we proceed to calculate them.

7) $R_{\theta\theta}$:

$$R_{\theta\theta} = \frac{\partial}{\partial q^\lambda}\Gamma^\lambda_{\underline{\theta\theta}} - \frac{\partial}{\partial\theta}\Gamma^\lambda_{\underline{\theta\lambda}} + \Gamma^\lambda_{\lambda\sigma}\Gamma^\sigma_{\underline{\theta\theta}} - \Gamma^\lambda_{\underline{\theta\sigma}}\Gamma^\sigma_{\underline{\theta\lambda}}$$

$$= \frac{\partial}{\partial r}\left(\frac{-r}{A}\right) - \frac{\partial}{\partial\theta}\left(\frac{\cos\theta}{\sin\theta}\right) + \left(\frac{-r}{A}\right)\Gamma^\lambda_{\lambda r}$$

$$- \left[\left(\frac{-r}{A}\right)\Gamma^\theta_{\theta r} + \left(\frac{1}{r}\right)\Gamma^r_{\theta\theta} + \cot\theta\,\Gamma^\phi_{\phi\theta}\right]$$

$$= -\frac{1}{A} + \frac{r}{A^2}A' - \left[-\frac{\sin\theta}{\sin\theta} - \frac{\cos^2\theta}{\sin^2\theta}\right] - \frac{r}{A}\left[\frac{A'}{2A} + \frac{1}{r} + \frac{1}{r} + \frac{B'}{2B}\right]$$

$$- \left[\left(\frac{-r}{A}\right)\frac{1}{r} + \left(\frac{1}{r}\right)\left(\frac{-r}{A}\right) + \cot^2\theta\right]$$

$$R_{\theta\theta} = 1 - \frac{1}{A} + \frac{r}{2A}\left(\frac{A'}{A} - \frac{B'}{B}\right) \qquad (7.66)$$

8) $\underline{R_{\phi\phi}}$:

$$R_{\phi\phi} = \frac{\partial}{\partial q^\lambda}\Gamma^\lambda_{\underline{\phi\phi}} - \frac{\partial}{\partial \phi}\Gamma^\lambda_{\underline{\phi\lambda}} + \Gamma^\lambda_{\lambda\sigma}\Gamma^\sigma_{\underline{\phi\phi}} - \Gamma^\lambda_{\underline{\phi}\sigma}\Gamma^\sigma_{\phi\lambda}$$

$$= \frac{\partial}{\partial r}\left(\frac{-r\sin^2\theta}{A}\right) + \frac{\partial}{\partial\theta}(-\sin\theta\cos\theta) + \left(-\frac{r\sin^2\theta}{A}\right)\Gamma^\lambda_{\lambda r}$$

$$+ (-\sin\theta\cos\theta)\Gamma^\lambda_{\lambda\theta}$$

$$- \left[\left(\frac{-r\sin^2\theta}{A}\right)\Gamma^\phi_{\phi r} - \sin\theta\cos\theta\Gamma^\phi_{\phi\theta} + \left(\frac{1}{r}\right)\Gamma^r_{\phi\phi} + \cot\theta\,\Gamma^\theta_{\phi\phi}\right]$$

$$= \frac{-\sin^2\theta}{A} + \frac{r\sin^2\theta A'}{A^2} - \cos^2\theta + \sin^2\theta - \frac{r\sin^2\theta}{A}\left[\frac{A'}{2A} + \frac{1}{r} + \frac{1}{r} + \frac{B'}{2B}\right]$$

$$- \sin\theta\cos\theta\left[\frac{\cos\theta}{\sin\theta}\right] + \left(\frac{r\sin^2\theta}{A}\right)\frac{1}{r} + \sin\theta\cos\theta\left[\frac{\cos\theta}{\sin\theta}\right]$$

$$- \frac{1}{r}\left(\frac{-r\sin^2\theta}{A}\right) - \cot\theta(-\sin\theta\cos\theta) \qquad (7.67)$$

Collection of terms gives a simple proportionality to $R_{\theta\theta}$.

$$R_{\phi\phi} = \sin^2\theta - \frac{\sin^2\theta}{A} + \left(\frac{rA'}{2A^2} - \frac{rB'}{2AB}\right)\sin^2\theta$$

$$= R_{\theta\theta}\sin^2\theta \qquad (7.68)$$

9) $\underline{R_{rr}}$:

$$R_{rr} = \frac{\partial}{\partial q^\lambda}\Gamma^\lambda_{\underline{rr}} - \frac{\partial}{\partial r}\Gamma^\lambda_{\underline{r\lambda}} + \Gamma^\lambda_{\lambda\sigma}\Gamma^\sigma_{\underline{rr}} - \Gamma^\lambda_{\underline{r}\sigma}\Gamma^\sigma_{r\lambda}$$

$$= \frac{\partial}{\partial r}\left(\frac{A'}{2A}\right) - \frac{\partial}{\partial r}\left[\frac{A'}{2A} + \frac{1}{r} + \frac{1}{r} + \frac{B'}{2B}\right] + \left(\frac{A'}{2A}\right)\Gamma^\lambda_{\lambda r}$$

$$- \left[\left(\frac{A'}{2A}\right)\Gamma^r_{rr} + \frac{1}{r}\Gamma^\theta_{\theta r} + \frac{1}{r}\Gamma^\phi_{\phi r} + \frac{B'}{2B}\Gamma^4_{4r}\right]$$

$$= \frac{2}{r^2} - \frac{B''}{2B} + \frac{(B')^2}{2B^2} + \frac{A'}{2A}\left[\frac{A'}{2A} + \frac{1}{r} + \frac{1}{r} + \frac{B'}{2B}\right]$$

$$- \left(\frac{A'}{2A}\right)^2 - \frac{1}{r^2} - \frac{1}{r^2} - \left(\frac{B'}{2B}\right)^2$$

$$R_{rr} = -\frac{B''}{2B} + \frac{1}{4}\left(\frac{B'}{B}\right)\left(\frac{B'}{B} + \frac{A'}{A}\right) + \frac{1}{r}\left(\frac{A'}{A}\right) \qquad (7.69)$$

10) $\underline{R_{44}}$:

$$R_{44} = \frac{\partial}{\partial q^\lambda}\Gamma^\lambda_{\underline{44}} - \frac{\partial}{\partial(ct)}\Gamma^\lambda_{\underline{4\lambda}} + \Gamma^\lambda_{\lambda\sigma}\Gamma^\sigma_{\underline{44}} - \Gamma^\lambda_{\underline{4\sigma}}\Gamma^\sigma_{4\lambda}$$

$$= \frac{\partial}{\partial r}\left(\frac{B'}{2A}\right) + \frac{B'}{2A}\left(\frac{A'}{2A} + \frac{1}{r} + \frac{1}{r} + \frac{B'}{2B}\right) - \left[\left(\frac{B'}{2A}\right)\Gamma^4_{r4} + \left(\frac{B'}{2B}\right)\Gamma^r_{44}\right]$$

$$= \frac{B''}{2A} - \frac{B'A'}{2A^2} + \frac{B'}{2A}\left(\frac{A'}{2A} + \frac{1}{r} + \frac{1}{r} + \frac{B'}{2B}\right)$$

$$- \left(\frac{B'}{2A}\right)\left(\frac{B'}{2B}\right) - \left(\frac{B'}{2B}\right)\left(\frac{B'}{2A}\right)$$

$$R_{44} = \frac{B''}{2A} - \frac{1}{4}\left(\frac{B'}{A}\right)\left(\frac{A'}{A} + \frac{B'}{B}\right) + \frac{1}{r}\left(\frac{B'}{A}\right) \tag{7.70}$$

In *summary*, the Ricci tensor for a spherically symmetric source, with the interval of Eq. (7.47) and corresponding metric of Eqs. (7.48), is given by

$$R_{\mu\nu} = 0 \qquad ; \mu \neq \nu$$

$$R_{\theta\theta} = 1 - \frac{1}{A} + \frac{r}{2A}\left(\frac{A'}{A} - \frac{B'}{B}\right)$$

$$R_{\phi\phi} = R_{\theta\theta}\sin^2\theta$$

$$R_{rr} = -\frac{B''}{2B} + \frac{1}{4}\left(\frac{B'}{B}\right)\left(\frac{A'}{A} + \frac{B'}{B}\right) + \frac{1}{r}\left(\frac{A'}{A}\right)$$

$$R_{44} = \frac{B''}{2A} - \frac{1}{4}\left(\frac{B'}{A}\right)\left(\frac{A'}{A} + \frac{B'}{B}\right) + \frac{1}{r}\left(\frac{B'}{A}\right) \tag{7.71}$$

The field equations *outside* the spherically symmetric source are

$$R_{\mu\nu} = 0 \tag{7.72}$$

These now provide four second-order, coupled, nonlinear differential equations for the two functions $[A(r), B(r)]$.

7.6 Solution in Vacuum

Let us look for a solution to the Einstein field equations that *everywhere* satisfies

$$R_{\mu\nu} = 0 \tag{7.73}$$

In *flat* space, the relevant solution from Eqs. (7.71) evidently is

$$B = \text{constant}$$
$$A = \text{constant}$$
$$1 - \frac{1}{A} = 0 \tag{7.74}$$

These imply

$$B = \text{constant} \qquad \text{; flat space}$$
$$A = 1 \tag{7.75}$$

Introduce *new coordinates* with $t \rightarrow t/\sqrt{B}$, which is simply a rescaling of the time t. The interval and metric then take the form

$$(d\mathbf{s})^2 = (d\vec{x})^2 - c^2(dt)^2 \qquad \text{; flat space}$$

$$\underline{g} = \begin{bmatrix} 1 & 0 & 0 & 0 \\ 0 & 1 & 0 & 0 \\ 0 & 0 & 1 & 0 \\ 0 & 0 & 0 & -1 \end{bmatrix} \tag{7.76}$$

This is the canonical form of the interval where $d\vec{x}$ is distance and dt is time, and the null interval now identifies the parameter c with the usual speed of light (see Prob. 6.6).

Consider next the solution to Eqs. (7.73) and (7.71) *outside* of a spherically symmetric source. Since one has four coupled, nonlinear differential equations in terms of just two functions, it is not at all clear *a priori* that such a solution exists; however, Schwarzschild found one. To obtain his solution, take

$$AB = \text{constant} \equiv \lambda \tag{7.77}$$

Differentiation of this relation implies

$$A'B + AB' = 0$$
$$\text{or ;} \qquad \frac{A'}{A} + \frac{B'}{B} = 0 \tag{7.78}$$

Substitute this relation in $R_{\theta\theta}$ and set $R_{\theta\theta} = 0$ in Eqs. (7.71), which also implies $R_{\phi\phi} = 0$. This gives

$$1 - \frac{1}{A} + \frac{rA'}{A^2} = 0 \tag{7.79}$$

Now make use of the relation

$$\frac{d}{dr}\left(\frac{r}{A}\right) = \frac{1}{A} - \frac{rA'}{A^2} \qquad (7.80)$$

to rewrite Eqs. (7.79) and (7.77) as

$$\frac{d}{dr}\left(\frac{r}{A}\right) = 1$$
$$AB = \lambda \qquad (7.81)$$

Next, use Eq. (7.78) in R_{rr} in Eqs. (7.71), and set $R_{rr} = 0$, which yields

$$-\frac{B''}{2B} + \frac{1}{r}\left(\frac{A'}{A}\right) = 0$$

or ;
$$-\frac{B''}{2B} = \frac{1}{r}\frac{B'}{B}$$
$$rB'' + 2B' = 0$$
$$\frac{d}{dr}\left(r^2 B'\right) = 0 \qquad (7.82)$$

The use of Eq. (7.78) in the equation for R_{44} in Eqs. (7.71) then reduces the relation $R_{44} = 0$ to

$$\frac{B''}{2A} + \frac{1}{r}\left(\frac{B'}{A}\right) = 0 \qquad (7.83)$$

But now this is identical to the third of Eqs. (7.82) that we got from R_{rr}! Thus the assumed form of solution in Eq. (7.77) has reduced the Einstein field equations in this case to the two relations

$$\frac{1}{\lambda}\frac{d}{dr}(rB) = 1 \qquad ; \text{Einstein's equations}$$
$$\frac{d}{dr}\left(r^2 B'\right) = 0 \qquad (7.84)$$

with A then determined from Eqs. (7.81) as

$$A = \frac{\lambda}{B} \qquad (7.85)$$

For a solution to the first of Eqs. (7.84), take

$$rB = \lambda(r - k) \qquad ; \; k = \text{constant}$$

$$\text{or} \; ; \qquad B = \lambda\left(1 - \frac{k}{r}\right)$$

$$B' = \frac{\lambda k}{r^2}$$

$$r^2 B' = \lambda k = \text{constant} \tag{7.86}$$

This then also provides a solution to the second of Eqs. (7.84).

Now again rescale the coordinate t according to $t \to t/\sqrt{\lambda}$. Also, redefine the constant k as (see later)

$$k \equiv \frac{2MG}{c^2} \tag{7.87}$$

The solution to the Einstein field equations then takes the form

$$A(r) = \frac{1}{1 - (2MG/c^2 r)} \qquad ; \; B(r) = 1 - \frac{2MG}{c^2 r} \tag{7.88}$$

The metric that we have found is thus

$$(ds)^2 = \frac{(dr)^2}{1 - (2MG/c^2 r)} + r^2(d\theta)^2 + r^2 \sin^2 \theta (d\phi)^2 - \left(1 - \frac{2MG}{c^2 r}\right)(cdt)^2$$
$$; \; \text{Schwarzschild metric} \tag{7.89}$$

This solution to the Einstein field Eqs. (7.72) outside a spherically symmetric source is due to Schwarzschild, and the metric is known as the *Schwarzschild metric* [Schwarzschild (1916)].

Consider the *newtonian limit* of these results. First, we are interested in distances far from the source so that $r \to \infty$. Then

$$A(r) \approx 1 + \frac{2MG}{c^2 r} \qquad ; \; r \to \infty \tag{7.90}$$

Now let $c^2 \to \infty$. The Schwarzschild metric then becomes

$$(ds)^2 \approx (d\vec{x})^2 - \left(1 - \frac{2MG}{c^2 r}\right)(cdt)^2 \quad ; \; \text{newtonian limit} \tag{7.91}$$

This is *identical* to Eq. (7.32), and thus we *reproduce newtonian physics.*[11] Furthermore, we can now identify the integration constant M as the mass

---

[11]Note that, in fact, only the dimensionless combination $2MG/c^2 r = R_s/r$ appears in the Schwarzschild metric, and this must be small to reproduce the newtonian limit.

of the source

$$M = \text{mass of source} \qquad (7.92)$$

We also reproduce the correct form of the gravitational potential Φ.[12]
   In summary:

We have found a solution to Einstein's field equations outside of a spherically symmetric source, and this solution reduces to the newtonian limit at large distances.

We define the *Schwarzschild radius* by

$$R_s \equiv \frac{2MG}{c^2} \qquad ; \text{Schwarzschild radius} \qquad (7.93)$$

The Schwarzschild metric is non-singular for $r > R_s$, and it is *singular* at $r = R_s$ (remember, this is the solution *outside* of the source). We shall have much more to say about this later.

We would expect that the solution we have found should also cover the vacuum case, since in both situations we are just solving the Einstein Eqs. (7.73) everywhere. Indeed, the Schwarzschild metric reduces to that in flat space if one simply sets the integration constant M equal to zero

$$M = 0 \qquad ; \text{vacuum case} \qquad (7.94)$$

To conclude these arguments:

We will take the reproduction of the newtonian limit outside of a spherically symmetric source as the basic justification of principles I-III of general relativity as we have stated them.

We proceed to investigate further implications of the theory of general relativity.

---

[12]We now use Φ for the gravitational potential to distinguish it from the azimuthal angle.

7.7 Interpretation of Schwarzschild Metric

Let us first find a way to *visualize* the Schwarzschild metric. Consider the situation where two of the coordinates (θ, ct) are held constant

$$\theta = \pi/2 = \text{constant} \qquad ; \text{ for visualization}$$
$$t = \text{constant} \tag{7.95}$$

In this case the metric takes the form

$$(ds)^2 = \frac{(dr)^2}{1 - R_s/r} + r^2 (d\phi)^2$$
$$= g_{rr}(r)(dr)^2 + g_{\phi\phi}(r)(d\phi)^2 \tag{7.96}$$

The configuration is reduced to that of a static, two-dimensional riemannian surface, and the generalized coordinates (r, ϕ) are used to locate a point on it.

Consider this surface in three-dimensional euclidian space (Fig. 7.9), and introduce cylindrical coordinates (r, z, ϕ). The interval in cylindrical coordinates is

$$(ds)^2 = (dr)^2 + (dz)^2 + r^2 (d\phi)^2 \tag{7.97}$$

There is a *tangent space* at each point on the surface, and in this tangent space the line element and interval are[13]

$$d\mathbf{s} = \mathbf{e}_r \, dr + \mathbf{e}_\phi \, d\phi$$
$$(d\mathbf{s})^2 = g_{rr}(r)(dr)^2 + g_{\phi\phi}(r)(d\phi)^2 \tag{7.98}$$

The surface is defined by the equation $z = z(r)$ in cylindrical coordinates. This function can be determined by making Eqs. (7.98) and (7.96) coincide. First, rewrite Eq. (7.97) as

$$(ds)^2 = \left[1 + \left(\frac{dz}{dr} \right)^2 \right] (dr)^2 + r^2 (d\phi)^2 \tag{7.99}$$

Then identify

$$g_{rr}(r) = 1 + \left(\frac{dz}{dr} \right)^2 = \frac{1}{1 - R_s/r} \tag{7.100}$$

---

[13]See chap. 2. Note the basis vectors here are orthogonal.

Now solve for $dz(r)/dr$

$$\left(\frac{dz}{dr}\right)^2 = g_{rr} - 1$$

$$= \frac{1}{1 - R_s/r} - 1 = \frac{R_s/r}{1 - R_s/r} = \frac{R_s}{r - R_s}$$

$$\frac{dz}{dr} = \pm \left(\frac{R_s}{r - R_s}\right)^{1/2} \tag{7.101}$$

Integration of this relation, with the choice of coordinates $z(R_s) = 0$, then gives

$$z(r) = \pm 2[R_s(r - R_s)]^{1/2} \qquad ; \text{ defines surface} \tag{7.102}$$

The plus sign gives the surface illustrated in Fig. 7.9. The minus sign gives that surface reflected in the (x, y) plane.

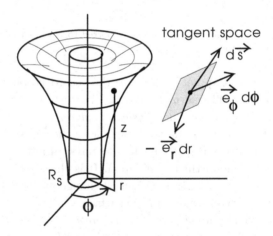

Fig. 7.9 The two-dimensional riemannian surface defined by the Schwarzschild metric with the two coordinates in Eq. (7.95) held constant, as viewed in three-dimensional euclidian space. The generalized coordinates used to locate a point on the surface are (r, ϕ). The tangent space at the point (r, ϕ) is indicated, as well as the basis vectors and line element at that point. The surface is described by Eq. (7.102).

Suppose one *lives in the surface.* A trip around the axis allows one to measure the circumference $\mathcal{C}$, and hence the radial distance r can be obtained from $\mathcal{C} = 2\pi r$. Thus

$$r = \frac{\mathcal{C}}{2\pi} \qquad ; \text{ coordinates in surface}$$
$$\phi = \phi \qquad\qquad\qquad (7.103)$$

These are an actual distance and angle; there is no ambiguity about these.

Now communicate with someone we shall call the "record keeper." This is a person who keeps track of position in coordinate space. He or she has a screen, or sheet of paper, in front of them and simply plots points in (r, ϕ) as they are reported (Fig. 7.10).

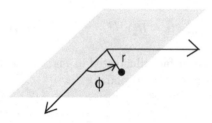

Fig. 7.10 Location on the surface as reported to, and plotted by, the "record keeper" in the space of the coordinates (r, ϕ).

The point-to-point motion on the record keeper's screen corresponds to the actual physical motion along the surface. The line element ds and the interval $(ds)^2$ are the *physical displacements* in the surface. It is the *metric* that provides the key that allows the record keeper to convert his or her coordinate displacements into physical intervals. This is the role of the metric (and of the coordinates). It is clear, for example, from Fig. 7.9 that just outside of R_s, a very small change in r at given ϕ corresponds to a very large displacement in the surface.

To provide a *physical interpretation* of the implications of the Schwarzschild metric, we introduce three reference frames (Fig. 7.11):

Reference frame I. This is the *global inertial laboratory system*. The coordinates are (r, θ, ϕ, ct). Note that *we can all know and agree on these coordinates*. For example, we can communicate r by the measurement of an equatorial circumference, or area of a sphere, and we can communicate t by synchronizing clocks when the position on an orbit has a given orientation with respect to the fixed stars. Together with these coordinates, we have the Schwarzschild metric to define the physical interval.

<u>Reference frame II</u>. This is a frame that is very far away from the source ($r \to \infty$). Here we have the newtonian limit and free, flat space. The interval and metric in this frame are

$$(ds)^2 = (d\vec{x})^2 - c^2(dt)^2 \qquad ; \text{ flat space}$$

$$g = \begin{bmatrix} 1 & 0 & 0 & 0 \\ 0 & 1 & 0 & 0 \\ 0 & 0 & 1 & 0 \\ 0 & 0 & 0 & -1 \end{bmatrix} \tag{7.104}$$

Here $d\vec{x}$ is distance and dt is time. *The coordinate t is the time in this frame.*

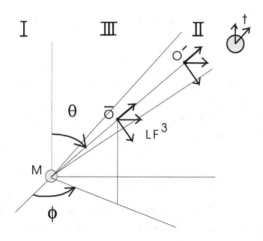

Fig. 7.11 Three reference frames introduced to obtain a physical interpretation of the implications of the Schwarzschild metric. Frame I is the global inertial laboratory system with coordinates (r, θ, ϕ, ct). Frame II is one very far away from the source, and frame III is the local freely falling frame (LF^3).

<u>Reference frame III</u>. This is the *local freely falling frame LF^3*. The interval and metric in this frame are (Fig. 7.12)

$$(ds)^2 = (d\vec{l})^2 - c^2(d\bar{t})^2 \qquad ; \text{ flat space}$$

$$g = \begin{bmatrix} 1 & 0 & 0 & 0 \\ 0 & 1 & 0 & 0 \\ 0 & 0 & 1 & 0 \\ 0 & 0 & 0 & -1 \end{bmatrix} \tag{7.105}$$

In the LF^3 one again has just free, flat space. There is no force of gravity, and no inertial force. Here $d\vec{l}$ is distance and $d\bar{t}$ is time. No matter where we are outside of the source, we always know what we are doing in this frame.

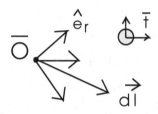

Fig. 7.12   Reference frame III, the local freely falling frame LF^3.

7.8   A Few Applications

Suppose one sits at the origin in the frame LF^3. This frame has *finite acceleration* but *zero velocity*. Thus

$$(\dot{r}, \dot{\theta}, \dot{\phi}) = 0 \qquad\qquad ; \text{origin of } LF^3$$
$$\Rightarrow \quad (dr, d\theta, d\phi) = (\dot{r}, \dot{\theta}, \dot{\phi})dt = 0 \qquad\qquad (7.106)$$

Now physics lies in the *interval*, which is independent of the coordinate system in which we choose to express it. At the origin in LF^3 the interval takes the form[14]

$$(d\mathbf{s})^2 = -c^2(d\bar{t})^2 \qquad\qquad ; \text{proper time interval} \quad (7.107)$$

In the global laboratory frame I, this same physical interval is given by

$$(d\mathbf{s})^2 = -\left(1 - \frac{R_s}{r}\right)(cdt)^2 \qquad\qquad ; \text{relation to clock in I} \quad (7.108)$$

This expresses the interval in terms of the radial coordinate r and the time dt in frame I. Upon equating these two expressions for the interval one finds

$$dt = \frac{d\bar{t}}{(1 - R_s/r)^{1/2}} \qquad\qquad ; \text{time dilation} \qquad (7.109)$$

---

[14]Note that here the LF^3 in general relativity plays a role analogous to that of the rest frame in special relativity.

There is again a *time dilation*, where the clock at the origin in LF^3 runs slower than the clock in the global laboratory frame. Note the effect depends on

$$\frac{R_s}{r} = -\frac{2\Phi}{c^2} \tag{7.110}$$

As a second application, suppose one has a *standard meter stick* of length $d\bar{l}$ oriented along the radial direction in LF^3 (see Fig. 7.12). The length of the stick is measured at a given instant in time in LF^3 so that $d\bar{t} = 0$. The physical interval, as expressed in LF^3, is then

$$(d\mathbf{s})^2 = (d\bar{l})^2 \qquad \text{; proper length interval} \tag{7.111}$$

Now let us measure the length in the global laboratory frame I. We again measure at a given instant in time, so that $dt = 0$. The same physical interval is then expressed in frame I as

$$(d\mathbf{s})^2 = \frac{(dl)^2}{1 - R_s/r} \qquad \text{; relation to length in I} \tag{7.112}$$

If these two expression for the interval are now equated one finds

$$dl = (1 - R_s/r)^{1/2} \, d\bar{l} \qquad \text{; radial contraction} \tag{7.113}$$

There is a contraction of the length in the radial direction when measured in the laboratory frame.

These two effects, time dilation and radial length contraction, are direct analogs of the time dilation and Lorentz contraction that we found in special relativity. Here, however, the effects rise from the relative *acceleration* of the frames, rather than from their relative *velocity*.[15]

---

[15]Recall that the LF^3 is accelerating relative to the inertial laboratory frame.

Chapter 8

Precession of Perihelion

The bound orbits in newtonian physics in the gravitational potential $\Phi = -MG/r$ are conic sections (circles or ellipses) with the central mass at a focus.[1] We proceed to calculate the orbits with the Schwarzschild metric. The orbits are the geodesics in that metric, and we will find that they are more complicated than simple conic sections.

fixed stars

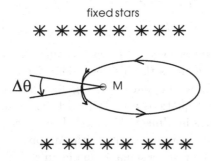

Fig. 8.1   Precession, with respect to the fixed stars, of the perihelion of an elliptical orbit in the Schwarzschild metric. The precession rate is $\Delta\theta$/revolution.

The *perihelion* of an orbit is the point of closest approach to the central mass. In an inertial frame, in newtonian physics with the potential $\Phi = -MG/r$, the perihelion of an elliptical orbit maintains a fixed orientation with respect to the fixed stars. In general relativity with the Schwarzschild metric, the orientation of the perihelion moves — it *precesses* with respect to the fixed stars. One can compute, and measure, the precession rate $\Delta\theta$/revolution (Fig. 8.1). There is no ambiguity in angle measurements

---

[1]See [Fetter and Walecka (2003)] chap. 1.

with the Schwarzschild metric, and *this is a well-defined piece of physics.* In fact, the predicted precession rate of the perihelion of the orbit of the planet mercury about the sun provided one of the first tests of the theory of general relativity (see later).

We shall attack this problem by finding an equivalent lagrangian for particle motion in the Schwarzschild metric and then simply deal with the orbit as an application of lagrangian mechanics.

8.1 Lagrangian

In chapter 6 we developed a relativistic Hamilton's principle in special relativity [see Eq. (6.24)]

$$\delta \int_{\tau_1}^{\tau_2} \bar{L} \, d\tau = 0 \qquad \text{; fixed endpoints} \qquad (8.1)$$

Given the lagrangian

$$\bar{L} = m \, g_{\mu\nu} \frac{dq^\mu}{d\tau} \frac{dq^\nu}{d\tau}$$
$$(d\mathbf{s})^2 = -c^2 (d\tau)^2 \qquad \text{; along actual path} \qquad (8.2)$$

this principle leads to the correct relativistic equations of motion. Here the proper time τ parameterizes the distance along the actual path, and the coordinates are varied about that path (Fig. 6.3). It was demonstrated in Eq. (6.49) that these equations are identical to those for the *geodesics* in curvilinear coordinates in Minkowski space. These arguments are immediately carried over to general relativity with an appropriate generalization of the metric to the four-dimensional riemannian space.

Let us again change variables from proper time τ to laboratory time t in Eq. (8.1) (see Fig. 8.2)

$$\delta \int_{\tau_1}^{\tau_2} \bar{L} \, d\tau = \delta \int_{t_1}^{t_2} \left(\bar{L} \frac{d\tau}{dt} \right) dt$$
$$= \delta \int_{t_1}^{t_2} L \, dt \qquad \text{; fixed endpoints} \qquad (8.3)$$

We now have the *usual Hamilton's principle* with the lagrangian

$$L = \bar{L} \frac{d\tau}{dt} \qquad (8.4)$$

Now t parameterizes the path, and we let everything else vary.

The general interval and lagrangian $\bar{L}$ are here given by

$$(ds)^2 = g_{\mu\nu} dq^\mu dq^\nu \qquad\qquad ; \bar{L} = m \left(\frac{ds}{d\tau}\right)^2 \qquad (8.5)$$

Therefore one has for the lagrangian L

$$L = \left[m \left(\frac{ds}{d\tau}\right)^2 \right] \left(\frac{d\tau}{dt}\right) = m \left(\frac{ds}{dt}\right)^2 \left(\frac{dt}{d\tau}\right) \qquad (8.6)$$

The second expression is just a rewriting of the first.

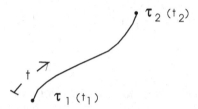

Fig. 8.2   Change of variables from proper time to time $\tau \to t$ in the relativistic Hamilton's principle. The endpoints are fixed.

Now express the lagrangian L in terms of the appropriate generalized coordinates $L(r, \theta, \phi; \dot{r}, \dot{\theta}, \dot{\phi})$ where the dot indicates the ordinary time derivative. We first use Schwarzschild's solution in Eq. (7.89) for the interval and factor out $(c\,dt)^2$

$$(ds)^2 = -\left\{ \left[1 + \frac{2\Phi(r)}{c^2}\right] - \frac{1}{c^2}\left[\frac{\dot{r}^2}{1 + 2\Phi(r)/c^2} + r^2\dot{\theta}^2 + r^2\sin^2\theta\,\dot{\phi}^2\right] \right\} c^2 (dt)^2 \qquad (8.7)$$

Since everything now varies in Hamilton's principle, the proper time is obtained from the interval *everywhere* by

$$(ds)^2 = -c^2 (d\tau)^2 \qquad\qquad ; \text{everywhere now} \qquad (8.8)$$

Thus the lagrangian for particle motion in the Schwarzschild metric is given by

$$L = m \left(\frac{d\mathbf{s}}{dt} \right)^2 \left(\frac{dt}{d\tau} \right) = -mc^2 \left(\frac{d\tau}{dt} \right)$$

$$= -mc^2 \left\{ \left[1 + \frac{2\Phi(r)}{c^2} \right] - \frac{1}{c^2} \left[\frac{\dot{r}^2}{1 + 2\Phi(r)/c^2} + r^2 \dot{\theta}^2 + r^2 \sin^2\theta \, \dot{\phi}^2 \right] \right\}^{1/2}$$

$$(8.9)$$

Here $\Phi(r) = -MG/r$ is the gravitational potential.

Note that since we now have $(d\mathbf{s}) = -c^2(d\tau)^2$ everywhere in our variational principle, this lagrangian is exactly the same L that one gets by simply writing[2]

$$L = m \left(\frac{d\mathbf{s}}{d\tau} \right)^2 \left(\frac{d\tau}{dt} \right) = -mc^2 \left(\frac{d\tau}{dt} \right) \qquad (8.10)$$

In summary:

We have an exact expression in Eq. (8.9) for the lagrangian $L(r, \theta, \phi \,; \dot{r}, \dot{\theta}, \dot{\phi})$ for particle motion in the Schwarzschild metric. We can now just do ordinary lagrangian mechanics!

We observe that[3]

- Expansion of the lagrangian L up to $O(1/c^2)$ gives the correct newtonian (NRL) limit;
- Expansion up to $O(1/c^4)$ gives the precession of the perihelion, which we will proceed to calculate.

Let us carry out this expansion in $1/c^2$. Use

$$(1 + \varepsilon)^{1/2} = 1 + \frac{1}{2}\varepsilon - \frac{1}{8}\varepsilon^2 + \cdots \qquad (8.11)$$

First expand inside the square root, and define the square of the usual velocity by

$$(\vec{v})^2 = v^2 = \dot{r}^2 + r^2\dot{\theta}^2 + r^2 \sin^2\theta \, \dot{\phi}^2 \qquad ; \text{ usual velocity} \quad (8.12)$$

---

[2]Compare the result in Eq. (6.36). When in free space with $M = 0$, Eq. (8.9) reproduces that result.

[3]Note the more precise terminology used here and henceforth: expansion *up to* a given order neglects all contributions of that order; expansion *through* a given order retains all of them.

Then

$$L \approx -mc^2 \left[1 + \frac{1}{c^2} \left(2\Phi - v^2 + \frac{2\Phi \dot{r}^2}{c^2} \right) \right]^{1/2}$$

$$\approx -mc^2 - \frac{m}{2} \left(2\Phi - v^2 + \frac{2\Phi \dot{r}^2}{c^2} \right) + \frac{m}{8c^2} (2\Phi - v^2)^2 \qquad (8.13)$$

With an organization of terms, the lagrangian is given by

$$L = -mc^2 + \frac{mv^2}{2} \left(1 + \frac{v^2}{4c^2} \right) - m\Phi \left(1 + \frac{\dot{r}^2}{c^2} + \frac{v^2}{2c^2} - \frac{\Phi}{2c^2} \right) + O\left(\frac{1}{c^4} \right) \qquad (8.14)$$

The first term of $O(c^2)$ is a constant and is irrelevant as far as Hamilton's principle is concerned; we discard it. The terms of $O(c^0)$ give the usual non-relativistic lagrangian

$$L_{\text{NRL}} = \frac{mv^2}{2} - m\Phi \qquad (8.15)$$

The terms of $O(1/c^2)$ provide the leading correction to the non-relativistic lagrangian for particle motion in the gravitational potential $\Phi = -MG/r$ in general relativity[4]

$$\delta L_{\text{GR}} = \frac{m}{c^2} \left[\frac{v^4}{8} - \frac{\Phi}{2} \left(2\dot{r}^2 + v^2 - \Phi \right) \right] \qquad (8.16)$$

8.2 Equations of Motion

We retain the terms in the lagrangian through order $O(1/c^2)$ in Eq. (8.14) and write Lagrange's equations for the coordinates (r, θ, ϕ). The configuration is indicated in Fig. 8.3.

$\underline{\phi \text{ equation}}$: Lagrange's equation for ϕ reads

$$\frac{d}{dt} \frac{\partial L}{\partial \dot{\phi}} = \frac{\partial L}{\partial \phi} \qquad (8.17)$$

---

[4]Of course, one must expand physical results in terms of a small *dimensionless parameter*, and we will always eventually identify that parameter. In all of our applications, that quantity will involve a factor of $1/c^2$, so that an explicit expansion in $1/c^2$ is the most direct and illuminating way of arriving at the results.

Since the coordinate ϕ is *cyclic* (*i.e.* it does not appear in L), then the momentum conjugate to ϕ is a constant of the motion

$$p_\phi = \frac{\partial L}{\partial \dot{\phi}} = \text{constant} \qquad (8.18)$$

Now, for orientation, recall the newtonian limit (see Prob. 8.8). In that limit, p_ϕ is the component of the angular momentum along the z-axis. One can choose to put the total angular momentum vector in the (x, y) plane, and then $p_\phi = 0$, as illustrated in Fig. 8.3. The motion then takes place in a plane perpendicular to the angular momentum. We use these arguments as *motivation* for the following.

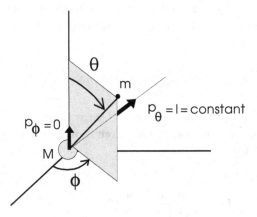

Fig. 8.3   Configuration for particle motion in the Schwarzschild metric. The motion takes place in the indicated (shaded) plane with constant $\phi = \phi_0$. In the newtonian limit, p_ϕ is the component of the angular momentum along the z-axis, and p_θ the component in the ϕ-direction.

In the full problem, let us again try to choose our coordinate system so that $p_\phi = 0$. Since the lagrangian depends quadratically on $\dot{\phi}$, one has

$$L = L(\dot{\phi}^2)$$
$$p_\phi = \frac{\partial L}{\partial \dot{\phi}} = 2\dot{\phi}\, \frac{\partial L}{\partial \dot{\phi}^2} \qquad (8.19)$$

Thus $p_\phi = 0$ implies $\dot{\phi} = 0$, and hence the angle ϕ does not change. The result is that the full motion is *again confined to a plane* as indicated in

Fig. 8.3.

$$\dot{\phi} = 0$$

$$\phi = \phi_0 = \text{constant} \qquad ; \text{ planar motion} \qquad (8.20)$$

We redraw the resulting planar configuration in Fig. 8.4. The condition in Eq. (8.20) certainly provides a solution to Lagrange's equations.

The square of the velocity in Eq. (8.12) now becomes

$$v^2 = \dot{r}^2 + r^2 \dot{\theta}^2 \qquad (8.21)$$

This is the familiar expression for the planar motion in Fig. 8.4. With this configuration, the lagrangian in Eq. (8.14) takes the form

$$L \doteq \frac{m}{2}(\dot{r}^2 + r^2\dot{\theta}^2) + \frac{m}{8c^2}(\dot{r}^4 + 2\dot{r}^2 r^2 \dot{\theta}^2 + r^4 \dot{\theta}^4) + $$
$$\frac{mMG}{r} + \frac{mMG}{r}\frac{\dot{r}^2}{c^2} + \frac{mMG}{r}\frac{(\dot{r}^2 + r^2\dot{\theta}^2)}{2c^2} + \frac{m}{2c^2}\left(\frac{MG}{r}\right)^2 \quad (8.22)$$

The symbol $\doteq$ indicates that the constant term $-mc^2$ has been dropped. This expression is then correct through $O(1/c^2)$.

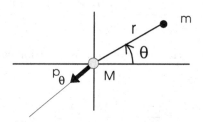

Fig. 8.4 Redrawing of Fig. 8.3 to illustrate the planar motion in the Schwarzschild metric, where the total angular momentum $p_\theta = l$ is a constant of the motion. In the newtonian limit, the angular momentum is a vector perpendicular to the plane of the motion.

We proceed to develop the equations of motion that follow from this L:

θ equation: The variable θ is now cyclic in the lagrangian of Eq. (8.22), and hence the conjugate momentum p_θ is a constant of the motion.

$$p_\theta = \frac{\partial L}{\partial \dot{\theta}} = l = \text{constant} \qquad ; \text{ angular momentum} \qquad (8.23)$$

This is now the total angular momentum. It is computed to be

$$l = \frac{\partial L}{\partial \dot\theta} = mr^2\dot\theta + \frac{m}{2c^2}r^2\dot{r}^2\dot\theta + \frac{m}{2c^2}r^4\dot\theta^3 + \frac{mMG}{c^2}r\dot\theta \qquad (8.24)$$

which may be rewritten as

$$l = mr^2\dot\theta\left(1 + \frac{\dot{r}^2}{2c^2} + \frac{r^2\dot\theta^2}{2c^2} + \frac{MG}{c^2 r}\right) \quad ; \text{ angular momentum} \quad (8.25)$$

This expression provides the angular momentum through order $(1/c^2)$.

r equation: Lagrange's equation for the coordinate r is

$$\frac{d}{dt}\frac{\partial L}{\partial \dot{r}} = \frac{\partial L}{\partial r} \qquad (8.26)$$

This gives

$$\frac{d}{dt}\left(m\dot{r} + \frac{m\dot{r}^3}{2c^2} + \frac{m\dot{r}r^2\dot\theta^2}{2c^2} + \frac{2mMG\dot{r}}{rc^2} + \frac{mMG\dot{r}}{rc^2}\right) = mr\dot\theta^2 +$$

$$\frac{mr\dot{r}^2\dot\theta^2}{2c^2} + \frac{mr^3\dot\theta^4}{2c^2} - \frac{mMG}{r^2} - \frac{3mMG\dot{r}^2}{2r^2c^2} + \frac{mMG\dot\theta^2}{2c^2} - \frac{m(MG)^2}{c^2 r^3} \quad (8.27)$$

With the cancelation of the factor of m, this equation can be rearranged to read

$$\frac{d}{dt}\left[\dot{r}\left(1 + \frac{\dot{r}^2}{2c^2} + \frac{r^2\dot\theta^2}{2c^2} + \frac{3MG}{rc^2}\right)\right] =$$

$$r\dot\theta^2\left(1 + \frac{\dot{r}^2}{2c^2} + \frac{r^2\dot\theta^2}{2c^2} + \frac{MG}{2rc^2}\right) - \frac{MG}{r^2}\left(1 + \frac{3\dot{r}^2}{2c^2} + \frac{MG}{rc^2}\right) \quad (8.28)$$

This expression is again correct through $O(1/c^2)$.

The angular momentum in Eq. (8.25) is a constant of the motion, and that equation can then be used to re-express the angular velocity $\dot\theta$ in terms of r. This in turn allows one to reduce Eq. (8.28) to a one-dimensional radial equation in an effective potential $V_{\text{eff}}(r)$. Since we are only working consistently through $O(1/c^2)$, the easiest way to achieve this is to just solve

Eq. (8.25) for $\dot{\theta}$ and then iterate. An expansion gives

$$\dot{\theta} = \frac{l}{mr^2}\left(1 + \frac{\dot{r}^2}{2c^2} + \frac{MG}{c^2 r} + \frac{r^2\dot{\theta}^2}{2c^2}\right)^{-1}$$

$$= \frac{l}{mr^2}\left[1 - \frac{\dot{r}^2}{2c^2} - \frac{MG}{c^2 r} - \frac{r^2\dot{\theta}^2}{2c^2} + \cdots\right] \tag{8.29}$$

Iteration then yields

$$\dot{\theta} = \frac{l}{mr^2}\left[1 - \frac{\dot{r}^2}{2c^2} - \frac{MG}{c^2 r} - \frac{r^2}{2c^2}\left(\frac{l}{mr^2}\right)^2\right] \tag{8.30}$$

which is correct through $O(1/c^2)$. Now substitute this expression in the radial Eq. (8.28), again working consistently through this order. The r.h.s. of that equation becomes

r.h.s. =
$$r\left(\frac{l}{mr^2}\right)^2\left[1 - \frac{\dot{r}^2}{c^2} - \frac{2MG}{rc^2} - \frac{r^2}{c^2}\left(\frac{l}{mr^2}\right)^2 + \frac{\dot{r}^2}{2c^2} + \frac{r^2}{2c^2}\left(\frac{l}{mr^2}\right)^2 + \frac{MG}{2rc^2}\right]$$

$$- \frac{MG}{r^2}\left[1 + \frac{3\dot{r}^2}{2c^2} + \frac{MG}{rc^2}\right] + O\left(\frac{1}{c^4}\right) \tag{8.31}$$

Hence the radial equation takes the form

$$\frac{d}{dt}\left\{\dot{r}\left[1 + \frac{\dot{r}^2}{2c^2} + \frac{r^2}{2c^2}\left(\frac{l}{mr^2}\right)^2 + \frac{3MG}{rc^2}\right]\right\} =$$

$$\frac{l^2}{m^2 r^3}\left[1 - \frac{\dot{r}^2}{2c^2} - \frac{r^2}{2c^2}\left(\frac{l}{mr^2}\right)^2 - \frac{3MG}{2rc^2}\right] - \frac{MG}{r^2}\left[1 + \frac{3\dot{r}^2}{2c^2} + \frac{MG}{rc^2}\right]$$

$$\tag{8.32}$$

The equation is exact through $O(1/c^2)$. Given the square of the angular momentum l^2, a constant of the motion, this expression is now in the form of a one-dimensional radial equation in an effective potential.

8.3 Equilibrium Circular Orbits

Let us look for a solution to these equations that describes equilibrium circular orbits

$$r = a = \text{constant} \qquad ; \text{ circular orbit}$$
$$\dot{r} = \ddot{r} = 0 \qquad\qquad\qquad (8.33)$$

The l.h.s. of Eq. (8.32) then vanishes, and the r.h.s gives

$$\frac{l^2}{m^2 a^3}\left[1 - \frac{l^2}{2c^2 m^2 a^2} - \frac{3MG}{2c^2 a}\right] = \frac{MG}{a^2}\left[1 + \frac{MG}{c^2 a}\right] \qquad (8.34)$$

Division by the factor in square brackets on the l.h.s. and expansion in $1/c^2$ then gives

$$\frac{l^2}{m^2 a^3} = \frac{MG}{a^2}\left[1 + \frac{MG}{c^2 a} + \frac{3MG}{2c^2 a} + \frac{l^2}{2c^2 m^2 a^2}\right] + O\left(\frac{1}{c^4}\right) \qquad (8.35)$$

An iteration then gives

$$\frac{l^2}{m^2 a^3} = \frac{MG}{a^2}\left(1 + \frac{3MG}{c^2 a}\right) \qquad (8.36)$$

This relation between the angular momentum and the radius of the circular orbit is correct through $O(1/c^2)$.

Equation (8.30) relates the angular velocity to the angular momentum, and it can now be used to determine $\dot{\theta}_0$, the angular velocity in the circular orbit. Substitution of the above result and an expansion and iteration yield

$$\dot{\theta}_0 = \left(\frac{MG}{a^3}\right)^{1/2}\left(1 + \frac{3MG}{2c^2 a} - \frac{MG}{c^2 a} - \frac{MG}{2c^2 a}\right) + \cdots$$
$$= \left(\frac{MG}{a^3}\right)^{1/2} \qquad\qquad ; \text{ angular velocity} \qquad (8.37)$$

There is *no* correction to this result through $O(1/c^2)$.[5]

8.4 Small Oscillations About Circular Orbits

Let us now consider *small oscillations* about the equilibrium circular orbit. Suppose the particle is given a small *transverse kick* where $\delta\vec{p} \propto \vec{r}$. At least in newtonian physics, this leaves the angular momentum unchanged

$$\delta\vec{l} = \vec{r} \times \delta\vec{p} = 0 \qquad ; \text{ transverse kick} \qquad (8.38)$$

---

[5] Compare Prob. 7.5.

With this observation as motivation, we here consider small oscillations about the circular orbit at *fixed angular momentum l*. We look for a solution to the equations of motion of the form (Fig. 8.5)

$$r = a\,[1 + \eta(t)]$$
$$\dot{r} = a\dot{\eta} \qquad ; \ddot{r} = a\ddot{\eta} \qquad (8.39)$$

We assume that η is very small, and we work *through first order in* η.[6] The equations of motion can then be linearized about the circular solution and terms such as $\dot{r}^2/c^2$ dropped.

It is convenient to now go over to *dimensionless quantities*. Introduce

$$\lambda \equiv \frac{MG}{c^2 a} \ll 1 \qquad (8.40)$$

With the use of Eq. (8.36), one obtains an alternate expression for λ

$$\lambda = \frac{l^2}{c^2 m^2 a^2} + O\left(\frac{1}{c^4}\right) \qquad (8.41)$$

We are working consistently *through first order in the small parameter* λ, *which characterizes the corrections coming from general relativity*. From Eq. (8.36) and (8.37) one has

$$\frac{l^2}{m^2 a^4} = \dot{\theta}_0^2(1 + 3\lambda) \qquad (8.42)$$

through $O(1/c^2)$.

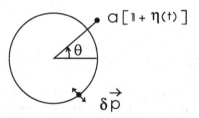

Fig. 8.5 Small oscillations about the equilibrium circular orbit at fixed total angular momentum l in the Schwarzschild metric. In the newtonian limit, a transverse kick leaves the angular momentum vector unchanged.

Now consider the radial Eq. (8.32) describing small oscillations about the equilibrium circular orbit. We neglect the terms in $\dot{r}^2/c^2$ in that equa-

---

[6]One can always make η small enough so that the corrections of $O(\eta^2)$ are negligible.

tion because they are of $O(\eta^2)$. Since it is already of $O(\eta)$, the l.h.s of that relation becomes

$$\text{l.h.s.} = a\ddot{\eta}\left(1 + \frac{1}{2}\lambda + 3\lambda\right) \qquad (8.43)$$

Here Eqs. (8.40) and (8.41) have been employed in the coefficient of η, which is now evaluated at equilibrium and all terms of $O(\lambda)$ retained.

For the r.h.s. of Eq. (8.32), we initially retain the full radial dependence, which will then be expanded about the equilibrium value of $\eta = 0$. Equations (8.37), (8.40), and (8.42) can be used to rewrite the r.h.s in terms of $\dot{\theta}_0^2$. With the retention of all terms of $O(\lambda)$ it takes the form

$$\text{r.h.s.} = a\frac{\dot{\theta}_0^2(1+3\lambda)}{(1+\eta)^3}\left[1 - \frac{1}{2}\frac{\lambda}{(1+\eta)^2} - \frac{3}{2}\frac{\lambda}{(1+\eta)}\right] -$$
$$a\frac{\dot{\theta}_0^2}{(1+\eta)^2}\left[1 + \frac{\lambda}{1+\eta}\right] \qquad (8.44)$$

This expression is now analyzed as follows:

- First retain only the terms linear in λ, since this is the order through which we are working;
- Then expand in η. The constant term should vanish since we are expanding about the equilibrium circular orbits defined by r.h.s. = 0; it is the *linear* term in η that is of interest here.

Thus, first

$$\frac{\text{r.h.s.}}{a\dot{\theta}_0^2} = \frac{1}{(1+\eta)^3}\left[1 + 3\lambda - \frac{\lambda}{2}\frac{1}{(1+\eta)^2} - \frac{3\lambda}{2}\frac{1}{(1+\eta)}\right]$$
$$- \frac{1}{(1+\eta)^2}\left[1 + \frac{\lambda}{1+\eta}\right] + O(\lambda^2) \qquad (8.45)$$

Now equate the l.h.s. and r.h.s. of the radial equation from Eqs. (8.43) and (8.45), cancel the factor of a, and rearrange the result in a slightly more transparent manner. This gives

$$\ddot{x}\left(1 + \frac{7}{2}\lambda\right) = \dot{\theta}_0^2\left\{\frac{1}{x^3} - \frac{1}{x^2} + \lambda\left[\frac{2}{x^3} - \frac{3}{2x^4} - \frac{1}{2x^5}\right]\right\} + O(\lambda^2)$$
$$x \equiv 1 + \eta \qquad (8.46)$$

If $\lambda = 0$, this reproduces the newtonian limit (NRL), where the expression on the right acts as an effective potential for the radial oscillations.

The term in λ on the r.h.s. gives the leading modification of this effective potential due to general relativity.[7]

The expansion in η then gives[8]

$$\ddot{\eta}\left(1 + \frac{7}{2}\lambda\right) = \dot{\theta}_0^2\left\{1 - 3\eta - (1 - 2\eta) + \right.$$

$$\left.\lambda\left[2(1 - 3\eta) - \frac{3}{2}(1 - 4\eta) - \frac{1}{2}(1 - 5\eta)\right]\right\} + \cdots$$

$$= -\dot{\theta}_0^2\left(1 - \frac{5}{2}\lambda\right)\eta + O(\eta^2) \tag{8.47}$$

Upon neglect of terms of $O(\eta^2)$, division by the factor on the left, and a final expansion in λ, this yields

$$\ddot{\eta} = -\Omega^2\eta$$
$$\Omega^2 = \dot{\theta}_0^2(1 - 6\lambda) \tag{8.48}$$

This result is correct through $O(1/c^2)$ and then exact through $O(\eta)$. We evidently have found

Stable simple harmonic oscillations, with an angular frequency Ω, of the radial coordinate about its value in the equilibrium circular orbit [recall Eq. (8.39)].

The solution to Eq. (8.48) with initial conditions

$$\eta(0) = -\eta_0 \qquad ; \dot{\eta}(0) = 0 \tag{8.49}$$

is elementary

$$\eta(t) = -\eta_0\cos(\Omega t) \tag{8.50}$$

The radial coordinate is then given by

$$r = a[1 - \eta_0\cos(\Omega t)] \tag{8.51}$$

This result is sketched in Fig. 8.6. The *time* taken from perihelion to perihelion (minimum r) is evidently given by

$$t = \frac{2\pi}{\Omega} \qquad ; \text{ perihelion to perihelion} \tag{8.52}$$

---

[7]The force is $-dV_{\text{eff}}(x)/dx$ and the one-dimensional radial equation has the form $d(m^\star\dot{x})/dt = -dV_{\text{eff}}(x)/dx$. The term in λ on the l.h.s. provides the $m^\star$.

[8]Use $(1 + \eta)^{-n} = 1 - n\eta + n(n + 1)\eta^2/2 + \cdots$.

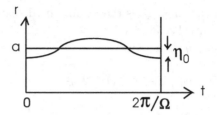

Fig. 8.6 Sketch of the radial coordinate for small oscillations about the equilibrium circular orbit as given by Eq. (8.51). The time from perihelion to perihelion (minimum r) is $t = 2\pi/\Omega$.

What is the increase in the angle θ over this time period? The rate of increase in θ is given by Eq. (8.30)

$$\dot{\theta} = \frac{l}{mr^2}\left[1 - \frac{\dot{r}^2}{2c^2} - \frac{MG}{c^2 r} - \frac{r^2}{2c^2}\left(\frac{l}{mr^2}\right)^2\right] \tag{8.53}$$

This is rewritten in terms of $\dot{\theta}_0$ and λ exactly as in Eqs. (8.44-8.46)

$$\dot{\theta} = \dot{\theta}_0\left[\frac{1}{x^2} + \lambda\left(\frac{3}{2x^2} - \frac{1}{x^3} - \frac{1}{2x^4}\right)\right] + O(\lambda^2)$$
$$x = 1 + \eta \tag{8.54}$$

Here the term in $\dot{r}^2/c^2$ has again already been neglected as quadratic in η. An expansion of Eq. (8.54) in η then gives

$$\dot{\theta} = \dot{\theta}_0[1 - 2(1 - \lambda)\eta(t)] + O(\eta^2) \tag{8.55}$$

Now neglect the terms of $O(\eta^2)$, and integrate Eq. (8.55) over one period of the radial oscillation

$$\int_0^{2\pi/\Omega} \frac{d\theta}{dt}dt = \theta\left(\frac{2\pi}{\Omega}\right) - \theta(0) = \left(\frac{2\pi}{\Omega}\right)\dot{\theta}_0 \tag{8.56}$$

Here we have used the fact that

$$\int_0^{2\pi/\Omega} \cos\left(\Omega t\right) dt = 0 \tag{8.57}$$

Consequently, the change in the angular velocity proportional to $\eta(t)$ averages out over one period, and we can simply use $\dot{\theta}_0$ as the angular velocity over that time. Thus

$$\theta\left(\frac{2\pi}{\Omega}\right) - \theta(0) = \left(\frac{2\pi}{\Omega}\right)\dot{\theta}_0$$
$$= \frac{2\pi}{(1 - 6\lambda)^{1/2}}$$
$$= 2\pi(1 + 3\lambda) + O(\lambda^2) \tag{8.58}$$

Hence the angle θ has increased by more than 2π as the radial coordinate goes from perihelion to perihelion, and there is an *advance of the perihelion* given by

$$\Delta\theta = 6\pi\lambda = 6\pi\left(\frac{MG}{c^2 a}\right) \qquad ; \text{ advance of perihelion} \tag{8.59}$$

The precession *rate* $\Omega_{\text{perihelion}}$ is $\Delta\theta$/revolution

$$\Omega_{\text{perihelion}} = \Delta\theta/\text{rev} \qquad ; \text{ precession rate} \tag{8.60}$$

This result is correct through $O(\lambda)$ and then exact as $\eta \to 0$. Note that in the newtonian limit $c^2 \to \infty$ one has $\Delta\theta = 0$, which confirms our observation that *the newtonian orbits are closed*.

8.5 Some Numbers and Comparison with Experiment

We recall two fundamental constants

$$G = 6.673 \times 10^{-20} \text{ km}^3/\text{kg-sec}^2$$
$$c = 2.998 \times 10^5 \text{ km/sec} \tag{8.61}$$

For the *sun*

$$M_\odot = 1.99 \times 10^{30} \text{ kg} \qquad ; \text{ sun}$$
$$R_\odot = 6.96 \times 10^5 \text{ km} \tag{8.62}$$

This gives a value of the parameter λ at the *surface of the sun* of

$$\lambda_\odot = \left(\frac{GM}{c^2 a}\right)_\odot = 2.12 \times 10^{-6} \qquad ; \text{ surface of sun} \tag{8.63}$$

Substitution into Eq. (8.59) then gives a value for the rate of advance of the perihelion at the surface of the sun of

$$\Omega_{\text{perihelion}} = 4.00 \times 10^{-5} \text{ rad/rev} \qquad \text{; surface of sun} \qquad (8.64)$$

The frequency of such an orbit is obtained from Eq. (8.37) as[9]

$$\nu_\odot = \frac{1}{2\pi} \left(\frac{MG}{a^3} \right)^{1/2}_\odot$$
$$= 3150 \text{ rev/year} \qquad (8.65)$$

Thus, with a change of units, the rate of advance of the perihelion for an orbit at the surface of the sun is

$$\Omega_{\text{perihelion}} = 0.126 \, \text{rad/year} = 7.22°/\text{year} \qquad \text{; surface of sun} \qquad (8.66)$$

Even as close as possible to the largest M in our solar system, this is a very small effect.

In this discussion we have focused on small oscillations about a circular orbit. If the orbit is actually elliptical to start with in the newtonian limit, the analysis is somewhat more involved. Weinberg gives the result in this case as [Weinberg (1972)]

$$\Delta\theta = 6\pi \left[\frac{MG}{c^2 a(1 - \varepsilon^2)} \right] \qquad \text{; } \varepsilon = \text{eccentricity of ellipse} \qquad (8.67)$$

Since the eccentricity of the ellipse is $\varepsilon = O(\eta)$ where η is the deformation parameter used in our analysis, we see that the correction to Eq. (8.59) is indeed of second order in η, as advertised.

For the planet mercury, Weinberg gives the numbers

$$a = 57.91 \times 10^6 \text{ km} \qquad \text{; planet mercury}$$
$$\text{rev/century} = 415$$
$$\varepsilon = 0.2056 \qquad (8.68)$$

These numbers give for the advance of the perihelion for mercury[10]

$$\Omega_{\text{perihelion}} = 43.0''/\text{century} \qquad \text{; planet mercury} \qquad (8.69)$$

---

[9]Recall that the frequency ν is related to the angular velocity ω by $\omega = 2\pi\nu$. We use 1 yr=3.156 $\times$ 10[7] sec.

[10]Note that for mercury $1/(1 - \varepsilon^2) = 1.044$ so the second-order correction in η is a 4.4% effect.

This is a very small number, but these astronomical quantities can be measured very accurately. Weinberg gives the actual comparison between the theoretical prediction and the experimental measurement of the advance of the perihelion for mercury as [Weinberg (1972)]

$$\Omega_{\text{perihelion}} = 43.03^{''}/\text{century} \qquad ; \text{ theory}$$
$$\Omega_{\text{perihelion}} = 43.11 \pm 0.45^{''}/\text{century} \qquad \text{experiment} \qquad (8.70)$$

Weinberg makes the statement at the time of publication of his book that[11]

"This is by far the most important experimental verification of general relativity"

This comparison is all the more striking when one realizes that the *observed* advance of the perihelion of mercury is *two orders of magnitude larger than the above result!*

$$\Omega_{\text{perihelion}} = 5600.73 \pm 0.41^{''}/\text{century} \qquad ; \text{ observed for mercury} \quad (8.71)$$

There are a whole host of corrections that have to be applied before one can come up with the experimental result quoted in Eq. (8.70), especially the *perturbation of the other planets.* The reader can start thinking up his or her own list of possible additional corrections:

• Quadrupole deformation of sun (both *inside* and as observed);
• Tidal forces and dynamic coupling;
• Interplanetary environment (meteors, dark matter);
• Magnetic fields, etc.

8.6 Deflection of Light

A related, but still substantially different, orbit problem is that of the deflection of a light ray in the Schwarzschild metric.[12] Since the observed quantity again involves an angle, it is an unambiguous physical effect and provides one of the "classic" tests of the theory of general relativity. Instead of carrying out this calculation in detail in the text, we guide the reader through an elementary (and approximate) calculation of the leading order deflection in Probs. 8.6 and 8.7.[13] This approach, through the

---

[11]The quote is from [Weinberg (1972)] p. 198.

[12]The existence of a rest mass and rest frame for a particle has played a crucial role in our development of lagrangian mechanics in the Schwarzschild metric.

[13]A proper calculation is carried out in [Weinberg (1972)].

effective index of refraction, has the advantage that it illustrates that a strong gravitational field acts as a gigantic *lens* for a distant light source [Einstein (1936)] — an imaging effect that is now seen regularly in modern astronomy.[14]

From Prob. 8.7 and Eq. (8.63), the prediction of general relativity for the deflection of light from the sun at an impact parameter equal to the sun's radius is given by

$$\Theta_{\text{deflect}} = 4\lambda_\odot \left(\frac{1+\gamma}{2} \right) = 1.75'' \left(\frac{1+\gamma}{2} \right) \qquad ; \text{ at } b = R_\odot \quad (8.72)$$

where $\gamma = 1$. This theoretical result compares favorably with many experimental observations of this quantity (see Table 8.1 in [Weinberg (1972)]).

---

[14]See [Falco (2005)] for a recent review of gravitational lensing. For some spectacular images, see [Hubble (2006)].

Chapter 9

Gravitational Redshift

We next turn to the phenomenon of the change in frequency of an oscilla-tor, or change in lifetime of a decaying source, as a function of its radial position in the gravitational field described by the Schwarzschild metric. The configuration is indicated in Fig. 9.1.

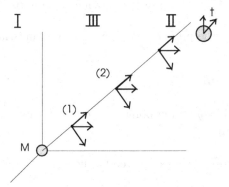

Fig. 9.1   Two points (r, θ, ϕ, ct), labeled by (1) and (2), in the global inertial laboratory frame I of the gravitational field described by the Schwarzschild metric. The frequency of an oscillator, or the lifetime of a source, is observed at those points. At each of those points there is a local freely falling frame III, and at the origin of the LF^3 one measures a number of oscillations (or decays) dN, proper time interval $d\tau$, and proper frequency $\bar{\nu} = dN/d\tau$. The observed laboratory frequency then changes because of the dependence of laboratory time on radial position. The notation here is that of Fig. 7.11.

We can all agree on how many oscillations have taken place, or just how many decays have occured. Thus we start from the following basic observation:

9.1   Basic Observation

The number of oscillations, or number of decays, has an invariant meaning.

The proper time is defined in terms of the invariant interval by

$$(d\mathbf{s})^2 = -c^2(d\tau)^2 \tag{9.1}$$

The proper time is identical to the clock time if one is

- at rest in the global inertial laboratory frame far away from the source (frame II);
- at rest in any local freely falling frame $(LF)^3$ (frame III).

Go to the origin of any LF^3. Let dN be the number of oscillations in a proper time $d\tau = d\bar{t}$. The *proper frequency* of the oscillations is then given by

$$\bar{\nu} = \frac{dN}{d\tau} \quad ; \text{ proper frequency } - \text{ an invariant} \tag{9.2}$$

From its definition, the proper frequency is also an invariant.[1]

9.2   Frequency Shift

The *observed frequency* at any point (r, θ, ϕ, ct) in the global inertial laboratory frame I is given by

$$\nu = \frac{dN}{dt} \quad ; \text{ observed frequency} \tag{9.3}$$

This now changes as a function of the radial position r, because the relation between proper time and laboratory time depends on that position.

Suppose a system undergoes a given set of oscillations dN at two different points (1) and (2) as indicated in Fig. 9.1. Since we all agree on the *number dN*, it is an *invariant*, one has the following relation between the observed laboratory frequencies and laboratory time elements at these two positions

$$dN = \nu_1 \, dt_1 = \nu_2 \, dt_2 \tag{9.4}$$

---

[1]The proper decay frequency (decay rate) for a collection of N particles is $\bar{\nu}_d = -(1/N)(dN/d\tau)$. The associated lifetime is $1/\bar{\nu}_d$. The subsequent analysis is immediately applied to the decay rate.

For a given dN, the corresponding invariant proper time interval $d\tau$ will also be the same at (1) and (2). The *key* is that at any given radial position r, the proper time interval $d\tau = d\bar{t}$ is related to the global laboratory time element dt by Eq. (7.109)

$$d\tau = \left(1 + \frac{2\Phi}{c^2}\right)^{1/2} dt \qquad ; \; \Phi = -\frac{MG}{r} \qquad (9.5)$$

Hence for a given element of proper time $d\tau$, the following relation exists between the radial coordinates and elements of laboratory time

$$d\tau = \left(1 + \frac{2\Phi_1}{c^2}\right)^{1/2} dt_1 = \left(1 + \frac{2\Phi_2}{c^2}\right)^{1/2} dt_2 \qquad (9.6)$$

It follows from Eqs. (9.4) and (9.6) that

$$\frac{\nu_1}{\nu_2} = \frac{dt_2}{dt_1} = \left(\frac{1 + 2\Phi_1/c^2}{1 + 2\Phi_2/c^2}\right)^{1/2} \qquad ; \text{ observed frequency of oscillator} \quad (9.7)$$

This is the observed frequency of the oscillator as a function of position to all orders in $1/c^2$. Through $O(1/c^2)$ it is given by

$$\frac{\nu_1}{\nu_2} = 1 + \frac{\Phi_1 - \Phi_2}{c^2} \qquad ; \text{ expansion through } O(1/c^2) \qquad (9.8)$$

These relations tell us what one observes regarding frequencies and the implied periods, or lifetimes, as a function of the radial position r in the Schwarzschild metric.

We make two comments:

- This is what is observed at two given stationary points (1) and (2) in the global inertial laboratory frame I in Fig. 9.1 — these points have neither velocity nor acceleration in this frame;
- The observed frequencies or lifetimes are reported to the *record keeper*, who keeps track of the coordinate position (r, θ, ϕ, ct) at which they occur. This is what the record keeper knows.

Let us now write

$$\nu_2 = \nu_1 + \Delta\nu$$
$$\Phi_2 = \Phi_1 + \Delta\Phi$$
$$r_2 = r_1 + \Delta r \qquad (9.9)$$

Then Eq. (9.8) takes the form

$$\frac{\Delta\nu}{\nu} = \frac{\Delta\Phi}{c^2} \tag{9.10}$$

This relation holds through $O(1/c^2)$, and through this order, one can use either $\nu = \nu_1$ or $\nu = \nu_2$ in the denominator on the l.h.s. The quantity in the numerator on the r.h.s. is given by

$$\Delta\Phi = -MG\left(\frac{1}{r_2} - \frac{1}{r_1}\right) = MG\left(\frac{r_2 - r_1}{r_1 r_2}\right)$$
$$\approx MG\frac{\Delta r}{r^2} \tag{9.11}$$

The last relation holds provided $|\Delta r| \ll r$.

In *summary*, the observed frequency of an oscillator, or of a decaying source, changes in the following manner as it is taken from point (1) to point (2) in the gravitational field described by the Schwarzschild metric

$$\frac{\Delta\nu}{\nu} = \frac{\Delta\Phi}{c^2} \qquad ; \text{ observed frequency of oscillator } 1 \to 2$$
$$\approx \frac{MG}{c^2}\frac{\Delta r}{r^2} \tag{9.12}$$

The first relation holds through $O(1/c^2)$ and the second assumes $|\Delta r| \ll r$. As one moves *out*, $\Delta r > 0$ and the gravitational potential *increases*; the observed frequency then also *increases*, $\Delta\nu > 0$. As one moves *in*, $\Delta r < 0$ and the gravitational potential *decreases*; the observed frequency then also *decreases*, $\Delta\nu < 0$.

9.3   Propagation of Light

Let us now consider the propagation of a light signal in the gravitational field described by the Schwarzschild metric. To examine this problem, we make use of the *conservation of energy*. We use the fact that we know from quantum theory that light is composed of *photons*, each of which has energy $\varepsilon = h\nu$, where h is Planck's constant.[2] The photon is created by some atomic or nuclear transition of energy ε, and $\nu = \varepsilon/h$ is the laboratory frequency of the photon (the one logged by the record keeper). Since energy and mass are equivalent through Einstein's celebrated relation from special

---

[2]It is interesting to note that Einstein was also responsible for the concept of a *photon*, introduced in his explanation of the photoelectric effect.

relativity, one can define an equivalent photon mass at the energy ε by $\varepsilon \equiv m_\gamma c^2$. Thus[3]

$$\varepsilon = h\nu = m_\gamma c^2 \qquad ; \text{photon} \tag{9.13}$$

Now suppose the photon is created by a transition at the point (1) in the gravitational field that arises from a spherically symmetric source, travels to the point (2), and is then detected at that second point as indicated in Fig. 9.2.

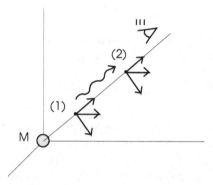

Fig. 9.2   A photon is created at point (1) and then observed at point (2) in the gravitational field of a spherically symmetric source. We write the photon energy as $h\nu = m_\gamma c^2$ and assume that the photon's total energy $m_\gamma c^2 + m_\gamma \Phi$ is conserved through $O(c^0)$.

There will now also be a gravitational *potential* energy of the mass m_γ in the potential Φ. We assume that the photon's *total* energy $h\nu + m_\gamma \Phi = m_\gamma c^2 + m_\gamma \Phi$ is conserved through $O(c^0)$. The familiar energy conservation relation should hold in general relativity to this order (compare Prob. 8.2). The statement of energy conservation as the photon travels from point (1) to point (2) therefore implies[4]

$$(m_\gamma c^2 + m_\gamma \Phi)_1 = (m_\gamma c^2 + m_\gamma \Phi)_2 \qquad ; \text{energy conservation} \tag{9.14}$$

This relation holds through $O(c^0)$.

---

[3]Note that m_γ is *not* the rest mass of the photon, which vanishes (as far as we know).
[4]Note this is now something that *happens to the light* as it passes from point (1) to point (2).

Substitution of Eq. (9.13) into (9.14) then leads to the *frequency* relation

$$h\nu_1 + h\nu_1 \frac{\Phi_1}{c^2} = h\nu_2 + h\nu_2 \frac{\Phi_2}{c^2}$$

$$\text{or ;} \quad \nu_1 \left(1 + \frac{\Phi_1}{c^2}\right) = \nu_2 \left(1 + \frac{\Phi_2}{c^2}\right)$$

$$\frac{\nu_1}{\nu_2} = \frac{1 + \Phi_2/c^2}{1 + \Phi_1/c^2} \tag{9.15}$$

The last line can now be expanded through $O(1/c^2)$ to give

$$\frac{\nu_1}{\nu_2} = 1 + \frac{\Phi_2 - \Phi_1}{c^2} \tag{9.16}$$

Note that Planck's constant h has disappeared from this relation.

Thus, with the definitions in Eqs. (9.9),

$$\frac{\Delta\nu}{\nu} = -\frac{\Delta\Phi}{c^2} \quad ; \text{ observed shift of photon frequency } 1 \rightarrow 2 \tag{9.17}$$

This relation holds through $O(1/c^2)$. Note the sign. This is the *observed shift in the photon frequency* as it travels from point (1) to point (2). As the photon moves to a region of *increased* gravitational potential $\Delta\Phi > 0$, its frequency will *decrease*, $\Delta\nu < 0$. One observes lower and lower frequencies as light moves outward from a gravitational source. Thus the light emitted by an atom at the surface of a star is *red-shifted* when it gets to us.[5]

We observe that the decrease in frequency of the *light* in going from (1) $\rightarrow$ (2) in Eq. (9.17) is just equal to the decrease in frequency of an *oscillator* in going in the opposite direction from (2) $\rightarrow$ (1) in Eq. (9.12). If the frequency of the light did *not* decrease in this way, then the following cyclic process would produce energy, in violation of the first law of thermodynamics.[6]

9.4 A Cyclic Process

Consider the following process as illustrated in Fig. 9.3.

---

[5]Problems 9.4–9.8 examine the radial propagation of light in the gravitational field of the Schwarzschild metric.

[6]It is a conclusion from these arguments that, to this order, when one observes the gravitational redshift of a photon coming out from a massive gravitating body, one is equivalently observing the slowing of the oscillating frequency of an oscillator lying deep in that field.

(1) Start with, say, an excited atom at position (2) in the gravitational field of a mass M. Write its total energy as

$$E_2 = m_A c^2 + \varepsilon \equiv m_A^\star c^2 \qquad (9.18)$$

Let the atom fall to point (1). Gravity does work $m_A^\star \Delta\Phi$;

(2) Let the atom radiate a photon at point (1). After radiating the photon with $h\nu_1 = \varepsilon$, the energy of the atom is $m_A c^2$;
(3) Now take the atom back to point (2). Gravity does work $-m_A \Delta\Phi$;
(4) If the photon frequency is unchanged when it gets to (2), then let it excite an identical atom.

The net result is a *cyclic process in which work is done and energy is produced*

$$\oint dW = (m_A^\star - m_A)\Delta\Phi = \varepsilon\frac{\Delta\Phi}{c^2} \qquad (9.19)$$

This violates the first law of thermodynamics.

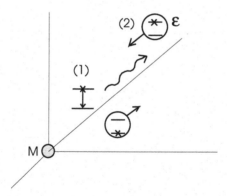

Fig. 9.3   A cyclic process, described in the text, that is a potential source of energy.

To *prevent* this from happening, the energy of the photon must just *decrease* by an equivalent amount

$$h(\nu_1 - \nu_2) = \varepsilon\frac{\Delta\Phi}{c^2} \qquad (9.20)$$

Since $\varepsilon = h\nu_1$, this gives[7]

$$\frac{\Delta\nu}{\nu} = -\frac{\Delta\Phi}{c^2} \tag{9.21}$$

for the frequency shift of the photon in going from (1) to (2), which holds through $O(1/c^2)$. This is precisely Eq. (9.17).

One can argue, therefore, that Eq. (9.21) is not at all a unique prediction of general relativity, but is merely a consequence of the nature of the photon, work done on an atom in a gravitational potential, and the first law of thermodynamics. General relativity does, however, provide a means of relating in leading order the gravitational redshift to the slowing of the oscillating frequency of an oscillator in the gravitational field. It also, of course, predicts the shift in oscillator frequency (or lifetime) to all orders in $1/c^2$ [see Eq. (9.7)].

While no attempt at all is made in this text to cite the primary experimental literature, we do want to single out the very clever experiment of Pound and Rebka. They made a laboratory measurement of the gravitational redshift in Eq. (9.21) using a ladder and the turntable Doppler shift of a Mössbauer line, which provided an amazingly accurate frequency determination [Pound and Rebka (1960)].

---

[7]In this argument, $\Delta\Phi = \Phi_2 - \Phi_1 > 0$, while $\Delta\nu = \nu_2 - \nu_1 < 0$. The photon frequency must *decrease* in going from (1) → (2). Note that all expressions of the form $\Delta\nu/\nu = \pm\Delta\Phi/c^2$ are exact through $O(1/c^2)$ in these arguments (see also Prob. 9.8).

Chapter 10

Neutron Stars

After this discussion of the classic tests of general relativity, we turn
to other interesting applications of the theory. Here we discuss *neutron stars*. These are stellar objects of nuclear densities held together
by gravity, and kept from collapsing by the short-distance repulsion of
the strong nuclear force. We will find that eventually, if the object is
large enough, the nuclear force will not be sufficient to overcome the
gravitational attraction, and the object will continue to collapse un-
til it lies *within its Schwarzschild radius*. At this point, the object
forms a *black hole* with many unusual properties. We will derive the
Tolman-Oppenheimer-Volkoff (TOV) equations that allow one to calculate the structure of a neutron star in general relativity [Tolman (1939);
Oppenheimer and Volkoff (1939)]. The required input is an *equation of
state* for neutron matter, and for that we will introduce relativistic mean
field theory (RMFT), which provides a simple and effective description of
the main features of nuclear structure [Walecka (2004)].

To start the analysis, we must first extend our discussion of hydrodynamics from special to general relativity.

10.1 Hydrodynamics in General Relativity

Start in the flat tangent space at the point $\bar{O}$ in Fig. 10.1. Within this
tangent space there is a particular frame (coordinate system), the LF^3, in
which one has a simple physical interpretation of the theory. There is *only
special relativity*

- There is no gravity;
- There are no inertial forces.

This is a *local* statement.

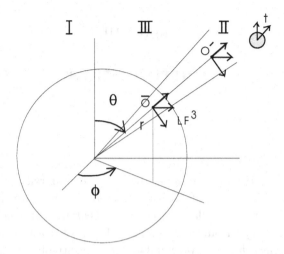

Fig. 10.1   Configuration for discussion of local hydrodynamics in general relativity (after Fig. 7.11).

Assume that in the LF^3 (frame III), one has an isotropic fluid with no shear forces and a velocity field $\vec{v}$ as studied in chap. 6.[1] The situation is illustrated in Fig. 10.2. Then from Eqs. (6.108) and (6.103)

$$\bar{T}^{\mu\nu} = P\,\bar{g}^{\mu\nu} + \left(\rho + \frac{P}{c^2}\right)\bar{u}^{\mu}\bar{u}^{\nu}$$

$$\frac{\bar{u}^{\mu}}{c} = (\beta,0,0,1)\frac{1}{(1-\beta^2)^{1/2}}$$

$$\bar{g}^{\mu\nu} = \begin{bmatrix} 1 & 0 & 0 & 0 \\ 0 & 1 & 0 & 0 \\ 0 & 0 & 1 & 0 \\ 0 & 0 & 0 & -1 \end{bmatrix} = \bar{g}_{\mu\nu} \qquad (10.1)$$

Here we use a bar over a symbol to indicate a quantity in the LF^3, and we write the coordinates as $(\bar{z}, \bar{x}, \bar{y}, c\bar{t})$. The combination $\bar{u}^{\mu}\bar{u}_{\mu}$ is an invariant

$$\bar{u}^{\mu}\bar{u}_{\mu} = -c^2 \qquad ; \text{ invariant} \qquad (10.2)$$

---

[1]We assume an arbitrary $\vec{v}$ to start with. Later we will set $\vec{v} = 0$ to describe the static situation.

The quantities (P, ρ) are the pressure and mass density defined in the rest frame of the fluid, and what is required as input is the *equation of state* $P(\rho)$ for the fluid

$$P = P(\rho) \qquad \text{; equation of state} \qquad (10.3)$$

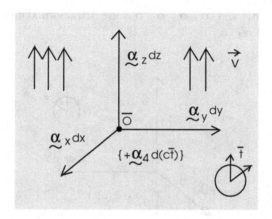

Fig. 10.2 Configuration in the LF^3 with origin at $\bar{O}$ in Fig. 10.1. One has an isotropic fluid with no shear forces and a velocity field $\vec{v}$ (after Fig. 6.12). The displacement in the fourth direction $\alpha_4 d(c\bar{t})$ is suppressed. We will later choose to put this z-axis in the radial direction and x-axis in the θ-direction in Fig. 10.1.

Now *transform to the global inertial laboratory frame I* where we will employ the spherical coordinates $q^\mu = (r, \theta, \phi, ct)$ as illustrated in Figs. 10.1 and 10.3. We work with the simple case of the *static, spherically symmetric metric* given by Eq. (7.48)

$$g_{\mu\nu} = \begin{bmatrix} A(r) & 0 & 0 & 0 \\ 0 & r^2 & 0 & 0 \\ 0 & 0 & r^2\sin^2\theta & 0 \\ 0 & 0 & 0 & -B(r) \end{bmatrix}$$

$$= \mathbf{e}_\mu \cdot \mathbf{e}_\nu \qquad (10.4)$$

The affine connection and Ricci tensor for this metric were computed in chap. 7. To transform to the laboratory frame I, we can now use the results from our *general tensor analysis*. If the z-direction in Fig. 10.2 is placed along the radial direction in Figs. 10.1 and 10.3, and the x-axis along the

θ-direction, then one can identify

$$\mathbf{e}_r = \sqrt{A(r)}\,\hat{\mathbf{e}}_r = \sqrt{A(r)}\,\boldsymbol{\alpha}_z$$
$$\mathbf{e}_\theta = r\,\hat{\mathbf{e}}_\theta = r\,\boldsymbol{\alpha}_x$$
$$\mathbf{e}_\phi = r\sin\theta\,\hat{\mathbf{e}}_\phi = r\sin\theta\,\boldsymbol{\alpha}_y$$
$$\mathbf{e}_4 = \sqrt{B(r)}\,\hat{\mathbf{e}}_4 = \sqrt{B(r)}\,\boldsymbol{\alpha}_4 \qquad (10.5)$$

Here $(\hat{\mathbf{e}}_r, \hat{\mathbf{e}}_\theta, \hat{\mathbf{e}}_\phi, \hat{\mathbf{e}}_4)$ and $(\boldsymbol{\alpha}_z, \boldsymbol{\alpha}_x, \boldsymbol{\alpha}_y, \boldsymbol{\alpha}_4)$ are orthonormal unit vectors with $\hat{\mathbf{e}}_4 \cdot \hat{\mathbf{e}}_4 = -1$ and $\boldsymbol{\alpha}_4 \cdot \boldsymbol{\alpha}_4 = -1$.

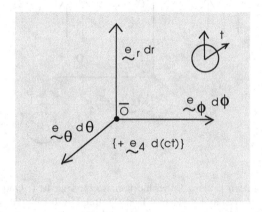

Fig. 10.3 Spherical coordinates in the global inertial laboratory frame I (see Fig. 10.1).

We are now in a position to identify the *coordinate transformation matrix* $a^\nu{}_\mu$ from the relation

$$\boldsymbol{\alpha}_\mu = a^\nu{}_\mu\,\mathbf{e}_\nu \qquad (10.6)$$

With the matrix definition $[\underline{a}]_{\nu\mu} \equiv a^\nu{}_\mu$ one has

$$[\underline{a}]_{\nu\mu} = a^\nu{}_\mu$$

$$\underline{a} = \begin{bmatrix} 1/\sqrt{A(r)} & 0 & 0 & 0 \\ 0 & 1/r & 0 & 0 \\ 0 & 0 & 1/r\sin\theta & 0 \\ 0 & 0 & 0 & 1/\sqrt{B(r)} \end{bmatrix} \qquad (10.7)$$

Here the rows are labeled by $\nu = (r, \theta, \phi, 4)$ where the fourth coordinate is (ct), while the columns are labeled by $\mu = (z, x, y, 4)$ corresponding to the coordinates $(\bar{z}, \bar{x}, \bar{y}, c\bar{t})$. This coordinate transformation takes us from the LF^3 frame III to the global inertial laboratory frame I (Fig. 10.1).

The metric is transformed according to the general tensor transformation law

$$g^{\mu\nu} = a^{\mu}{}_{\mu'} a^{\nu}{}_{\nu'} \bar{g}^{\mu'\nu'} \tag{10.8}$$

This relation may be rewritten with the introduction of matrix notation $a^{\mu}{}_{\nu} = [\underline{a}]_{\mu\nu}$ and $g^{\mu\nu} = [\underline{g}^{-1}]_{\mu\nu}$ as

$$g^{\mu\nu} = a^{\nu}{}_{\nu'} \left[a^{\mu}{}_{\mu'} \bar{g}^{\mu'\nu'} \right] = a^{\nu}{}_{\nu'} \left[\underline{a}(\bar{\underline{g}}^{-1}) \right]_{\mu\nu'} = a^{\nu}{}_{\nu'} \left[\underline{a}(\bar{\underline{g}}^{-1}) \right]_{\nu'\mu}$$
$$= \left[\underline{a}\,\underline{a}(\bar{\underline{g}}^{-1}) \right]_{\nu\mu} = \left[\underline{a}\,\underline{a}(\bar{\underline{g}}^{-1}) \right]_{\mu\nu} \tag{10.9}$$

Here the symmetry of the resulting matrices (explicitly verified below) has been used in obtaining the last equality in each line.

Thus one simply has to carry out the indicated matrix multiplications[2]

$$\underline{a}(\bar{\underline{g}}^{-1}) = \begin{bmatrix} 1/\sqrt{A(r)} & 0 & 0 & 0 \\ 0 & 1/r & 0 & 0 \\ 0 & 0 & 1/r\sin\theta & 0 \\ 0 & 0 & 0 & -1/\sqrt{B(r)} \end{bmatrix}$$

$$\underline{a}\,\underline{a}(\bar{\underline{g}}^{-1}) = \begin{bmatrix} 1/A(r) & 0 & 0 & 0 \\ 0 & 1/r^2 & 0 & 0 \\ 0 & 0 & 1/r^2\sin^2\theta & 0 \\ 0 & 0 & 0 & -1/B(r) \end{bmatrix} \tag{10.10}$$

Both matrices are symmetric, as advertised, and the second is indeed the inverse of the metric matrix in Eq. (10.4).

The four-velocity of the fluid is similarly transformed according to

$$\frac{u^{\mu}}{c} = a^{\mu}{}_{\mu'} \frac{\bar{u}^{\mu'}}{c}$$

$$\frac{1}{c}\underline{u} = \gamma \begin{bmatrix} 1/\sqrt{A(r)} & 0 & 0 & 0 \\ 0 & 1/r & 0 & 0 \\ 0 & 0 & 1/r\sin\theta & 0 \\ 0 & 0 & 0 & 1/\sqrt{B(r)} \end{bmatrix} \begin{bmatrix} \beta \\ 0 \\ 0 \\ 1 \end{bmatrix}$$

$$\gamma = \frac{1}{(1-\beta^2)^{1/2}} \tag{10.11}$$

---

[2] Recall the last of Eqs. (10.1).

For a *fluid at rest* in the static, spherically symmetric metric of Eq. (10.4) this gives

$$\frac{u^\mu}{c} = \left(0, 0, 0, \frac{1}{\sqrt{B(r)}}\right) \tag{10.12}$$

Note that

$$\frac{u_\mu}{c} = \left(0, 0, 0, -\sqrt{B(r)}\right)$$

$$\frac{u_\mu u^\mu}{c^2} = -1 \qquad ; \text{ invariant} \tag{10.13}$$

Let us now compute the *source term* in Einstein's field equations in a medium with an isotropic fluid with no shear forces at rest and in the case of a static, spherically symmetric metric. The general form of the stress tensor in this case in the LF^3 is given by Eq. (6.108)

$$\bar{T}_{\mu\nu} = P\,\bar{g}_{\mu\nu} + \left(\rho + \frac{P}{c^2}\right)\bar{u}_\mu\bar{u}_\nu \qquad ; \text{ in } LF^3 \tag{10.14}$$

Now make the above coordinate transformation to the global inertial laboratory frame I

$$T_{\mu\nu} = Pg_{\mu\nu} + \left(\rho + \frac{P}{c^2}\right)u_\mu u_\nu \qquad ; \text{ laboratory frame I} \tag{10.15}$$

Compute $T \equiv T^\mu{}_\mu$

$$T = T^\mu{}_\mu = 4P - \left(\rho + \frac{P}{c^2}\right)c^2 = 3P - \rho c^2 \tag{10.16}$$

In *summary*, the source term in the Einstein field equations for an isotropic fluid with no shear forces and a static, spherically symmetric metric is given by

$$T_{\mu\nu} - \frac{1}{2}T\,g_{\mu\nu} = \frac{1}{2}(\rho c^2 - P)\,g_{\mu\nu} + \left(\rho + \frac{P}{c^2}\right)u_\mu u_\nu \tag{10.17}$$

Here, now using the explicit matrix notation for the metric, one has

$$g_{\mu\nu} = [\underline{g}]_{\mu\nu} \qquad ; g^{\mu\nu} = [\underline{g}^{-1}]_{\mu\nu} \tag{10.18}$$

with metric matrices given by

$$\underline{g} = \begin{bmatrix} A(r) & 0 & 0 & 0 \\ 0 & r^2 & 0 & 0 \\ 0 & 0 & r^2 \sin^2 \theta & 0 \\ 0 & 0 & 0 & -B(r) \end{bmatrix}$$

$$\underline{g}^{-1} = \begin{bmatrix} 1/A(r) & 0 & 0 & 0 \\ 0 & 1/r^2 & 0 & 0 \\ 0 & 0 & 1/r^2 \sin^2 \theta & 0 \\ 0 & 0 & 0 & -1/B(r) \end{bmatrix} \tag{10.19}$$

The four-velocity in Eq. (10.17) for the fluid at rest is given by

$$\frac{u_\mu}{c} = \left(0, 0, 0, -\sqrt{B(r)} \right) \tag{10.20}$$

It is then a simple matter to combine the above results to obtain the components with $\mu = (r, \theta, \phi, 4)$ that are needed for the r.h.s. of the Einstein field Eqs. (7.14) and (7.71) when one is *inside* the source. The components of $(T_{\mu\nu} - T g_{\mu\nu}/2)$ are

$$
\begin{aligned}
T_{\mu\nu} - \frac{1}{2} T g_{\mu\nu} &= 0 && ; \underline{\mu \neq \nu} \\
&= \frac{1}{2}(\rho c^2 - P) A(r) && ; \underline{rr} \\
&= \frac{1}{2}(\rho c^2 - P) r^2 && ; \underline{\theta\theta} \\
&= \frac{1}{2}(\rho c^2 - P) r^2 \sin^2 \theta && ; \underline{\phi\phi} \\
&= \frac{1}{2}(\rho c^2 + 3P) B(r) && ; \underline{44}
\end{aligned}
\tag{10.21}
$$

To recapitulate, the Einstein field equations inside the source are

$$R_{\mu\nu} = \kappa \left(T_{\mu\nu} - \frac{1}{2} T g_{\mu\nu} \right) \tag{10.22}$$

where the metric is given by Eq. (10.19), the Ricci tensor is given in Eqs. (7.71), and the source terms on the r.h.s. by Eqs. (10.21). The following assumptions have been made:

• Spherical symmetry;

- Static case;[3]
- Isotropic fluid with no shear forces;
- Equation of state $P = P(\rho)$ available.

In the following, we require the physical volume element. Recall that the determinant of the metric is here given by

$$g \equiv \det \underline{g}$$
$$= -A(r)B(r)r^4 \sin^2 \theta \qquad (10.23)$$

The *negative* of the determinant satisfies

$$-g = -\det \underline{g}$$
$$= A(r)B(r)r^4 \sin^2 \theta \geq 0 \qquad (10.24)$$

The pesky minus sign arises from the structure of the Lorentz metric, and to obtain a *real* expression, we employ $\sqrt{-g}$ in the volume element

$$\sqrt{-g} = \sqrt{A(r)B(r)} \, r^2 \sin \theta \qquad (10.25)$$

The invariant volume element then follows from Eq. (5.152) as

$$d^4\tau = \sqrt{-g} \, dr \, d\theta \, d\phi \, d(ct)$$
$$= \sqrt{A(r)B(r)} \, r^2 \sin \theta \, dr \, d\theta \, d\phi \, d(ct) \qquad (10.26)$$

With the Schwarzschild metric, which holds *outside* of the spherically symmetric source, Eqs. (7.88) imply that this expression simplifies to

$$d^4\tau = r^2 \sin \theta \, dr \, d\theta \, d\phi \, d(ct) \qquad ; \text{ Schwarzschild metric} \quad (10.27)$$

In the following we also require the affine connection with a pair of contracted indices $\Gamma^\lambda_{\mu\lambda}$, and from Eq. (5.170) this can be rewritten as

$$\Gamma^\lambda_{\mu\lambda} = \frac{1}{2g}\frac{\partial g}{\partial q^\mu} \equiv \frac{1}{2(-g)}\frac{\partial(-g)}{\partial q^\mu}$$
$$= \frac{\partial \ln\sqrt{-g}}{\partial q^\mu} \qquad (10.28)$$

Recall that in moving to a neighboring point the components of a four-vector change according to

$$\mathcal{D}v^\mu = dv^\mu + \Gamma^\mu_{\mu'\lambda}v^{\mu'} dq^\lambda \qquad (10.29)$$

---

[3]It should be mentioned that the metric outside of a *rotating* sphere is presented in [Kerr (1963)]; however, we shall have nothing more to say on this topic.

The *covariant derivative* and *covariant divergence* of the four-vector then follow as

$$v^{\mu}{}_{;\nu} = \frac{\partial v^{\mu}}{\partial q^{\nu}} + \Gamma^{\mu}{}_{\mu'\nu} v^{\mu'}$$

$$v^{\mu}{}_{;\mu} = \frac{\partial v^{\mu}}{\partial q^{\mu}} + \Gamma^{\mu}{}_{\mu'\mu} v^{\mu'} \tag{10.30}$$

With the aid of the last of Eqs. (10.28), the covariant divergence can be rewritten as[4]

$$v^{\mu}{}_{;\mu} = \frac{1}{\sqrt{-g}} \frac{\partial}{\partial q^{\mu}} (\sqrt{-g}\, v^{\mu}) \qquad ; \text{ covariant divergence} \tag{10.31}$$

Since one just has special relativity in the LF^3, we know from chap. 6 that conservation of the baryon current and of the energy-momentum tensor implies that both have vanishing covariant divergence

$$\bar{N}^{\mu}{}_{;\mu} = 0$$

$$\bar{T}^{\mu\nu}{}_{;\nu} = 0 \tag{10.32}$$

We remind the reader that a bar above a symbol indicates a quantity in the LF^3 (frame III). One now simply carries out a *coordinate transformation* to the global inertial laboratory frame I and makes use of general tensor transformation properties.

The physical statement that the *baryon current is conserved* is independent of coordinate system since evidently upon the coordinate transformation

$$N^{\mu}{}_{;\mu} = \bar{N}^{\mu}{}_{;\mu} = 0 \tag{10.33}$$

This is an *invariant* statement.[5]

Recall that in moving to a neighboring point the components of a second-rank tensor are changed by

$$\mathcal{D}T^{\mu\nu} = dT^{\mu\nu} + \Gamma^{\mu}{}_{\mu'\lambda} T^{\mu'\nu} dq^{\lambda} + \Gamma^{\nu}{}_{\nu'\lambda} T^{\mu\nu'} dq^{\lambda} \tag{10.34}$$

---

[4]Note that $\Gamma^{\mu}{}_{\mu'\mu} v^{\mu'} \equiv \Gamma^{\lambda}{}_{\mu\lambda} v^{\mu}$.

[5]The relevant particle current in nuclear physics is the *baryon current*, since it is the number of baryons that is conserved. The nucleons (p, n) are baryons.

Hence the covariant derivative and covariant divergence of this tensor are

$$T^{\mu\nu}{}_{;\lambda} = \frac{\partial T^{\mu\nu}}{\partial q^\lambda} + \Gamma^\mu_{\mu'\lambda} T^{\mu'\nu} + \Gamma^\nu_{\nu'\lambda} T^{\mu\nu'}$$

$$T^{\mu\nu}{}_{;\nu} = \frac{\partial T^{\mu\nu}}{\partial q^\nu} + \Gamma^\mu_{\mu'\nu} T^{\mu'\nu} + \Gamma^\nu_{\nu'\nu} T^{\mu\nu'} \qquad (10.35)$$

With the aid of the last of Eqs. (10.28), the covariant divergence can be rewritten as

$$T^{\mu\nu}{}_{;\nu} = \frac{1}{\sqrt{-g}} \frac{\partial}{\partial q^\nu} (\sqrt{-g}\, T^{\mu\nu}) + \Gamma^\mu_{\lambda\nu} T^{\lambda\nu} \quad ; \text{ covariant divergence} \quad (10.36)$$

The physical statement that the *energy-momentum tensor is conserved* is independent of coordinate system since upon the coordinate transformation

$$T^{\mu\nu}{}_{;\nu} = \bar{T}^{\mu\nu}{}_{;\nu} = 0 \qquad (10.37)$$

This is also an *invariant* statement.

10.2 Tolman-Oppenheimer-Volkoff (TOV) Equations

We assume

- Spherical symmetry;
- Static situation;
- Isotropic fluid with no shear forces;

The Einstein field equations are

$$G_{\mu\nu} = R_{\mu\nu} - \frac{1}{2} R\, g_{\mu\nu} = \kappa T_{\mu\nu}$$

$$\text{or ;} \qquad R_{\mu\nu} = \kappa \left(T_{\mu\nu} - \frac{1}{2} T\, g_{\mu\nu} \right) \qquad (10.38)$$

The metric is given in Eq. (10.19). The general form of the Ricci tensor is given in Eqs. (7.71). The general form of the energy-momentum tensor is given in Eq. (10.15), and source terms on the r.h.s. of the second equation are then given in Eqs. (10.21). These results can be combined to yield the

Einstein field equations in the form

$$-\frac{B''}{2B} + \frac{1}{4}\left(\frac{B'}{B}\right)\left(\frac{B'}{B} + \frac{A'}{A}\right) + \frac{1}{r}\left(\frac{A'}{A}\right) = \frac{\kappa}{2}(\rho c^2 - P)A \qquad ; \underline{rr}$$

$$\frac{B''}{2A} - \frac{1}{4}\left(\frac{B'}{A}\right)\left(\frac{A'}{A} + \frac{B'}{B}\right) + \frac{1}{r}\left(\frac{B'}{A}\right) = \frac{\kappa}{2}(\rho c^2 + 3P)B \qquad ; \underline{44}$$

$$1 - \frac{1}{A} + \frac{r}{2A}\left(\frac{A'}{A} - \frac{B'}{B}\right) = \frac{\kappa}{2}(\rho c^2 - P)r^2 \qquad ; \underline{\theta\theta}$$

$$\sin^2\theta\,[R_{\theta\theta}] = \sin^2\theta\left[\kappa\left(T_{\theta\theta} - \frac{1}{2}Tg_{\theta\theta}\right)\right] \qquad ; \underline{\phi\phi}$$

$$0 = 0 \qquad ; \underline{\mu \neq \nu} \qquad (10.39)$$

These equations are augmented with the equation of state[6]

$$P = P(\rho) \qquad ; \text{equation of state} \qquad (10.40)$$

Now note that if the $\theta\theta$ equation is satisfied, the $\phi\phi$ equation is also satisfied automatically, and the equations with $\mu \neq \nu$ are satisfied identically. Thus one is left with the first three coupled, nonlinear, second-order differential equations in the three unknowns $[A(r), B(r), \rho(r)]$. We assume that these equations have a solution. We know that they have a solution outside of $\rho(r)$, it is the Schwarzschild metric. One can envision integrating the above equations in from this. Let us then derive some more useful representations of this solution.

Recall that we have established the fact that both sides of the Einstein field equations have vanishing covariant divergence

$$G^{\mu\nu}{}_{;\nu} = 0$$

$$T^{\mu\nu}{}_{;\nu} = 0 \qquad (10.41)$$

We make use of the second relation to get one useful combination of the previous equations

$$T^{\mu\nu} = P g^{\mu\nu} + (P + \rho c^2)\frac{u^\mu u^\nu}{c^2}$$

$$T^{\mu\nu}{}_{;\nu} = \underbrace{[P g^{\mu\nu}]_{;\nu}}_{P g^{\mu\nu}{}_{;\nu} + g^{\mu\nu} P_{;\nu}} + \left[(P + \rho c^2)\frac{u^\mu u^\nu}{c^2}\right]_{;\nu} \qquad (10.42)$$

---

[6]Note that we are working at zero temperature, $T = 0$. One should not get confused with the notation here. The equation of state $P = P(\rho)$ implies that there is a *functional relation* (assumed invertible) between pressure and mass density. The *radial dependence* of these quantities is then related according to $P(r) = P[\rho(r)]$.

In the last line, the covariant divergence of the metric tensor vanishes, and the covariant derivative of a scalar is just the partial derivative. One can then employ Eq. (10.36) to reduce the covariant divergence of the energy-momentum tensor to

$$T^{\mu\nu}{}_{;\nu} = g^{\mu\nu}\frac{\partial P}{\partial q^\nu} +$$

$$\left\{ \frac{1}{\sqrt{-g}}\frac{\partial}{\partial q^\nu}\left[\sqrt{-g}\,(P+\rho c^2)\frac{u^\mu u^\nu}{c^2}\right] + \Gamma^\mu_{\lambda\nu}(P+\rho c^2)\frac{u^\lambda u^\nu}{c^2} \right\} \quad (10.43)$$

In the static case there is only a radial dependence left in $[P(r), u^4(r)]$, and u^4 is given by Eq. (10.12). Then

$$\frac{\partial}{\partial q^4}\left[\sqrt{-g}(P+\rho c^2)\frac{u^\mu u^4}{c^2}\right] = 0 \quad (10.44)$$

The affine connection is given in Table 7.1. It follows that the only non-zero element in Eq. (10.43) comes from $\mu = r$, with the result that

$$g^{rr}\frac{dP(r)}{dr} + \Gamma^r_{44}(P+\rho c^2)\frac{u^4 u^4}{c^2} = 0$$

$$\text{or ;} \qquad \frac{1}{A}\frac{dP}{dr} + \frac{B'}{2A}(P+\rho c^2)\frac{1}{B} = 0 \quad (10.45)$$

Thus

$$\frac{dP}{dr} = -\frac{1}{2}(P+\rho c^2)\frac{B'}{B} \quad (10.46)$$

This equation relates the pressure $P(r)$ [or $\rho(r)$ — they are equivalent by the equation of state] to the function $B(r)$ in the metric in Eq. (10.19).

Another useful relation describing the solution is obtained by taking the combination $(rr)/2A + (44)/2B + (\theta\theta)/r^2$ in Eqs. (10.39). This leads to

$$\frac{A'}{2A^2 r} + \frac{B'}{2ABr} + \frac{1}{r^2} - \frac{1}{Ar^2} + \frac{1}{2Ar}\left(\frac{A'}{A} - \frac{B'}{B}\right) =$$

$$\frac{\kappa}{4}(\rho c^2 - P) + \frac{\kappa}{4}(\rho c^2 + 3P) + \frac{\kappa}{2}(\rho c^2 - P)$$

$$\text{or ;} \qquad \frac{1}{r^2} + \frac{A'}{A^2 r} - \frac{1}{Ar^2} = \kappa\rho c^2$$

$$\frac{1}{r^2}\left[1 - \frac{d}{dr}\left(\frac{r}{A}\right)\right] = \kappa\rho c^2 \quad (10.47)$$

Hence

$$\frac{d}{dr}\left(\frac{r}{A}\right) = 1 - \kappa\rho c^2 r^2 \tag{10.48}$$

This equation relates $P(r)$ [or $\rho(r)$] to the function $A(r)$ in the metric. This equation can be integrated using the boundary condition

$$\frac{r}{A} = 0 \quad ; \text{ at } r = 0 \tag{10.49}$$

That is, at the origin inside the star, $A(0)$ should be *finite*. Define[7]

$$\mathcal{M}(r) \equiv \int_0^r 4\pi s^2 \rho(s)\, ds \tag{10.50}$$

Then the integration of Eq. (10.48) out from the origin gives

$$\frac{r}{A} = r - \frac{\kappa c^2}{4\pi}\mathcal{M}(r) \tag{10.51}$$

Hence $A(r)$ is determined in terms of the mass density $\rho(r)$ according to

$$A(r) = \left[1 - \left(\frac{\kappa c^4}{8\pi}\right)\frac{2\mathcal{M}(r)}{c^2 r}\right]^{-1} \tag{10.52}$$

A third useful relation between $[A(r), B(r), P(r)]$ is obtained from the $\theta\theta$ expression in Eqs. (10.39)

$$1 - \frac{1}{A} + \frac{r}{2A}\left(\frac{A'}{A} - \frac{B'}{B}\right) = \frac{\kappa}{2}(\rho c^2 - P)r^2 \tag{10.53}$$

Make use of the following relation from Eq. (10.46)

$$\frac{B'}{B} = -\frac{2}{P + \rho c^2}\frac{dP}{dr} \tag{10.54}$$

Now define[8]

$$\kappa \equiv \frac{8\pi G}{c^4} \tag{10.55}$$

Equation (10.52) then takes the form

$$\frac{1}{A} = 1 - \frac{2G\mathcal{M}(r)}{c^2 r} \tag{10.56}$$

---

[7]Note the volume element $4\pi s^2\, ds$ used in this definition (here s is simply an integration variable).

[8]Note this is the first time we have actually had to use κ!

Differentiate this relation

$$-\frac{A'}{A^2} = \frac{2G}{c^2 r^2}\mathcal{M} - \frac{2G}{c^2 r}[4\pi r^2 \rho] \tag{10.57}$$

The last term in square brackets is just $d\mathcal{M}/dr$.

Now substitute Eqs. (10.54)-(10.57) into Eq. (10.53), with the result that

$$\frac{2G\mathcal{M}}{c^2 r} - \frac{r}{2}\left[\frac{2G\mathcal{M}}{c^2 r^2} - \frac{8\pi G r \rho}{c^2}\right] + \frac{r}{2}\left(1 - \frac{2G\mathcal{M}}{c^2 r}\right)\left(\frac{2}{P + \rho c^2}\frac{dP}{dr}\right) =$$
$$\frac{4\pi G}{c^4}(\rho c^2 - P)r^2$$

or ;
$$\frac{dP}{dr} = \frac{(P + \rho c^2)}{(1 - 2G\mathcal{M}/c^2 r)}\frac{1}{r}\left[-\frac{4\pi G P r^2}{c^4} - \frac{G\mathcal{M}}{c^2 r}\right] \tag{10.58}$$

With a rewriting of this expression and collection of results, we thus have

$$\frac{dP}{dr} = -\frac{G\mathcal{M}(r)\rho(r)}{r^2}\left[\frac{1 + P(r)/\rho(r)c^2}{1 - 2G\mathcal{M}(r)/c^2 r}\right]\left[1 + \frac{4\pi P(r)r^3}{\mathcal{M}(r)c^2}\right]$$
$$\mathcal{M}(r) = \int_0^r 4\pi s^2 \rho(s)\,ds$$
$$P = P(\rho) \qquad\qquad ; \text{ equation of state} \tag{10.59}$$

These equations provide a first-order, non-linear, integro-differential equation for the pressure $P(r)$, or equivalently the mass density $\rho(r)$. They form the Tolman-Oppenheimer-Volkoff (TOV) equations for stellar structure.

The functions $[A(r), B(r)]$ in the metric in Eq. (10.19) are given in terms of the solution to the TOV equations by Eq. (10.56), and the integration of Eq. (10.54), respectively.

The newtonian limit $c^2 \to \infty$ of Eqs. (10.59) takes the familiar form

$$\frac{dP}{dr} = -\frac{G\mathcal{M}(r)\rho(r)}{r^2} \qquad ; \text{ newtonian limit}$$
$$\mathcal{M}(r) = \int_0^r 4\pi s^2 \rho(s)\,ds \tag{10.60}$$

Here the pressure just balances the gravitational force, which is obtained with the aid of Gauss' law (see Prob. 10.1).

We must discuss the *boundary conditions* that accompany the TOV equations. We assume (see Fig. 10.4):

1) $\rho(0), A(0)$; finite

2) $P(R) = \rho(R) = 0$; defines surface

3) $P(r) = \rho(r) = 0$; $r > R$

4) $B(\infty) = 1$; newtonian limit   (10.61)

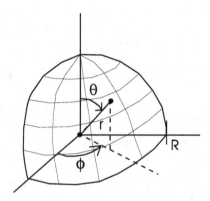

Fig. 10.4   Visualization of boundary conditions for TOV equations.

The TOV equations are directly applicable to condensed stellar objects such as neutron stars, where general relativistic effects are important. Let us investigate some consequences of these results.

1) Since the angular part of the metric is just that of spherical coordinates, the circumference of the star is given by $2\pi R$ and the area of the surface by $4\pi R^2$. This provides an unambiguous way of determining the coordinate R, the radius of the star.

$$\text{circumference of star} = 2\pi R$$
$$\text{area of surface} = 4\pi R^2 \qquad (10.62)$$

2) The mass of the star is defined by

$$M \equiv \mathcal{M}(R) = \int_0^R 4\pi s^2 \rho(s)\, ds \qquad ; \text{ mass} \qquad (10.63)$$

This follows from Eq. (10.56), since we have the identification with the Schwarzschild metric outside the star

$$A(r) = \frac{1}{1 - 2MG/c^2 r} \qquad\qquad ; \ r \geq R \qquad (10.64)$$

3) To get $B(r)$ we must still integrate Eq. (10.54)

$$\frac{B'}{B} = \frac{d\ln B}{dr} = -\frac{2}{P + \rho c^2} \frac{dP}{dr} \qquad\qquad (10.65)$$

This equation is integrated from r to ∞ with the aid of the last boundary condition in Eqs. (10.61) to give

$$\int_r^\infty d\ln B = -\ln B(r) = -2 \int_r^\infty \frac{dP}{P + \rho c^2} \qquad\qquad (10.66)$$

The differential of the pressure follows from the TOV Eq. (10.59)

$$\frac{dP}{P + \rho c^2} = -\frac{G\,dr}{c^2\,r^2} \left[\frac{\mathcal{M}(r) + 4\pi r^3 P(r)/c^2}{1 - 2G\mathcal{M}(r)/c^2 r} \right] \qquad\qquad (10.67)$$

Substitution into the above, and exponentiation, then yields an expression for $B(r)$

$$B(r) = \exp\left\{ -\frac{2G}{c^2} \int_r^\infty \frac{dr}{r^2} \left[\frac{\mathcal{M}(r) + 4\pi r^3 P(r)/c^2}{1 - 2G\mathcal{M}(r)/c^2 r} \right] \right\} \qquad (10.68)$$

This expression holds for all r. For $r \geq R$, the pressure vanishes and one has

$$\mathcal{M}(r) = \int_0^R 4\pi s^2 \rho(s)\,ds = M \qquad\qquad ; \ r \geq R \qquad (10.69)$$

Thus outside of the star

$$B(r) = \exp\left\{ -\frac{2G}{c^2} \int_r^\infty \frac{dr}{r^2} \frac{M}{(1 - 2GM/c^2 r)} \right\} \qquad ; \ r \geq R \qquad (10.70)$$

Note that we have assumed here that

$$\frac{2GM}{c^2 R} = \frac{R_s}{R} < 1 \qquad\qquad (10.71)$$

so that there are no singularities in the integral in Eq. (10.68). This inequality implies that the Schwarzschild radius lies *inside* of the star.

Introduce a change of variable in the integral in Eq. (10.70)

$$u = \frac{1}{r} \qquad ; \ du = -\frac{dr}{r^2} \qquad\qquad (10.72)$$

Then

$$B(r) = \exp\left\{ \frac{2GM}{c^2} \int_{1/r}^{0} \frac{du}{(1 - 2GMu/c^2)} \right\}$$

$$= \exp\left\{ \left[-\ln\left(1 - \frac{2GM}{c^2}u\right) \right]_{1/r}^{0} \right\}$$

$$= \exp\left\{ \ln\left(1 - \frac{2GM}{c^2 r}\right) \right\} \qquad ; r \geq R \qquad (10.73)$$

Hence, when one is outside of the star with $r \geq R$, the functions $[A(r), B(r)]$ are given by

$$B(r) = \left(1 - \frac{2GM}{c^2 r}\right) \qquad ; r \geq R$$

$$A(r) = \frac{1}{(1 - 2GM/c^2 r)} \qquad \text{Schwarzschild metric} \qquad (10.74)$$

This is just the *Schwarzschild metric*. Thus

- M is indeed the mass of the star;
- G is the gravitational constant, and the quantity κ that appears in the Einstein field equations is identified as

$$\kappa = \frac{8\pi G}{c^4} \qquad (10.75)$$

- All of the physics *outside* of the star follows from our discussion of the Schwarzschild metric!

In *summary*, the TOV equations are

$$\frac{dP}{dr} = -\frac{G\mathcal{M}(r)\rho(r)}{r^2} \left[\frac{1 + P(r)/\rho(r)c^2}{1 - 2G\mathcal{M}(r)/c^2 r} \right] \left[1 + \frac{4\pi P(r)r^3}{\mathcal{M}(r)c^2} \right]$$

$$\mathcal{M}(r) = \int_{0}^{r} 4\pi s^2 \rho(s)\, ds$$

$$P = P(\rho) \qquad ; \text{ equation of state} \qquad (10.76)$$

These provide a first-order, non-linear, integro-differential equation for the pressure $P(r)$, or equivalently the mass density $\rho(r)$. The appropriate boundary conditions are those in Eqs. (10.61). The TOV equations are directly applicable to condensed stellar objects such as neutron stars, where general relativistic effects are important. In the newtonian limit, they reduce to the newtonian description of stellar structure.

It is instructive to expand the TOV differential equation through $O(1/c^2)$, consistently obtaining all the corrections to the newtonian limit

$$\frac{dP}{dr} = -\frac{G\mathcal{M}(r)\rho(r)}{r^2}\left[1 + \frac{P(r)}{c^2\rho(r)} + \frac{4\pi r^3 P(r)}{c^2\mathcal{M}(r)} + \frac{2G\mathcal{M}(r)}{c^2 r}\right] \quad (10.77)$$

The first term is the newtonian limit. The terms in $P(r)$ arise because the pressure appears in the stress tensor and is thus an additional source of gravity in the Einstein field equations. The last term arises from the modification of the radial part of the metric.[9] Note that all the $O(1/c^2)$ corrections in the square brackets are positive and serve to *increase the magnitude* of the slope dP/dr.

Consider a *numerical solution* to the TOV Eqs. (10.76). First, convert the differential equation to a finite difference equation

$$P_{n+1} = (r_{n+1} - r_n)F(P_n, \rho_n, \mathcal{M}_n, r_n) + P_n$$
$$P_n = P(\rho_n)$$
$$\mathcal{M}_n = \int_0^{r_n} 4\pi s^2 \rho(s)\, ds \quad (10.78)$$

Then impose the initial conditions

$$\rho(0) = \rho_0 \qquad\qquad ; \text{ finite}$$
$$P(\rho_0) = P_0 \qquad\qquad\qquad\qquad (10.79)$$

We study the star for different initial mass densities at the origin (Fig. 10.5)

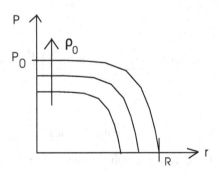

Fig. 10.5 Numerical integration of the TOV equations for different initial mass densities at the origin $\rho(0) = \rho_0$ and corresponding pressures $P(\rho_0) = P_0$.

---

[9]We leave it as an exercise for the dedicated reader to explain just why the last term takes this form.

As $r \to 0$, the mass $\mathcal{M}$, and corresponding pressure gradient in the TOV Eqs. (10.76) can be expanded as

$$\mathcal{M} \to \frac{4\pi}{3}\rho_0\, r^3 \qquad ; r \to 0$$

$$\frac{dP}{dr} \to -\frac{4\pi G}{3}\rho_0^2\, r\left(1 + \frac{P_0}{\rho_0 c^2}\right)\left(1 + \frac{3P_0}{\rho_0 c^2}\right) \qquad (10.80)$$

Since the last expression is proportional to r, one concludes that

$$\frac{dP}{dr} \to 0 \qquad ; r \to 0 \qquad (10.81)$$

Hence the slope of the pressure is *flat* at the origin.[10]

Now simply iterate the finite difference equation and step out in r. The slope dP/dr is negative for r greater than zero, and eventually one will cross the axis and find a value of $r = R$ for which the pressure vanishes (Fig. 10.5). This defines the radius of the star. In this way, one generates a series of stars with $[M(\rho_0), R(\rho_0)]$. If $2GM(\rho_0)/c^2R(\rho_0) < 1$, there is no difficulty with these arguments.

Suppose, however, that in stepping out in r one finds a value of r for which

$$1 - \frac{2G\mathcal{M}(r)}{c^2 r} \to 0 \qquad (10.82)$$

It follows from Eqs. (10.76) that at this point $dP/dr \to -\infty$, and the pressure clearly crosses the axis at that value of r (Fig. 10.6). At that point one has found both (M, R), and they evidently satisfy

$$\frac{2GM}{c^2 R} = 1 \qquad ; \text{ Schwarzschild radius} \qquad (10.83)$$

The neutron star then lies just within its Schwarzschild radius.[11] *Outside* of this radius, the star has the Schwarzschild metric with $R_s/r < 1$.

The *interval* $(d\mathbf{s})^2$ generated by the TOV equations takes the form

$$(d\mathbf{s})^2 = A(r)(dr)^2 + r^2(d\theta)^2 + r^2\sin^2\theta(d\phi)^2 - B(r)c^2(dt)^2 \qquad (10.84)$$

Here $[A(r), B(r)]$ are given for all r by Eqs. (10.56) and (10.68), respectively.

---

[10]There is no *cusp* in the pressure at the origin.

[11]It forms a *black hole* — see later.

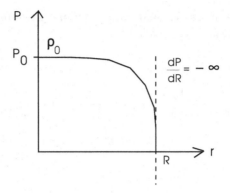

Fig. 10.6   Numerical integration of the TOV equations where $1 - 2G\mathcal{M}(r)/(c^2 r) \to 0$ at some r and $dP/dr \to -\infty$.

The *lagrangian* for the motion of a massive point particle in this metric (only) follows from Eq. (8.10) as

$$L = -mc^2 \frac{d\tau}{dt}$$

$$= -mc^2 \left\{ B(r) - \frac{1}{c^2} \left[A(r)\dot{r}^2 + r^2\dot{\theta}^2 + r^2 \sin^2\theta\,\dot{\phi}^2 \right] \right\}^{1/2} \quad (10.85)$$

This equation holds everywhere and provides a means for probing the metric for $r < R$. For $r > R$, one recovers motion in the Schwarzschild metric. Note that as $r \to 0$, one has

$$\mathcal{M}(r) \to \frac{4\pi}{3} \rho_0\, r^3 \qquad ;\, r \to 0$$

$$A(0) = 1$$

$$B(0) = \exp\left\{ -\frac{2G}{c^2} \int_0^\infty \frac{dr}{r^2} \left[\frac{\mathcal{M}(r) + 4\pi r^3 P(r)/c^2}{1 - 2\mathcal{M}(r)G/c^2 r} \right] \right\} \quad (10.86)$$

Since $g_{rr}(0) = 1$, there is no spatial distortion arising from the metric at the origin, but since B is minimized at the origin, there is maximal time dilation there.

We close this section by collecting some relevant physical constants and physical quantities, given in c.g.s. units to three-figure accuracy (from

[Weinberg (1972)])[12]

$$c = 3.00 \times 10^{10}\,\text{cm/sec}$$
$$G = 6.67 \times 10^{-8}\,\text{cm}^3/\text{gm-sec}^2$$

$$M_\odot = 1.99 \times 10^{33}\,\text{gm} \qquad ; \ R_\odot = 6.96 \times 10^{10}\,\text{cm}$$
$$M_e = 5.98 \times 10^{27}\,\text{gm} \qquad ; \ R_e = 6.38 \times 10^8\,\text{cm} \qquad (10.87)$$

These allow us to compute the Schwarzschild radii of the sun and the earth

$$R_\odot^s = \frac{2M_\odot G}{c^2} = 2.95 \times 10^5\,\text{cm} = 2.95\,\text{km}$$
$$R_e^s = \frac{2M_e G}{c^2} = 0.886\,\text{cm} \qquad ; \ \text{Schwarzschild radii} \qquad (10.88)$$

We can also compute the dimensionless ratio of Schwarzschild radius to observed radius R_s/R_{obs} for these bodies

$$\frac{2M_\odot G}{c^2 R_\odot} = 4.24 \times 10^{-6} \qquad ; \ \text{Schwarzschild radius/observed radius}$$
$$\frac{2M_e G}{c^2 R_e} = 1.39 \times 10^{-9} \qquad (10.89)$$

We note that this is a *very small dimensionless parameter* for both the earth and the sun. This ratio only comes close to unity for objects such as neutron stars. When it reaches unity, one has a black hole.

To proceed, one requires an equation of state $P(\rho)$ which holds at observed nuclear densities and well beyond. To provide this, we turn to the relativistic mean field theory of nuclear matter.

10.3 Relativistic Mean Field Theory of Nuclear Matter (RMFT)

In this section we present the relativistic mean field theory (RMFT) of nuclear matter, originally introduced to provide a covariant source term in the Einstein field equations. Subsequent developments showed that RMFT provides a framework for understanding much of nuclear structure, including nuclear densities, the nuclear shell model, and the spin observables in intermediate-energy nucleon scattering. The success of the model is currently understood within the framework of an effective field theory for

---

[12]One also has $\hbar = h/2\pi = 1.05 \times 10^{-27}$ erg-sec.

quantum chromodynamics (QCD), the underlying theory of the strong interactions, and density functional theory. This material is developed in detail in [Walecka (2004)]. For clarity and ease of presentation, we use the conventions of that reference in this section (only), taking care to write the final results in explicitly dimensionally correct form.[13]

We start from a simple model relativistic quantum field theory of the nuclear many-body system based on the following fields:

- A baryon field ψ of mass M_b, with the proton and neutron as components

$$\psi = \begin{pmatrix} p \\ n \end{pmatrix} \tag{10.90}$$

- A neutral scalar field ϕ of mass m_s, coupled to the scalar baryon density $\bar{\psi}\psi$ with coupling constant g_s;
- A neutral vector field V_μ of mass m_v, coupled to the conserved baryon current $i\bar{\psi}\gamma_\mu\psi$ with coupling constant g_v.

The scalar field provides the long-range nuclear attraction, and the vector field provides the short-range repulsion.[14]

The field equations are

$$\frac{\partial}{\partial x_\nu} V_{\mu\nu} + m_v^2 V_\mu = ig_v \bar{\psi}\gamma_\mu\psi$$

$$\left(\Box - m_s^2\right)\phi = -g_s\bar{\psi}\psi$$

$$\left[\gamma_\mu\left(\frac{\partial}{\partial x_\mu} - ig_v V_\mu\right) + (M_b - g_s\phi)\right]\psi = 0 \tag{10.91}$$

Here $\Box \equiv \nabla^2 - \partial^2/\partial(ct)^2$ is the wave operator, and the vector field tensor is defined by

$$V_{\mu\nu} \equiv \frac{\partial V_\nu}{\partial x_\mu} - \frac{\partial V_\mu}{\partial x_\nu} \tag{10.92}$$

---

[13]These conventions include $\hbar = c = 1$, in which case all masses, energies, and momenta are inverse lengths. Four-vectors are written $x_\mu = (\vec{x}, ix_0)$, and a scalar product of four-vectors takes the form $a \cdot b = \vec{a} \cdot \vec{b} - a_0 b_0$ (see Prob. 6.12). The Dirac gamma matrices are hermitian and satisfy $\gamma_\mu\gamma_\nu + \gamma_\nu\gamma_\mu = 2\delta_{\mu\nu}$.

[14]Since the electromagnetic current is simply replaced by the baryon current in the vector field equations, like baryon charges will repel, and the massive nature of the vector field implies a short-range interaction. The ratio of masses to coupling constants will later be adjusted [see Eqs. (10.105)] so that the scalar attraction dominates at low density and the vector repulsion at high density.

The first of Eqs. (10.91) is just the set of Maxwell's equations with the vector potential (photon field) replaced by a massive vector meson field and the electromagnetic current replaced by the conserved baryon current as the source. The second of Eqs. (10.91) is the Klein-Gordon equation for the scalar field with the scalar baryon density as source. The third of Eqs. (10.91) is the Dirac equation for the baryon field with the scalar and vector fields introduced in a minimal fashion.

If the fields (ψ, ϕ, V_μ) are quantized in the canonical fashion, one has all the complexities of an interacting relativistic quantum field theory, with the further difficulty of large (strong) coupling constants.

Imagine, however, a box of volume V containing a uniform distribution of B baryons (nucleons) where one is free to increase the baryon density $\rho_B \equiv B/V$ (Fig. 10.7).

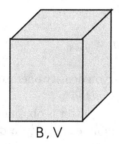

B, V

Fig. 10.7   A box of volume V containing a uniform distribution of B baryons (nucleons) where one is free to increase the baryon density $\rho_B \equiv B/V$.

A major simplification of the field equations occurs when one is able to replace the meson fields and sources by their expectation values.[15] Then

$$\langle \phi \rangle = \phi_0 \qquad ; \text{ RMFT}$$
$$\langle V_\lambda \rangle = i\delta_{\lambda 4} V_0 \qquad \text{constants for nuclear matter} \qquad (10.93)$$

The isotropy of the medium implies that only the fourth component of the vector field can develop an expectation value. Furthermore, the quantities on the r.h.s. of these equations are *constants* for uniform nuclear matter.

The hamiltonian density and baryon density operators take a very sim-

---

[15] Just as in classical E&M.

ple form in this RMFT[16]

$$\hat{\mathcal{H}}_{\mathrm{MFT}} = \frac{1}{2} m_s \phi_0^2 - \frac{1}{2} m_v^2 V_0^2 + g_v V_0 \, \hat{\rho}_{\mathrm{B}} +$$

$$\frac{1}{V} \sum_{\vec{k}\lambda} \left(\vec{k}^2 + M_b^{\star 2} \right)^{1/2} \left(A_{\vec{k}\lambda}^\dagger A_{\vec{k}\lambda} + B_{\vec{k}\lambda}^\dagger B_{\vec{k}\lambda} \right)$$

$$\hat{\rho}_{\mathrm{B}} = \frac{1}{V} \sum_{\vec{k}\lambda} \left(A_{\vec{k}\lambda}^\dagger A_{\vec{k}\lambda} - B_{\vec{k}\lambda}^\dagger B_{\vec{k}\lambda} \right) \qquad (10.94)$$

Here $(A^\dagger, A)$ are creation and destruction operators for the baryons (fermions) while $(B^\dagger, B)$ are those of the anti-baryons. The sum over $\vec{k}$ goes over all wave numbers satisfying periodic boundary conditions in the big box, and the sum over λ goes over the helicity $(\uparrow, \downarrow)$ and the z-component of isospin (p, n) of the nucleon. In the first of Eqs. (10.94), $M_b^\star$ is the *effective mass* of the nucleon defined by

$$M_b^\star \equiv M_b - g_s \phi_0 \qquad ; \text{ effective mass} \qquad (10.95)$$

Both the hamiltonian and baryon density operators are diagonal, and hence this RMFT has been *solved exactly*.

The ground state of nuclear matter is obtained by filling the momentum states up to the Fermi wavenumber k_{F} with a degeneracy factor of $\gamma = 4$ for $(p \uparrow, p \downarrow, n \uparrow, n \downarrow)$ (see Fig. 10.8).

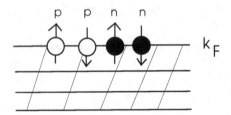

Fig. 10.8   Ground state of nuclear matter in RMFT. The states are filled to the Fermi level with wavenumber k_{F}, and there is a degeneracy factor of $\gamma = 4$ for $(p \uparrow, p \downarrow, n \uparrow, n \downarrow)$.

For large V, the baryon density is related to the Fermi wavenumber by

---

[16]There is an additional vacuum zero-point energy that is discussed in [Walecka (2004)].

the familiar expression

$$\rho_{\rm B} = \frac{\gamma}{(2\pi)^3} \int^{k_{\rm F}} d^3k = \frac{\gamma k_{\rm F}^3}{6\pi^2} \qquad ; \text{ baryon density} \qquad (10.96)$$

This is a conserved quantity. The mean vector meson field is related to $\rho_{\rm B}$ through the expectation value of the vector meson field equation

$$V_0 = \frac{g_v}{m_v^2} \rho_{\rm B} \qquad\qquad (10.97)$$

The energy density follows as the expectation value of the hamiltonian density

$$\varepsilon(\rho_{\rm B}, \phi_0) = \frac{1}{2} \frac{g_v^2}{m_v^2} \rho_{\rm B}^2 + \frac{1}{2} \frac{m_s^2}{g_s^2} (M_b - M_b^\star)^2 + \frac{\gamma}{(2\pi)^3} \int^{k_{\rm F}} (\vec{k}^2 + M_b^{\star 2})^{1/2} d^3k$$
$$; \text{ energy density} \qquad (10.98)$$

Here Eq. (10.97) has been used to replace V_0, and Eq. (10.95) relates ϕ_0 to $M_b^\star$.

The expression for the energy density in Eq. (10.98) still depends on the mean value of the scalar field ϕ_0, which can be obtained with the aid of thermodynamics — the system will minimize its energy at a given baryon density

$$\left(\frac{\partial \varepsilon}{\partial \phi_0}\right)_{\rho_{\rm B}} = 0 \qquad\qquad ; \text{ equilibrium} \qquad (10.99)$$

This yields the expression

$$\frac{m_s^2}{g_s^2}(M_b - M_b^\star) = \frac{\gamma}{(2\pi)^3} \int^{k_{\rm F}} \frac{M_b^\star}{(\vec{k}^2 + M_b^{\star 2})^{1/2}} d^3k$$
$$; \text{ self-consistency} \qquad (10.100)$$

This is a transcendental *self-consistency* relation for the scalar field ϕ_0 (or, equivalently, for the effective mass $M_b^\star$), which must be satisfied at each baryon density.

The pressure can also be obtained with the aid of thermodynamics; it follows from the first law

$$P = -\left(\frac{\partial E}{\partial V}\right)_{\rm B}$$

$$\text{or } ; \qquad \mathcal{P} = \rho_{\rm B}^2 \frac{\partial}{\partial \rho_{\rm B}}\left(\frac{\varepsilon}{\rho_{\rm B}}\right) \qquad\qquad (10.101)$$

We leave it as a problem to show that the pressure is then given by[17]

$$\mathcal{P}(\rho_B, \phi_0) = \frac{1}{2}\frac{g_v^2}{m_v^2}\rho_B^2 - \frac{1}{2}\frac{m_s^2}{g_s^2}(M_b - M_b^\star)^2 + \frac{\gamma}{(2\pi)^3}\frac{1}{3}\int^{k_F}\frac{\vec{k}^2}{(\vec{k}^2 + M_b^{\star 2})^{1/2}}d^3k$$

$$; \text{ pressure} \qquad (10.102)$$

If the coupling constants appearing here are dimensionless, and all masses are inverse Compton wavelengths $mc/\hbar$, then the dimensions of both $(\mathcal{P}, \varepsilon)$ are $[1/L^4]$. The proper units are therefore restored with a factor of $\hbar c$, which converts one factor of $mc/\hbar$ to mc^2. Thus the quantities with explicitly correct dimensions are related to those given above by[18]

$$\rho c^2 = \hbar c\, \varepsilon \qquad ; \text{ energy density}$$

$$P = \hbar c\, \mathcal{P} \qquad \text{pressure}$$

$$M_b = \frac{m_b c}{\hbar} \qquad \text{nucleon mass} \qquad (10.103)$$

We have evidently generated a parametric equation of state for nuclear matter

$$P = P(\rho_B) \qquad ; \text{ parametric equation of state}$$

$$\rho = \rho(\rho_B) \qquad\qquad\qquad (10.104)$$

The first of Eqs. (10.103) can be used to calculate the binding energy of nuclear matter as a function of density (Fig. 10.9). Interestingly enough, for an appropriate range of the two parameters appearing in the theory, this curve has a *minimum* and nuclear matter *saturates*.[19] The position of the minimum, and value there, are fit with the following choice of the two parameters in this RMFT

$$C_s^2 = g_s^2\left(\frac{M_b}{m_s}\right)^2 = 267.1$$

$$C_v^2 = g_v^2\left(\frac{M_b}{m_v}\right)^2 = 195.9 \qquad (10.105)$$

---

[17]This calculation is, in fact, carried out in appendix B.1 of [Walecka (2004)].

[18]A useful conversion factor to remember is $\hbar c = 197.3 \times 10^{-13}$ MeV-cm. The proton mass is $m_p c^2 = 938.3$ MeV, and the neutron is 1.3 MeV heavier; $m_b c^2$ is an appropriate mean value.

[19]This is a feature of RMFT. A static nucleon-nucleon potential built on scalar and vector meson exchange does not lead to nuclear saturation.

The results of the calculation are shown in Fig. 10.9. The ground-state energy is determined at all other values of ρ_B in RMFT.

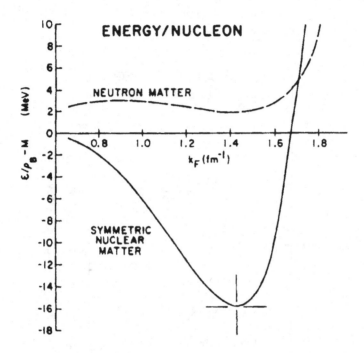

Fig. 10.9   Saturation curve for nuclear matter calculated in RMFT with baryons and neutral scalar and vector mesons. The binding-energy/nucleon $\hbar c(\varepsilon/\rho_B - M_b)$ is plotted against the Fermi wavenumber k_F. Note that $1\,\text{fm} = 10^{-13}$ cm, and the Fermi wavenumber is related to the baryon density by Eq. (10.96). The coupling constants are chosen to fit the value and position of the minimum. The prediction for neutron matter ($\gamma = 2$) is also shown. From [Serot and Walecka (1986)].

In order to obtain these curves, the self-consistency Eq. (10.100) must first be solved at each density. This result is shown in Fig. 10.10. The effective mass is driven to zero at high baryon density by the self-consistency relation. In nuclear matter, at the observed density, one finds $(M^\star/M)_b = 0.56$.

The properties of *neutron matter* are then determined in this RMFT, for it is simply necessary to change the degeneracy factor from $\gamma = 4$ to

$\gamma = 2$

$$\gamma = 2 \qquad ; \text{neutron matter} \qquad (10.106)$$

The results are included in Figs. 10.9 and 10.10. Pure neutron matter is unbound (fortunately, since we do not see large chunks of it around!), and the effective mass again goes to zero at high density.

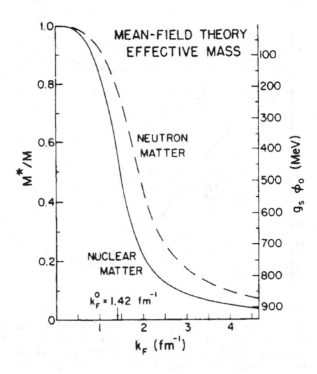

Fig. 10.10   Effective mass M_b^*/M_b as a function of Fermi wavenumber for nuclear ($\gamma = 4$) and neutron ($\gamma = 2$) matter. From [Serot and Walecka (1986)].

The calculated equation of state of neutron matter at all densities, in c.g.s. units, is then shown in Fig. 10.11. Note from Eqs. (10.98) and (10.102) that at high baryon density the pressure and energy density are dominated by the vector meson repulsion, and they become identical

$$\mathcal{P} = \varepsilon = \frac{1}{2}\frac{g_v^2}{m_v^2}\rho_{\mathrm{B}}^2 \qquad ; \rho_{\mathrm{B}} \to \infty$$

$$\text{or}\ ; \qquad P = \rho c^2 \qquad\qquad (10.107)$$

With this equation of state, the speed of sound becomes equal to the speed of light (see Prob. 10.7). The equation of state is as "stiff" as possible, while still remaining consistent with microscopic causality — the causal propagation of signals. An equation of state of this form was first envisioned by Zeldovitch.[20]

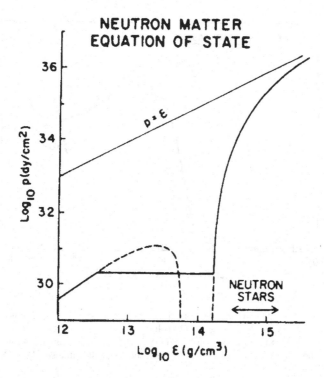

Fig. 10.11 RMFT equation of state for neutron matter in c.g.s. units based on Fig. 10.9. A Maxwell construction is used to determine the equilibrium curve in the region of the gas-liquid phase transition. The mass density regime relevant to neutron stars is indicated. From [Serot and Walecka (1986)].[21]

---

[20]See [Serot and Walecka (1986)] for relevant references.

[21]On the abscissa, once the units are restored, one has $\log \varepsilon = \log \rho c^2 = \log \rho + \log c^2 = \log \rho + 20.95$; it is $\log \rho$ that is indicated.

10.4   Neutron Stars and Black Holes

Now that one has the equation of state for neutron matter shown in Fig. 10.11, the TOV Eqs. (10.76) can be integrated numerically to find neutron star masses and radii as a function of their central density. The resulting masses are shown in Fig. 10.12.

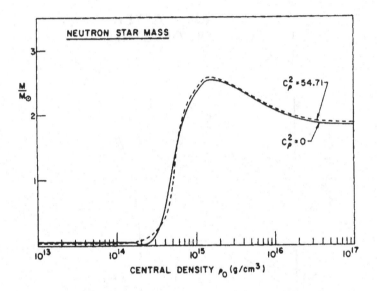

Fig. 10.12   Calculated neutron star mass in units of the solar mass $M_\odot$ as a function of the central density based on Fig. 10.11 (solid curve). The RMFT equation of state is given by Eqs. (10.98) and (10.102) and the TOV Eqs. (10.76) are integrated numerically. All stars to the right of the maximum are unstable against collapse. From [Serot and Walecka (1986)].

There is evidently a *maximum* neutron star mass in the RMFT

$$\left(\frac{M}{M_\odot}\right)_{max} = 2.57 \qquad ; \text{ neutron star in RMFT} \qquad (10.108)$$

Furthermore, in the region to the right of the maximum in Fig. 10.12 where $dM/d\rho_0 < 0$, the star is *unstable against collapse*. This is readily proven as follows. Take the variation in density $\delta\rho$ at a fixed number of baryons B as indicated in Fig. 10.13. This variation *decreases* the total mass M of

the star. The star is therefore unstable against this variation, which moves
mass from the surface to the interior and indicates collapse.

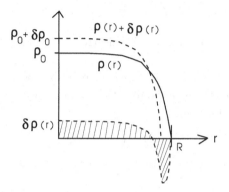

Fig. 10.13   A variation $\delta\rho$ at a fixed number of baryons B for which the total mass of
the star M decreases in the case $dM/d\rho_0 < 0$.

The equation of state used to compute Fig. 10.12 is about as "stiff"
as one can have and still be consistent with microscopic causality and the
observed properties of nuclear matter (and observed nuclei). Thus the value
in Eq. (10.108) should really be considered an *upper limit* to the mass of
a neutron star. The existence of such a maximum is very general — it is
due to the fact that gravity will eventually overcome the repulsion in the
nuclear force if enough mass is present.

Neutron stars are formed as supernovae remnants after an explosive
disappearance of the protons, through the capture of the accompanying
electrons, in the dense nuclear residue of the spent stellar fuel. The basic
reaction converts a proton to a neutron with the emission of a neutrino

$$e^- + p \to n + \nu_e \qquad (10.109)$$

Several comments concerning neutron stars are relevant:

- The neutron is heavier than the proton, and free neutrons β-decay
  to protons. The gravitational binding energy stabilizes neutron stars
  against this process (just as nuclear binding energies stabilize large
  classes of nuclei);
- There will always be *some* protons and electrons present in equilibrium

in neutral neutron stars due to the presence of the reactions

$$n \rightarrow p + e^- + \bar{\nu}_e$$
$$p + e^- \rightarrow n + \nu_e \tag{10.110}$$

It is the Fermi pressure of the electrons that keeps this component suppressed;

- There are many other possibilities for reactions, weak and strong, that can relieve the Fermi pressure of the neutrons and other baryons in neutron stars, for example

$$n + n \rightarrow n + \Lambda \qquad ; \text{weak}$$
$$\Lambda + \Lambda \rightarrow \Xi^0 + n \qquad \text{strong, etc.} \tag{10.111}$$

Many of these additional species will be present to some extent;

- There is also the possibility of other condensed meson fields in addition to (ϕ_0, V_0);
- Nucleons are built from quarks. They are held together by the strong gluon interactions of QCD. At very high baryon densities, one expects nuclear matter to undergo a transition to a state of *unconfined quarks and gluons*. One expects to find such a state in the deep interior of stars, which, due to the asymptotic freedom of QCD, is expected to have the equation of state of an ideal relativistic gas (see Prob. 10.5)

$$P = \frac{1}{3}\rho c^2 \qquad ; \text{ideal relativistic gas} \tag{10.112}$$

A much more extensive discussion of these matters is contained in [Walecka (2004)];

- The inside of a neutron star is evidently quite complex. For an expanded discussion of the theory of neutron star interiors, see [Glendenning (2000)];
- One can obtain indirect evidence concerning the structure of neutron stars from some of their observed properties (star quakes, slowing down of pulsars, etc.). It is difficult to imagine a method of directly sampling the deep interior of a neutron star (although a few years ago, one would have said the same thing about our sun!).

All of the effects mentioned above tend to *soften* the equation of state at high baryon densities. Several observed values of neutron star masses are shown in Fig. 10.14.

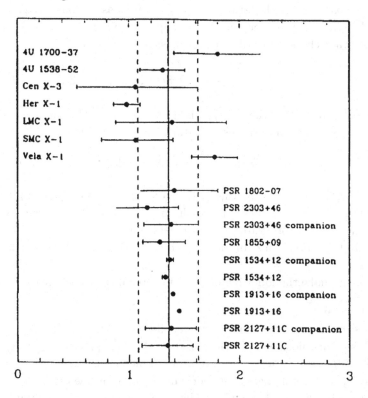

Fig. 10.14 Measured masses of neutron stars in units of $M_\odot$. From [Brown (1994)].

The value of the masses for this set of neutron stars is

$$\left(\frac{M}{M_\odot}\right)_{\text{obs}} \approx 1.4 \pm 0.2 \qquad ; \text{ neutron stars observed} \qquad (10.113)$$

Higher mass neutron stars may exist, but they have not yet been observed. They also may not be readily formed in the supernovae explosions that lead to neutron stars.[21]

---

[21] It is evident from our discussion that the actual value of $(M/M_\odot)_{\text{max}}$ depends on the true nature of the repulsive component of the nuclear force in a high-density, degenerate system of neutrons (for a more sophisticated analysis of this quantity within the effective-field theory approach, see [Müller and Serot (1996)]). The reader should note that the experimental number quoted in Eq. (10.113), while similar, bears no obvious relation to the Chandrasekhar limit on the mass of white dwarf stars, which follows from the Fermi pressure of a high-density, degenerate system of electrons in a neutralizing background of ionized ^{4}He.

What happens if the cold remnant of a supernovae explosion, after the exterior of the star has blown off, has a mass larger than $(M/M_\odot)_{\mathrm{max}}$? Nothing prevents its collapse to smaller and smaller radii. The Schwarzschild radius of a star of mass M is given by

$$ R_s = \frac{2MG}{c^2} \qquad ; \text{ Schwarzschild radius} \qquad (10.114) $$

If the end product of the collapse is a star of mass M that lies entirely *within* its Schwarzschild radius, the star forms a *black hole*

$$ R \leq R_s \qquad ; \text{ black hole} \qquad (10.115) $$

When a light signal propagating in the radial direction originates just outside of the Schwarzschild radius with such an object, its radial velocity dr/dt vanishes (Prob. 9.6). It never gets out.[22] This observation, and Eq. (10.115), provide a definition of a black hole that is sufficient for our purposes.

We can make the following observations regarding black holes:

- They have a given mass M;
- The mass lies inside the Schwarzschild radius in Eq. (10.114). Thus these black holes are characterized by the two quantities (M, R) with $R \leq R_s$. (They may have additional macroscopic properties such as charge and total angular momentum.) Note that one can, in principle, have a black hole of any mass. The arguments in this section imply that a massive cold stellar object, ultimately a neutron star, will collapse into a black hole if its mass is greater than $(M/M_\odot)_{\mathrm{max}}$;
- The cause of this collapse is that if the mass M is large enough, the long-range gravitational attraction of all of the particles will always overcome the short-range nuclear repulsion of neighboring constituents;
- *Inside* of the Schwarzschild radius, one foregoes any description in terms of a medium with a given equation of state;[23]
- *Outside* of the Schwarzschild radius, black holes are described by the previously discussed static, spherically symmetric Schwarzschild solution to the Einstein field equations;

---

[22]In fact, the null light geodesics are trapped at the Schwarzschild radius for a black hole. This is the reason for the name "black hole," and why the Schwarzschild radius is frequently referred to as the "event horizon."

[23]One is always free to speculate on just what happens *inside* of the Schwarzschild radius.

• Outside of the Schwarzschild radius, the black hole simply acts as a spherically symmetric gravitational source, and all the previous analysis of physics in the Schwarzschild metric applies. For example, in the newtonian limit, the equation of motion of a binary star with components of mass (m_1, m_2) and reduced mass $\mu = m_1 m_2 / (m_1 + m_2)$ separated by a distance $\vec{r} = \vec{r}_1 - \vec{r}_2$ is given by

$$\mu \frac{d^2 \vec{r}}{dt^2} = - \frac{G m_1 m_2}{r^3} \vec{r}$$

or ; $$\frac{d^2 \vec{r}}{dt^2} = - \frac{G(m_1 + m_2)}{r^3} \vec{r} \qquad (10.116)$$

The orbit dynamics thus measures the quantity $(m_1 + m_2)$ and provides a way of determining m_2 if m_1 is known by some other means;
• Black holes can accrete mass from the surrounding stellar medium, although angular momentum may prevent material from falling in.[24]

The experimental observation of a black hole lies in determining the pair of quantities (M, R) for a cold stellar object and verifying that $R \leq R_s$. (limits on the radius can be determined by observing phenomena in the vicinity of the object). Many candidates have been discovered. It is now believed that a *massive black hole* lies at the center of almost every galaxy [Sky and Telescope (2000)].

Black holes have many fascinating properties such as temperature and entropy. They are a source of Hawking radiation, by which black holes can eventually decay, albeit with extraordinarily long lifetimes. Problems 10.9-10.12 are intended to familiarize the reader with some of these concepts. There are many good books available, for example [Taylor and Wheeler (2000)], for further reading on this fascinating subject.

---

[24] Just as the angular momentum of the planets prevents them from falling into the sun.

Chapter 11

Cosmology

We next turn to the topic of *cosmology*. It is here that general relativity has found its most significant application—one that captures the imagination of one and all. Cosmology is the study of the time development, and current structure, of the universe we see around us. A primary reason for the great interest in cosmology is that it has now become an *experimental science*. With the advent of satellite instruments such as the Cosmic Background Explorer (COBE), or the Wilkinson Microwave Anisotropy Probe (WMAP), it is now possible to study in detail the uniformity and granularity of the black-body cosmic microwave background left over from the "big bang."[1] Extended observations of element abundances tell us about element formation at early times. In addition, thorough astronomical studies now exhibit large-scale anisotropies, "walls" and "filaments," in the mass distribution in the observed universe. Furthermore, studies of "standard candles" tell us about the current *rate of expansion* of the universe. There are many examples one can give. Cosmology is no longer an esoteric exercise in pure mathematics. It is now a true physics subject where theories must continually be constrained and guided by high-quality experimental data.

The ability of Einstein's theory of general relativity to provide an understanding of the universe we see around us, and its development from just after the "big bang," is truly awe inspiring.

The material in this text is meant as an *introduction* to cosmology. There are many good books available for further study of this fascinat-

---

[1]Up-to-date information on such experiments is best obtained from their websites, for example [COBE (2006)] and [WMAP (2006)]. The cosmic microwave background is discussed in Probs. 11.8-11.9.

ing topic, including [Weinberg (1972)], [Kolb and Turner (1990)], [Linde (1990)], [Peebles (1993)], [Peacock (1999)], and [Hartle (2002)].

In this chapter we study the simplest situation of a uniform mass distribution and a corresponding flat three-dimensional space characterized by a Robertson-Walker metric with $k = 0$. Interestingly enough, this appears to describe well what we currently see around us. Later, we shall return to the more general situation of curved space with $k \neq 0$, and we will discuss how we apparently got to where we are today.

11.1 Uniform Mass Background

If there is a finite, uniform mass density throughout the universe (Fig. 11.1), then general relativity implies that three-dimensional space *cannot be euclidian and static*. Here we are talking about *interstellar* space (which is most of the universe!). We will find that the required current mass density of interstellar space is given approximately by

$$\rho \approx 10^{-29} \, \text{gm/cm}^3 \qquad ; \text{ interstellar space} \qquad (11.1)$$

Recall that the mass of the proton is $m_p = 1.673 \times 10^{-24} \, \text{gm}$, so we are talking about a mass density of approximately

$$\rho \approx 1 \, \text{proton/m}^3 \qquad (11.2)$$

This is a good way to visualize and remember it.[2]

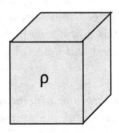

Fig. 11.1 Model of a finite, uniform mass density throughout the universe. The current required value is $\rho \approx 1 \, \text{proton/m}^3$.

---

[2]We will start our discussion of cosmology with the simple model of a uniform distribution of ordinary baryonic mass. Later, we will become more sophisticated and categorize the gravitational source in terms of ordinary matter, "dark matter," and "dark energy," with the latter two actually dominating.

Let us return to the arbitrary two-dimensional surface with which we started this text. Assume the surface is flat, and lay out a uniform orthogonal coordinate grid $q^\mu = (q^1, q^2)$. The position on the surface is then located by specifying the pair (q^1, q^2) as illustrated in Fig. 11.2. As before, the physical displacements on the surface are given in terms of the basis vectors by $\mathbf{e}_1 \, dq^1$ and $\mathbf{e}_2 \, dq^2$ (Fig. 11.3). Assume the surface is isotropic and the coordinate grid is a uniform one.

The *physical displacement* corresponding to an infinitesimal change in grid points on the surface can now evidently be parameterized in terms of a *single overall scale parameter* Λ according to

$$dl^i = \Lambda \, dq^i \qquad \text{; physical displacement}$$
$$i = 1, 2 \qquad\qquad\qquad (11.3)$$

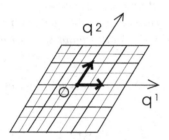

Fig. 11.2   Uniform orthogonal coordinate grid (q^1, q^2) on a flat, two-dimensional surface (perspective view). We will later visualize the surface as a sheet of rubber that stretches uniformly in all directions as a function of time.

The line element then takes the form

$$d\mathbf{s} = \mathbf{e}_i \, dq^i = \Lambda \, \hat{\mathbf{e}}_i \, dq^i \qquad\qquad (11.4)$$

Here $\hat{\mathbf{e}}_i$ are orthonormal *unit* vectors. The basis vectors are then identified as

$$\mathbf{e}_i = \Lambda \, \hat{\mathbf{e}}_i \qquad \text{; basis vectors} \qquad\qquad (11.5)$$

The *metric* on the surface follows as

$$g_{ij} = \mathbf{e}_i \cdot \mathbf{e}_j = \begin{pmatrix} \Lambda^2 & 0 \\ 0 & \Lambda^2 \end{pmatrix} \qquad\qquad (11.6)$$

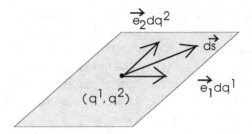

Fig. 11.3   Basis vectors and line element corresponding to the uniform orthogonal co-ordinate grid (q^1, q^2) on the flat surface in Fig. 11.2 (perspective view).

Now suppose the surface is actually a sheet of rubber that stretches uniformly in all directions as a function of time. The coordinate grid remains embedded in the surface, but the physical distance between pairs of coordinate points changes with time.

In this case, the parameter Λ becomes a *function of time* and

$$dl^i = \Lambda(t)\, dq^i$$
$$\mathbf{e}_i = \Lambda(t)\, \hat{\mathbf{e}}_i$$
$$g_{ij} = \begin{bmatrix} \Lambda^2(t) & 0 \\ 0 & \Lambda^2(t) \end{bmatrix} \qquad (11.7)$$

We will work with a spatial metric of this form, and we now extend the arguments to four-dimensional space-time.

11.2   Robertson-Walker Metric with $k = 0$

We assume that space is homogeneous and isotropic so that all three-dimensional directions are equivalent. We further assume that one has the *same time* everywhere, and we assume that the metric describes an undistorted Minkowski space with respect to time. Thus we look for a metric of the form

$$g_{\mu\nu} = \begin{bmatrix} \Lambda^2(t) & 0 & 0 & 0 \\ 0 & \Lambda^2(t) & 0 & 0 \\ 0 & 0 & \Lambda^2(t) & 0 \\ 0 & 0 & 0 & -1 \end{bmatrix}$$

$$g^{\mu\nu} = \begin{bmatrix} 1/\Lambda^2(t) & 0 & 0 & 0 \\ 0 & 1/\Lambda^2(t) & 0 & 0 \\ 0 & 0 & 1/\Lambda^2(t) & 0 \\ 0 & 0 & 0 & -1 \end{bmatrix} \qquad (11.8)$$

This is known as a Robertson-Walker metric with $k = 0$ [Robertson (1935); Walker (1936)], and this form of the metric makes things particularly simple for us. Interestingly enough, it also corresponds well with what is actually observed today. The situation is illustrated in Fig. 11.4.

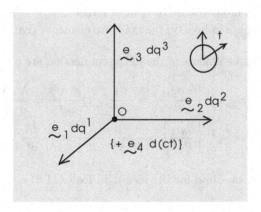

Fig. 11.4   Illustration of physical spatial displacements in the Robertson-Walker metric with $k = 0$. The fourth coordinate displacement $e_4\, d(ct)$ is suppressed.

The *interval* is now given by[3]

$$(ds)^2 = g_{\mu\nu} dq^\mu dq^\nu = \Lambda^2(t)\, dq^i dq^i - c^2 (dt)^2 \qquad \text{; interval} \quad (11.9)$$

---

[3]We postpone the discussion of the general Robertson-Walker metric with interval

$$(ds)^2 = \Lambda^2(t) \left[A(r)(dr)^2 + r^2(d\theta)^2 + r^2 \sin^2\theta (d\phi)^2 \right] - B(r)c^2 (dt)^2$$
$$A(r) = (1 - kr^2)^{-1} \qquad ; B(r) = 1$$

Here the repeated Latin indices are again summed from 1 to 3. The *proper time* element is related to the interval by the usual relation

$$(ds)^2 = -c^2(d\tau)^2 \tag{11.10}$$

The simple form of the metric in Eqs. (11.8) makes it relatively easy to compute the *affine connection*

$$\Gamma^\mu_{\lambda\nu} = \frac{1}{2} g^{\mu\rho} \left[\frac{\partial g_{\rho\nu}}{\partial q^\lambda} + \frac{\partial g_{\rho\lambda}}{\partial q^\nu} - \frac{\partial g_{\lambda\nu}}{\partial q^\rho} \right] \tag{11.11}$$

We make the following observations concerning this affine connection:

- The only non-zero derivative is with respect to $q^4 = ct$;
- The metric and its inverse are diagonal; thus one index must be 4;
- The time dependence occurs in the spatial part of the metric; hence the other two indices must be spatial indices;
- All pairs of space indices on the non-zero elements give the same result.

Hence the non-zero elements of the affine connection are all obtained from

$$\Gamma^4_{11} = \frac{1}{2} g^{44} \left[-\frac{\partial g_{11}}{\partial q^4} \right] = -\frac{1}{2c}(-1)\frac{d(\Lambda^2)}{dt} = \frac{\Lambda}{c}\frac{d\Lambda}{dt}$$

$$\Gamma^1_{41} = \frac{1}{2} g^{11} \left[\frac{\partial g_{11}}{\partial q^4} \right] = \frac{1}{2c}\frac{1}{\Lambda^2}\frac{d(\Lambda^2)}{dt} = \frac{1}{\Lambda c}\frac{d\Lambda}{dt} \tag{11.12}$$

Thus the affine connection for the metric in Eqs. (11.8) is given by[4]

$$\Gamma^4_{11} = \Gamma^4_{22} = \Gamma^4_{33} = \frac{\Lambda}{c}\frac{d\Lambda}{dt}$$

$$\Gamma^1_{41} = \Gamma^2_{42} = \Gamma^3_{43} = \frac{1}{\Lambda c}\frac{d\Lambda}{dt} \qquad ; \text{ all others vanish} \tag{11.13}$$

The *Ricci tensor* is then also calculated with relative ease

$$R_{\mu\nu} = \frac{\partial \Gamma^\lambda_{\mu\nu}}{\partial q^\lambda} - \frac{\partial \Gamma^\lambda_{\mu\lambda}}{\partial q^\nu} + \Gamma^\lambda_{\lambda\rho}\Gamma^\rho_{\mu\nu} - \Gamma^\lambda_{\nu\rho}\Gamma^\rho_{\lambda\mu} \tag{11.14}$$

All derivatives must again be with respect to time, and clearly $R_{11} = R_{22} = $

---

[4] As usual, the affine connection is also symmetric in its lower indices, $\Gamma^1_{14} = \Gamma^1_{41}$, etc.

R_{33}. We proceed to calculate the components of $R_{\mu\nu}$ in detail.

$$R_{11} = \frac{1}{c^2}\frac{d}{dt}\left(\Lambda\frac{d\Lambda}{dt}\right) + \Gamma^4_{11}\Gamma^\lambda_{\lambda 4} - \left[\Gamma^4_{11}\Gamma^1_{41} + \Gamma^1_{14}\Gamma^4_{11}\right]$$

$$= \frac{1}{c^2}\frac{d}{dt}\left(\Lambda\frac{d\Lambda}{dt}\right) + \frac{\Lambda}{c}\frac{d\Lambda}{dt}\frac{3}{\Lambda c}\frac{d\Lambda}{dt} - \frac{2}{c^2}\left(\frac{d\Lambda}{dt}\right)^2$$

$$= \frac{1}{c^2}\left[\dot\Lambda^2 + \Lambda\ddot\Lambda + 3\dot\Lambda^2 - 2\dot\Lambda^2\right]$$

$$R_{11} = R_{22} = R_{33} = \frac{1}{c^2}\left[2\dot\Lambda^2 + \Lambda\ddot\Lambda\right] \qquad ; \dot\Lambda \equiv \frac{d\Lambda}{dt} \qquad (11.15)$$

Furthermore

$$R_{44} = -\frac{1}{c}\frac{\partial}{\partial t}\Gamma^\lambda_{4\lambda} + \Gamma^\rho_{44}\Gamma^\lambda_{\rho\lambda} - 3\left[\Gamma^1_{41}\Gamma^1_{14}\right]$$

$$= -\frac{3}{c^2}\frac{d}{dt}\left(\frac{1}{\Lambda}\frac{d\Lambda}{dt}\right) - \frac{3}{c^2\Lambda^2}\left(\frac{d\Lambda}{dt}\right)^2$$

$$R_{44} = -\frac{3}{c^2}\frac{\ddot\Lambda}{\Lambda} \qquad\qquad (11.16)$$

Clearly, since all spatial indices are equivalent, $R_{12} = R_{13} = R_{23}$ and

$$R_{12} = \frac{\partial}{\partial q^\lambda}\Gamma^\lambda_{12} - \frac{\partial}{\partial q^2}\Gamma^\lambda_{1\lambda} + \Gamma^\rho_{12}\Gamma^\lambda_{\rho\lambda} - \left[\Gamma^4_{22}\Gamma^2_{41} + \Gamma^2_{24}\Gamma^4_{21}\right] = 0$$

$$R_{12} = R_{13} = R_{23} = 0 \qquad\qquad (11.17)$$

It also follows that $R_{14} = R_{24} = R_{34}$, and

$$R_{14} = \frac{\partial}{\partial q^\lambda}\Gamma^\lambda_{14} - \frac{\partial}{\partial q^4}\Gamma^\lambda_{1\lambda} + \Gamma^\rho_{14}\Gamma^\lambda_{\rho\lambda} - \left[\Gamma^1_{41}\Gamma^1_{11} + \Gamma^2_{42}\Gamma^2_{21} + \Gamma^3_{43}\Gamma^3_{31}\right] = 0$$

$$R_{14} = R_{24} = R_{34} = 0 \qquad\qquad (11.18)$$

In *summary*, the ten elements of the Ricci tensor corresponding to the metric in Eqs. (11.8) are given by

$$R_{ij} = \frac{1}{c^2}\left[2\dot\Lambda^2 + \Lambda\ddot\Lambda\right]\delta_{ij}$$

$$R_{44} = -\frac{3}{c^2}\frac{\ddot\Lambda}{\Lambda} \qquad ; R_{i4} = 0 \qquad (11.19)$$

Here (i,j) are spatial indices running from 1 to 3, and δ_{ij} is the usual Kronecker delta. Also $\dot\Lambda \equiv d\Lambda(t)/dt$ and $\ddot\Lambda \equiv d^2\Lambda(t)/dt^2$.

The source term in the Einstein field equations is the *energy-momentum tensor*, and for this very rare medium (Fig. 11.1), the pressure contribution will be negligible with respect to that of the mass

$$P \ll \rho c^2 \tag{11.20}$$

It follows from Eqs. (10.1) that the energy-momentum tensor here takes the approximate form

$$T^{\mu\nu} = \rho\, u^\mu u^\nu \qquad ; P \ll \rho c^2 \tag{11.21}$$

For a fluid at rest in the laboratory frame, the four-velocity now reduces to[5]

$$\frac{u^\mu}{c} = (0,0,0,1)$$

$$\frac{u_\mu}{c} = (0,0,0,-1)$$

$$\frac{u^\mu u_\mu}{c^2} = -1 \tag{11.22}$$

The quantity $T = T^\mu{}_\mu$ follows from the above as

$$T = T^\mu{}_\mu = -\rho c^2 \tag{11.23}$$

Thus, with the metric of Eqs. (11.8), the *source term* in the Einstein field equations takes the form

$$T_{\mu\nu} - \frac{1}{2} T\, g_{\mu\nu} = \rho c^2 \left(\frac{1}{2} g_{\mu\nu} + \frac{u_\mu u_\nu}{c^2} \right) \tag{11.24}$$

The components of this relation are given by

$$T_{\mu\nu} - \frac{1}{2} T\, g_{\mu\nu} = \frac{1}{2}\rho c^2 \qquad ; \underline{44}$$
$$= 0 \qquad ; \underline{i4}$$
$$= \frac{\Lambda^2}{2}\rho c^2 \delta_{ij} \qquad ; \underline{ij} \tag{11.25}$$

The Einstein field Eqs. (7.14) read

$$R_{\mu\nu} = \kappa \left(T_{\mu\nu} - \frac{1}{2} T\, g_{\mu\nu} \right)$$

$$\kappa = \frac{8\pi G}{c^4} \tag{11.26}$$

---

[5]Note that with this metric $g_{44} = -1$, and $B(r) = 1$ in our previous results. Nothing unusual now happens with the fourth component. Note also that the only non-zero element of the energy-momentum tensor here is $T^{44} = \rho c^2$.

The value of κ comes from Eq. (10.55). The Ricci tensor in Eqs. (11.19) can now be combined with the source terms in Eqs. (11.25) to yield the *Einstein field equations* for the system in Fig. 11.1 with the metric in Eqs. (11.8)

$$-\frac{3}{c^2}\frac{\ddot{\Lambda}}{\Lambda} = \frac{\kappa}{2}\rho c^2$$

$$\frac{1}{c^2}\left[\frac{\ddot{\Lambda}}{\Lambda} + 2\left(\frac{\dot{\Lambda}}{\Lambda}\right)^2\right] = \frac{\kappa}{2}\rho c^2 \qquad ; \text{ Einstein field equations} \quad (11.27)$$

These are a pair of coupled, non-linear, second-order differential equations for the function $\Lambda(t)$ which appears in the spatial part of the metric in Eqs. (11.8). Note that the r.h.s. of these equations are identical; they depend on $\rho(t)$, which, of course, is also a function of time.[6] Note also that a non-zero mass density ρ *precludes* the static euclidian spatial metric with $\dot{\Lambda} = \ddot{\Lambda} = 0$.

11.3 Solution to Einstein's Equations

The Einstein field equations for a universe with a uniform matter distribution as source, and the $k = 0$ Robertson-Walker metric of Eqs. (11.8), are given by Eqs. (11.27). Let us proceed to solve those coupled, non-linear differential equations for the scale function $\Lambda(t)$ appearing in the metric. Introduce the ratio

$$h(t) \equiv \frac{\dot{\Lambda}(t)}{\Lambda(t)} = \frac{1}{\Lambda}\frac{d\Lambda}{dt} \qquad (11.28)$$

Differentiate this expression with respect to time

$$\dot{h} = \frac{\ddot{\Lambda}}{\Lambda} - \left(\frac{\dot{\Lambda}}{\Lambda}\right)^2$$

$$\text{or ;} \qquad \frac{\ddot{\Lambda}}{\Lambda} = \dot{h} + h^2 \qquad (11.29)$$

Now substitute these relations and rewrite Eqs. (11.27) in terms of $h(t)$

$$-3\left(\frac{dh}{dt} + h^2\right) = \frac{\kappa}{2}\rho c^4$$

$$\frac{dh}{dt} + 3h^2 = \frac{\kappa}{2}\rho c^4 \qquad ; \text{ Einstein field equations} \quad (11.30)$$

---

[6]We presumably can determine by observation $\rho(t_p)$ where t_p is the present time.

The l.h.s. of these equations can now be equated with the result that[7]

$$2\frac{dh}{dt} + 3h^2 = 0 \tag{11.31}$$

This is a first-order non-linear differential equation, which is readily integrated in the following manner

$$\frac{1}{h^2}\frac{dh}{dt} = -\frac{3}{2}$$

$$\frac{d}{dt}\left(\frac{1}{h}\right) = \frac{3}{2}$$

$$\frac{1}{h(t)} = \frac{3}{2}(t - t_0) \tag{11.32}$$

Here t_0 is a constant of integration. Thus one has an *explicit solution* for the ratio $h(t)$ in Eq. (11.28)

$$h(t) = \frac{2}{3(t - t_0)} \tag{11.33}$$

We will refer to $(t - t_0)$ as the "age of the universe."

Note that whereas the static Schwarzschild solution to the Einstein field equations introduces a singularity in the radial coordinate, which has interesting consequences, the Robertson-Walker solution introduces a singularity in the time, which has its own fascinating implications.

A combination of the first of Eqs. (11.30) with the first of Eqs. (11.32) relates h to the strength of the source

$$-3\left(-\frac{3}{2}h^2 + h^2\right) = \frac{3}{2}h^2 = \frac{\kappa}{2}\rho c^4 \tag{11.34}$$

This gives

$$h(t) = \left[\frac{\kappa}{3}\rho(t)c^4\right]^{1/2} = \left[\frac{8\pi G\rho(t)}{3}\right]^{1/2} \tag{11.35}$$

A combination of Eq. (11.33) and Eq. (11.35) then yields

$$t - t_0 = \frac{2}{3h(t)} = \left[\frac{1}{6\pi G\rho(t)}\right]^{1/2} \quad ; \text{ age of universe} \tag{11.36}$$

This equation relates the age of the universe at the time t to the mass density at that time $\rho(t)$.

---

[7]Note that this relation no longer depends on $\rho(t)$!

Equation (11.28) can be written as

$$h(t) = \frac{1}{\Lambda(t)} \frac{d\Lambda(t)}{dt} = \frac{d}{dt} \ln \Lambda(t) \qquad (11.37)$$

Equation (11.33) can be then be integrated to actually find $\Lambda(t)$

$$\ln \Lambda(t) = \frac{2}{3} \ln (t - t_0) + \text{constant}$$
$$\Lambda(t) = \gamma^{1/3}(t - t_0)^{2/3} \qquad (11.38)$$

Here the constant of integration is parameterized as $\ln (\gamma)^{1/3}$. Thus the quantity $\Lambda^2(t)$ appearing in the spatial part of the Robertson-Walker metric in Eqs. (11.8) depends on $(t - t_0)$ according to

$$\Lambda^2(t) = \gamma^{2/3}(t - t_0)^{4/3} \qquad ; \gamma = \text{constant} \qquad (11.39)$$

The space "stretches" with time.

Let us expand the scale parameter $\Lambda(t)$ about the *present time* t_p

$$\Lambda(t) = \Lambda(t_p) + (t - t_p)\Lambda'(t_p) + \frac{1}{2}(t - t_p)^2\Lambda''(t_p) + \cdots$$
$$= \Lambda(t_p) \left[1 + (t - t_p)H_0 - \frac{1}{2}(t - t_p)^2 q_0 H_0^2 + \cdots \right] \qquad (11.40)$$

Here we have defined *Hubble's constant*

$$H_0 \equiv \left[\frac{1}{\Lambda} \frac{d\Lambda}{dt} \right]_{t_p} = h(t_p) \qquad (11.41)$$

where the last relation follows from Eq. (11.28). Thus, with the explicit solution for h in Eq. (11.33),

$$H_0 = h(t_p) = \frac{2}{3(t_p - t_0)} \qquad ; \text{Hubble's constant} \qquad (11.42)$$

In Eq. (11.40) we have also defined the *deceleration parameter*

$$-q_0 H_0^2 \equiv \left[\frac{1}{\Lambda} \frac{d^2\Lambda}{dt^2} \right]_{t_p} \qquad (11.43)$$

This is evaluated from Eq. (11.38) as

$$-q_0 H_0^2 = \frac{2}{3} \left(-\frac{1}{3} \right) \frac{1}{(t_p - t_0)^2} \qquad (11.44)$$

With the use of Eq. (11.42) this gives

$$q_0 = \frac{1}{2} \qquad ; \text{ deceleration parameter} \qquad (11.45)$$

We shall later discuss experimental measurements of (H_0, q_0). For now, we merely make some comments on the simple matter-dominated cosmology used in this section. The primary reason for the use of the model with a diffuse, uniform mass distribution and the spatially flat Robertson-Walker metric with $k = 0$ is that it provides a simple introduction to cosmology, where it is relatively easy to write and solve the Einstein field equations. As we shall see, however, it is the case that this spatially flat metric describes well what is actually observed. Moreover, as recently as 1972, Weinberg wrote[8]

> *"One is forced to the conclusion that a mass density of $\rho \approx 2 \times 10^{-29}$ gm/cm^3 must be found somewhere outside normal galaxies. $\cdots$ (It is) more conservative to suppose the missing mass takes the form of a tenuous hydrogen gas, ionized or neutral, filling all space."*

There have been great strides in experimental cosmology in recent years, and much more is now known about the nature of the source $T_{\mu\nu}$ governing cosmology — we shall later become more sophisticated with this analysis.

We close this section by augmenting Eqs. (10.87) with some commonly employed astronomical units of time and distance, again given to three-figure accuracy [Weinberg (1972)][9]

$$1 \, \text{yr} = 3.16 \times 10^7 \, \text{sec}$$
$$1 \, \text{light-yr} = 9.46 \times 10^{17} \, \text{cm}$$
$$1 \, \text{pc} = 3.26 \, \text{light-yrs}$$
$$1 \, \text{Mpc} = 10^6 \, \text{pc} = 3.08 \times 10^{24} \, \text{cm} = 3.08 \times 10^{19} \, \text{km} \qquad (11.46)$$

We further tabulate the results obtained in the present cosmology with a value of the mass density of $\rho(t_p) = 2 \times 10^{-29}$ gm/cm^3. The quantity

---

[8]The quotes are from [Weinberg (1972)] pp. 478, 481.

[9]Here "pc" stands for *parsec*, and "Mpc" for *megaparsec*.

$h(t_p)$ is then calculated from Eq. (11.35) as

$$h(t_p) = 3.34 \times 10^{-18} \sec^{-1}$$

$$= 3.34 \times 10^{-18} \left(\frac{3.08 \times 10^{19} \, \text{km}}{1 \, \text{Mpc}} \right) \sec^{-1}$$

$$= 103 \, (\text{km/sec})/\text{Mpc} \tag{11.47}$$

We shall see later why astronomers find this rather convoluted expression a convenient way of re-expressing the units of $\sec^{-1}$ in Hubble's constant. Thus, in *summary*, in the present simple cosmology with this value of $\rho(t_p)$[10]

$$\rho(t_p) = 2 \times 10^{-29} \, \text{gm/cm}^3$$

$$h(t_p) = 3.34 \times 10^{-18} \sec^{-1} = 103 \, (\text{km/sec})/\text{Mpc}$$

$$t_p - t_0 = 2.00 \times 10^{17} \sec = 0.632 \times 10^{10} \, \text{yrs}$$

$$\frac{1}{h(t_p)} = 9.47 \times 10^9 \, \text{yrs}$$

$$q_0 = \frac{1}{2} \tag{11.48}$$

11.4 Interpretation

Let us discuss the physical interpretation of the results obtained in the present cosmology. The line element corresponding to the Robertson-Walker metric with $k = 0$ is given by

$$d\mathbf{s} = \mathbf{e}_i \, dq^i + \mathbf{e}_4 \, d(ct)$$

$$= \hat{\mathbf{e}}_i \left[\Lambda(t) dq^i \right] + \mathbf{e}_4 \, d(ct) \tag{11.49}$$

Here $(\hat{\mathbf{e}}_1, \hat{\mathbf{e}}_2, \hat{\mathbf{e}}_3)$ are a set of orthonormal spatial unit vectors, and the *physical* displacement in the direction of these unit vectors corresponding to a *coordinate* displacement dq^i is given by $dl^i = \Lambda(t) dq^i$. This physical displacement increases with time.

Consider the two-dimensional rubber sheet analogy introduced at the beginning of this chapter. The situation is illustrated in Fig. 11.5. Suppose two events occur at a fixed coordinate position (q^1, q^2).[11]

---

[10]These values are readily scaled to other values of $\rho(t_p)$. Note that it is common practice to instead refer to $h(t_p)^{-1}$ as the present "age of the universe," since Hubble's constant [see Eq. (11.41)] has an experimental significance that goes beyond the present model.

[11]The coordinates now represent, say, a certain number of tick marks on the axes.

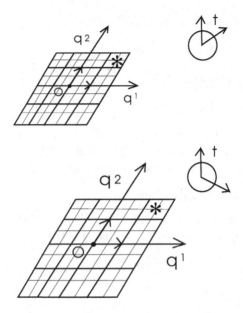

Fig. 11.5   Two-dimensional rubber sheet analogy for the Robertson-Walker metric with $k = 0$. Shown is the configuration at two times, the bottom one later than the top. The stars mark two events that occur at a fixed coordinate position (q^1, q^2).

Now nothing changes but the time t. The coordinates do not change. The record keeper notes only the moving clock. Nothing changes on the screen in front of the record keeper [Fig. (11.6)].

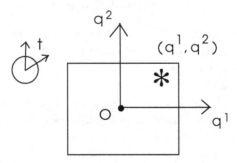

Fig. 11.6   Record keeper's screen corresponding to the situation in Fig. 11.5. Only the clock moves. It is the metric that describes how the physical separation between the events increases with time.

The *physical separation* of the events at (q^1, q^2), however, *changes with time*, and this physical separation is described by the metric. Again, the record keeper must appeal to the metric to determine what is actually going on.

The local freely falling frame LF^3 always provides the key to a physical interpretation of the theory, and we introduce the LF^3 at a given point in space-time. Since three-dimensional space is homogeneous and isotropic here, there is no preferred direction, and thus there is no gravitational field $\vec{g}$. Furthermore, for the same reason, we can take the LF^3 to be *global in space*. It is the fourth component (ct) that now varies, so the LF^3 must remain *local in time*. In the LF^3 one just has the laws of special relativity.

Write the coordinates in the LF^3 in a cartesian basis as $q^\mu = (\bar{x}, \bar{y}, \bar{z}, c\bar{t})$. Since the fourth component of the metric in Eq. (11.8) is just that of Minkowski space, there is no coordinate transformation in the time when one goes from the LF^3 to the global inertial laboratory frame, and hence the *time in the two frames is identical*

$$\bar{t} = t \qquad ; \text{ time in } LF^3 \text{ and laboratory} \qquad (11.50)$$

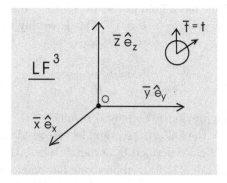

Fig. 11.7   The local freely falling frame LF^3 for the gravitational field described by the Robertson-Walker metric with $k = 0$. There is no $\vec{g}$. The LF^3 is *local in time* but *global in space*. The coordinate axes $(1, 2, 3)$ are here labeled by (x, y, z).

The situation is illustrated in Fig. 11.7. Just as before, the coordinate transformation from the LF^3 to the laboratory frame can immediately be obtained from the relation between the basis vectors, which can be taken as

$$\mathbf{e}_i = \Lambda(t)\,\hat{\mathbf{e}}_i \qquad ; \; i = 1, 2, 3$$
$$\boldsymbol{\alpha}_i = \hat{\mathbf{e}}_i$$
$$\boldsymbol{\alpha}_4 = \mathbf{e}_4 \tag{11.51}$$

The coordinate transformation matrix is obtained from

$$\boldsymbol{\alpha}_\mu = a^\nu_{\;\mu}\,\mathbf{e}_\nu \tag{11.52}$$

Hence, with $[\underline{a}]_{\nu\mu} \equiv a^\nu_{\;\mu}$ one has

$$[\underline{a}]_{\nu\mu} = a^\nu_{\;\mu}$$

$$\underline{a} = \begin{bmatrix} 1/\Lambda(t) & 0 & 0 & 0 \\ 0 & 1/\Lambda(t) & 0 & 0 \\ 0 & 0 & 1/\Lambda(t) & 0 \\ 0 & 0 & 0 & 1 \end{bmatrix} \tag{11.53}$$

This coordinate transformation takes us from the local freely falling frame LF^3 with coordinates $\bar{q}^\mu = (\bar{x}, \bar{y}, \bar{z}, c\bar{t}\,)$, where there is no gravity and only special relativity, to the global inertial laboratory frame I with coordinates $q^\mu = (q^1, q^2, q^3, ct)$, which in this case is an expanding universe with a uniform mass density.

For example, if one has a fluid at rest in the LF^3, then the four-velocity of the fluid is simply $\bar{u}^\mu/c = (0, 0, 0, 1)$. This is readily transformed to the lab with $\underline{a}$

$$\bar{u}^\mu = (0, 0, 0, c)$$
$$u^\mu = a^\mu_{\;\nu}\bar{u}^\nu = (0, 0, 0, c) \tag{11.54}$$

This reproduces our previously employed result.

In preparation for the following section, let us consider light propagation in the gravitational field described by a metric where the time coordinate is undistorted, as in Eqs. (11.8). Suppose one has an electromagnetic wave (many photons) with frequency $\bar{\nu}$ in the LF^3 undergoing $d\bar{N} = \bar{\nu}\,d\bar{t}$ oscillations in a time interval $d\bar{t}$. If $\bar{\lambda}$ is the wavelength of the light, then the physical spatial distance traveled by the light signal in the LF^3 in this time interval is $|d\vec{l}\,| = \bar{\lambda}\,d\bar{N} = \bar{\lambda}(\bar{\nu}\,d\bar{t})$. Since light follows the null interval in the LF^3 one has

$$(d\mathbf{s})^2 = (d\vec{l}\,)^2 - (c\,d\bar{t}\,)^2 = (\bar{\lambda}\,\bar{\nu}\,d\bar{t}\,)^2 - (c\,d\bar{t}\,)^2 = 0$$
$$\bar{\lambda}\,\bar{\nu} = c \tag{11.55}$$

A familiar result.

Now repeat this argument in the global inertial laboratory frame I. Since the time interval is unaltered $dt = d\bar{t}$, and since the *number* of oscillations is unaltered $dN = d\bar{N}$, the frequencies must be the same $\nu = \bar{\nu}$. Furthermore, since the interval is invariant under a coordinate transformation, it must again vanish in the laboratory frame. Hence a repetition of the above argument gives

$$(ds)^2 = (d\vec{l}\,)^2 - (c\,dt)^2 = (\lambda\nu\,dt)^2 - (c\,dt)^2 = 0$$
$$\lambda\nu = c \qquad\qquad (11.56)$$

Evidently the wavelengths must again be the same $\lambda = \bar{\lambda}$.

Now $\lambda\nu\,dt = c\,dt$ is the *physical spatial distance* dl traveled by the light signal in the time interval dt in the global inertial laboratory frame I. We therefore reach the central conclusion that

Light travels a physical spatial distance $dl = c\,dt$ in the time interval dt in the global inertial laboratory frame I in this cosmology.[12]

11.5 Cosmological Redshift

Let us return to the rubber sheet analogy. The physical separation of two events is given by the interval, which in the global inertial laboratory frame I in this cosmology is

$$(ds)^2 = \Lambda^2(t)\,dq^i dq^i - c^2(dt)^2 \qquad\qquad (11.57)$$

where the repeated Latin index is summed from 1 to 3. The physical spatial distance between any two points is defined by a measurement of this interval at a given instant in time, which implies $dt = 0$. In this case the interval reduces to

$$(ds)^2 = \Lambda^2(t)\,dq^i dq^i \qquad ; \text{ when } dt = 0 \qquad\qquad (11.58)$$

Thus the *physical spatial distance* between two points separated by the *coordinate distance* $d\vec{q}$ is given by

$$(ds)^2 = \Lambda^2(t)\,(d\vec{q})^2 \qquad\qquad (11.59)$$

---

[12]These same arguments lead to the conclusion that light travels a proper distance $dl = c\,dt$ in the time interval dt in the LF^3. For a further discussion of the LF^3 and light propagation in this cosmology, see Probs. 11.3-11.4.

One can extend the infinitesimal relation to a finite $\vec{q}$, and a corresponding finite distance l, since the spatial part of the metric is global (i.e. the same everywhere). Thus, with a subsequent square root,

$$l = \Lambda(t)\left(\vec{q}^{\,2}\right)^{1/2} \qquad ; \text{ physical spatial distance} \qquad (11.60)$$

Now $\vec{q}$ is a spatial vector in coordinate space running from the origin O to the point (q^1, q^2, q^3), or to the point (q^1, q^2) in our two-dimensional analogy, and that point will be held *fixed* in the following discussion.[13]

Suppose a *light signal* is emitted at the coordinate point $q = (q^1, q^2)$ at the time t and detected later at the origin O (Fig. 11.8). We have shown above that light travels with velocity c in the global inertial laboratory frame I since it follows the null geodesic $(ds)^2 = 0$. The time it takes to get to the origin, which lies a physical spatial distance l away at the instant of emission, is therefore[14]

$$\Delta t = \frac{l}{c} \qquad ; \text{ time to origin } O \qquad (11.61)$$

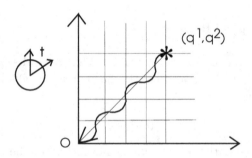

Fig. 11.8   Emission of a light signal at coordinate point (q^1, q^2) at the time t that then travels to the origin O and is detected there. The point (q^1, q^2) lies a physical spatial distance $l = \Lambda(t)\left(\vec{q}^{\,2}\right)^{1/2}$ away from the origin at the instant of emission. The time the light signal takes to get to the origin is $\Delta t = l/c$.

When the light signal is received at the origin, the *physical spatial separation* of the point (q^1, q^2) and the origin O has increased to (see Fig. 11.9)

$$l + \Delta l = \Lambda(t + \Delta t)(\vec{q}^{\,2})^{1/2} \qquad (11.62)$$

---

[13]The vector $\vec{q}$ is fixed on the record keeper's screen (Fig. 11.6).

[14]At the instant of emission, the spatial distance l is the same in the global inertial laboratory frame and in the LF^3 (see Probs. 11.3-11.4).

As shown above, both the time interval and frequency are the same in the global inertial laboratory frame and in the LF^3. Hence the number of oscillations of the source over the time interval Δt can be written as

$$\Delta N = \bar{\nu}\,\Delta \bar{t} = \nu\,\Delta t \qquad ;\ \text{oscillations of source} \qquad (11.63)$$

The wavelength of the light if the source had not moved is given by the physical distance the light has traveled divided by the number of oscillations

$$\lambda_0 = \frac{l}{\nu\,\Delta t} = \frac{\Lambda(t)\left(\vec{q}^{\,2}\right)^{1/2}}{\nu\Delta t} \qquad ;\ \text{wavelength for fixed source} \quad (11.64)$$

The *actual* wavelength of the light is given by the *actual* physical distance between the source and observer at the time of detection divided by the number of oscillations (Fig. 11.9)

$$\lambda = \frac{l + \Delta l}{\nu\,\Delta t} = \frac{\Lambda(t + \Delta t)\left(\vec{q}^{\,2}\right)^{1/2}}{\nu\Delta t} \qquad ;\ \text{actual wavelength} \qquad (11.65)$$

It follows that there is a shift in wavelength of the light given by

$$\frac{\lambda - \lambda_0}{\lambda_0} = \frac{\Lambda(t + \Delta t) - \Lambda(t)}{\Lambda(t)} \qquad ;\ \Delta t = \frac{l}{c} \qquad (11.66)$$

Here t is the time of emission of the light signal from a source that lies a physical spatial distance $l = c\,\Delta t$ away at the time of emission.

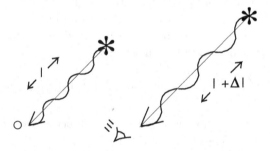

Fig. 11.9 Detection of a light signal at the origin O, which was emitted at the coordinate point (q^1, q^2) at the time t in a space that is stretching according to the present cosmology. At the time of emission t the physical spatial distance of the point from the origin is l, and at the time of detection $t + \Delta t$ the physical distance is $l + \Delta l$.

There is the shift in wavelength of the light due to the fact that the underlying space is stretching while the light signal is propagating from the point of emission to the point of detection. It is known as the *cosmological redshift*.

The full expression for $\Lambda(t)$ is given by Eq. (11.38)

$$\Lambda(t) = \gamma^{1/3}(t - t_0)^{2/3} \tag{11.67}$$

For the finite time interval $\Delta t = l/c$ considered above, Eq. (11.66) can be written as

$$\frac{\lambda - \lambda_0}{\lambda_0} = \frac{(t + \Delta t - t_0)^{2/3} - (t - t_0)^{2/3}}{(t - t_0)^{2/3}} \quad ; \; \Delta t = \frac{l}{c} \tag{11.68}$$

This is a general result in this cosmology. If the time of transit of the light is much less than the age of the universe at the time of emission, $\Delta t \ll (t - t_0)$, then this expression can be expanded as

$$\frac{\lambda - \lambda_0}{\lambda_0} = \frac{2}{3(t - t_0)}\Delta t = h(t)\Delta t \quad ; \; \Delta t \ll (t - t_0) \tag{11.69}$$

This can be rewritten as

$$\frac{\lambda - \lambda_0}{\lambda_0} = h(t)\frac{l}{c} \quad ; \; l \ll c(t - t_0) \tag{11.70}$$

This expression is exact through first order in $l/c(t - t_0)$.

Through leading order in $1/c$, Eq. (11.70) is just the *Doppler shift* of the wavelength of the light due to the fact that the source of the radiation is actually in motion away from the observer. Recall from freshman physics the expression that describes such a situation

$$\lambda = \lambda_0 \left(1 + \frac{v_r}{c}\right) \quad ; \; \text{Doppler shift through } O(v_r/c) \tag{11.71}$$

Here v_r is the relative velocity of source and observer. In our case $v_r(t)$ is given by the limit as $dt \to 0$ of

$$\begin{aligned}
v_r &= \frac{l(t + dt) - l(t)}{dt} = \left[\frac{l(t + dt) - l(t)}{l(t)dt}\right]l \\
&= \left[\frac{\Lambda(t + dt) - \Lambda(t)}{\Lambda(t)dt}\right]l \\
&\to \left[\frac{1}{\Lambda(t)}\frac{d\Lambda}{dt}\right]l \quad ; \; dt \to 0
\end{aligned} \tag{11.72}$$

Hence

$$v_r = h(t)\, l$$

$$\frac{\lambda - \lambda_0}{\lambda_0} = \frac{v_r}{c} = h(t)\frac{l}{c} \qquad ; \text{Doppler shift} \qquad (11.73)$$

Thus, through $O(1/c)$, the cosmological redshift is just the Doppler shift due to the relative motion of the source arising from the underlying time-dependent stretching of the space. Furthermore, since $t + \Delta t = t + l/c$, either the time of emission or the time of detection can be used in $h(t)$ through this order. It follows as in Eq. (11.70) that the dimensionless expansion parameter here is $l/c(t - t_0)$.

If Eq. (11.73) is evaluated at the present time t_p, then it becomes

$$\frac{\lambda - \lambda_0}{\lambda_0} = H_0 \frac{l}{c} \qquad ; \text{cosmological redshift} \qquad (11.74)$$

where *Hubble's constant* has been identified from Eq. (11.41). This is an extremely important relation, for it means that l, the *distance of an emitting star*, can be measured from the cosmological redshift! This result is of common use in astronomy. The present derivation applies to stars whose distance satisfies $l \ll c(t - t_0)$ where t is the time of emission; however, the analysis can be extended (see problems).

11.6 Horizon

We next introduce the concept of the *horizon*. We have demonstrated that light travels with a velocity c in this cosmology. Consider light that was emitted at the *birth* of the universe $t = t_0$ and has been traveling for the present *age* of the universe $\Delta t = t_p - t_0$ to get to us. This is "old light." The physical spatial distance that light has traveled is $D_H = c(t_p - t_0)$, and this distance is *as far as we can see* at the present time. This distance marks the location of our *horizon* (Fig. 11.10)[15]

$$D_H = c(t_p - t_0) \qquad ; \text{distance to horizon} \qquad (11.75)$$

The light from the horizon will be strongly red-shifted when it gets to us.[16]

---

[15] In the present model cosmology where there is a uniform mass distribution that fills *all* of space, there can be other disjoint universes, with their own horizons, that do not overlap ours!

[16] In fact, the redshift of the light from the horizon is infinite (see problems). Remember that there are also *other* sources of redshift.

Let us put in some numbers for D_H using Eqs. (11.46) and (11.48)

$$D_H = 0.632 \times 10^{10} \text{ light-yrs}$$

$$= 0.632 \times 10^{10} \text{ light-yrs} \left(\frac{1\,\text{Mpc}}{3.26 \times 10^6 \text{ light-yrs}} \right)$$

$$D_H = 1.94 \times 10^3 \text{ Mpc} \tag{11.76}$$

The distance to the horizon in our universe is approximately 2000 megaparsecs in this cosmology.

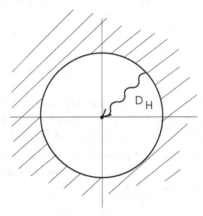

Fig. 11.10   Our horizon lies at a physical spatial distance $D_H = c(t_p - t_0)$. This is the distance that light that was emitted at the birth of the universe has traveled to get to us. It is as far as we can see at the present time. Here $D_H \approx 2000\,\text{Mpc}$.

Let us compute the *relative velocity* v_H of a point a physical spatial distance $l = D_H$ from us

$$l = D_H = c(t_p - t_0) \qquad ; \text{physical spatial distance} \tag{11.77}$$

The velocity is determined in terms of the change in this distance in an infinitesimal time dt according to

$$v_H = \left[\frac{l(t_p + dt) - l(t_p)}{l(t_p)\,dt} \right] l(t_p)$$

$$= \left[\frac{\Lambda(t_p + dt) - \Lambda(t_p)}{\Lambda(t_p)\,dt} \right] l(t_p)$$

$$\rightarrow \left[\frac{1}{\Lambda} \frac{d\Lambda}{dt} \right]_{t_p} D_H \tag{11.78}$$

Here the relation between distance and scale factor in Eq. (11.60) has been introduced in the second line. With the introduction of $\Lambda(t)$ from Eq. (11.67) this becomes

$$\frac{v_{\mathrm{H}}}{c} = \frac{2}{3} \frac{1}{(t_p - t_0)} (t_p - t_0) \tag{11.79}$$

Thus in our cosmology the horizon is always receding from us due to the stretching of space with a constant relative velocity of[17]

$$\frac{v_{\mathrm{H}}}{c} = \frac{2}{3} \qquad ; \text{ relative velocity of horizon} \tag{11.80}$$

11.7 Some Comments

Finally, let us calculate the *scalar curvature* of the four-dimensional riemannian space described by the Robertson-Walker metric with $k = 0$. From Eq. (11.26) and (11.24) one has

$$R = R^{\mu}{}_{\mu} = \kappa \rho c^2 \left(\frac{1}{2} g^{\mu}{}_{\mu} + \frac{u^{\mu} u_{\mu}}{c^2} \right)$$
$$= \kappa \rho c^2 \tag{11.81}$$

Even though three-dimensional space is flat, the four-dimensional riemannian space has *uniform curvature* $R(t)$, and since the mass density goes to zero at infinite time, one has

$$R(t) = \kappa \rho(t) c^2 \qquad ; \text{ space of uniform curvature}$$
$$\rightarrow 0 \qquad \qquad ; \text{ as } t \rightarrow \infty \tag{11.82}$$

The most recent measurement of the Hubble constant gives a value [WMAP (2006)][18]

$$H_0 = 70 \pm 3 \ (\mathrm{km/sec})/\mathrm{Mpc} \tag{11.83}$$

The corresponding time is

$$(H_0)^{-1} = 13.9 \times 10^9 \, \mathrm{yrs} \tag{11.84}$$

---

[17]Note that the relative velocity of the horizon satisfies $v_{\mathrm{H}}/c < 1$. There are points beyond the horizon that are clearly moving with a relative velocity greater than c, but they cannot be connected to us with a light signal.

[18]A compatible value of $H_0 = 72 \pm 8 \ (\mathrm{km/sec})/\mathrm{Mpc}$ was obtained by the Hubble Space Telescope Project to Measure the Hubble Constant [Freedman, *et al.* (2001)]; this paper contains a host of relevant references. See also [Hubble (2006)].

Challenging recent measurements now indicate that the deceleration parameter has the *opposite sign* from that in Eq. (11.45).[19] This necessitates the development of a more sophisticated cosmology than the simple Robertson-Walker $k = 0$ metric with a uniform mass distribution, as presented in this chapter. We shall return to this topic in chap. 13.

---

[19]It is actually an *acceleration* parameter! One would expect the expansion rate of a matter-dominated universe to be slowed down by the gravitational attraction between the constituents.

Chapter 12

Gravitational Radiation

So far we have considered the solution to Einstein's equations for the static spherically symmetric gravitational field produced by a massive body, as well as the dynamic cosmological solution for a uniform distribution of ordinary baryonic matter. The question arises as to whether there are source-free solutions corresponding to propagating oscillations of the metric, just as there are propagating electromagnetic waves in E&M. The Laser Interferometer Gravitational-Wave Observatory (LIGO) makes the direct detection of such gravitational waves one of the forefront experimental activities of modern physics [LIGO (2006)].

We proceed to study gravitational radiation in the weak-field approximation where we linearize the Einstein field equations about the free, spatially flat Minkowski metric.

12.1 Linearized Theory

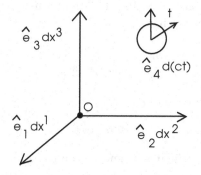

Fig. 12.1   Coordinates in pseudo-euclidian (Minkowski) space.

Start with "pseudo-euclidian" (*i.e.* Minkowski) space (Fig 12.1) with the following cartesian coordinates and metric

$$x^\mu = (x^1, x^2, x^3, ct)$$

$$g^0_{\mu\nu} = \begin{bmatrix} 1 & 0 & 0 & 0 \\ 0 & 1 & 0 & 0 \\ 0 & 0 & 1 & 0 \\ 0 & 0 & 0 & -1 \end{bmatrix} \tag{12.1}$$

To visualize what follows, consider the two-dimensional surface in the $(1,2)$ plane illustrated in Fig. 12.2.

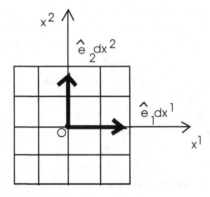

Fig. 12.2   Visualization of the two-dimensional surface in the $(1,2)$ plane in Fig. 12.1, with line elements.

Suppose there is now a small *distortion* of the space, as illustrated for the two-dimensional surface in Fig. 12.3, and a new set of generalized coordinates in the space so that

$$g_{\mu\nu} = g^0_{\mu\nu} + h_{\mu\nu}$$

$$q^\mu = (q^1, q^2, q^3, q^4) \tag{12.2}$$

Our operating principle will be to

Assume $h_{\mu\nu}$ is small and work to first order in $h_{\mu\nu}$.

It follows that the coordinates q^μ and x^μ can be assumed to differ by $O(h)$

$$q^\mu = x^\mu + O(h) \tag{12.3}$$

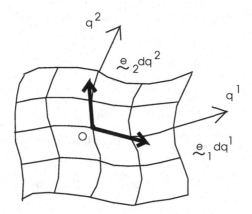

Fig. 12.3 Small distortion of the two-dimensional surface in Fig. 12.2 (exaggerated) with coordinate system (q^1, q^2) on the surface and basis vectors.

Since any derivative of the metric necessarily involves $h_{\mu\nu}$, the *affine connection* is also of $O(h)$

$$\Gamma^{\mu}_{\lambda\nu} \approx \frac{1}{2}g_0^{\mu\rho}\left[\frac{\partial h_{\rho\nu}}{\partial q^\lambda} + \frac{\partial h_{\rho\lambda}}{\partial q^\nu} - \frac{\partial h_{\lambda\nu}}{\partial q^\rho}\right] \qquad ; \text{ of order } h \qquad (12.4)$$

This simplifies things considerably, since the *Ricci tensor* is then given to $O(h)$ by the terms linear in the affine connection

$$R_{\mu\nu} = \frac{\partial}{\partial q^\lambda}\Gamma^{\lambda}_{\mu\nu} - \frac{\partial}{\partial q^\nu}\Gamma^{\lambda}_{\mu\lambda} + O(h^2) \qquad (12.5)$$

Substitution of Eq. (12.4) then leads to

$$R_{\mu\nu} = \frac{\partial}{\partial q^\lambda}\left\{\frac{1}{2}g_0^{\lambda\rho}\left[\frac{\partial h_{\rho\mu}}{\partial q^\nu} + \frac{\partial h_{\rho\nu}}{\partial q^\mu} - \frac{\partial h_{\mu\nu}}{\partial q^\rho}\right]\right\}$$
$$- \frac{\partial}{\partial q^\nu}\left\{\frac{1}{2}g_0^{\lambda\rho}\left[\frac{\partial h_{\rho\mu}}{\partial q^\lambda} + \frac{\partial h_{\rho\lambda}}{\partial q^\mu} - \frac{\partial h_{\mu\lambda}}{\partial q^\rho}\right]\right\} \qquad (12.6)$$

This expression is now correct through $O(h)$. We then develop this expression with the following series of steps:

(1) The metric is symmetric, and thus the additional term in the metric in Eq. (12.2) must satisfy

$$h_{\lambda\sigma} = h_{\sigma\lambda} \qquad ; \text{ symmetric} \qquad (12.7)$$

(2) Through $O(h)$, the metric g_0 can be used to raise and lower indices on h itself. Thus

$$g_0^{\lambda\rho} h_{\rho\nu} = g_0^{\lambda\rho} h_{\nu\rho} = h^\lambda{}_\nu = h_\nu{}^\lambda \qquad (12.8)$$

(3) The contraction of the additional term in the metric is defined as

$$h^\lambda{}_\lambda \equiv h \qquad (12.9)$$

Through $O(h)$, the Ricci tensor in Eq. (12.6) thus takes the form

$$R_{\mu\nu} = -\frac{1}{2} g_0^{\lambda\rho} \frac{\partial^2}{\partial q^\lambda \partial q^\rho} h_{\mu\nu} + \frac{1}{2} \frac{\partial^2}{\partial q^\lambda \partial q^\mu} h_\nu{}^\lambda + \frac{1}{2} \frac{\partial^2}{\partial q^\nu \partial q^\rho} h_\mu{}^\rho - \frac{1}{2} \frac{\partial^2}{\partial q^\nu \partial q^\mu} h$$
$$(12.10)$$

(4) Define a new tensor $\psi_\mu{}^\nu$ by

$$\psi_\mu{}^\nu \equiv h_\mu{}^\nu - \frac{1}{2} h \, \delta_\mu{}^\nu = \psi^\nu{}_\mu \qquad (12.11)$$

(5) Use

$$\frac{\partial}{\partial q^\mu} = \frac{\partial}{\partial x^\mu} + O(h) \qquad (12.12)$$

(6) Then note that up to $O(h)$

$$g_0^{\lambda\rho} \frac{\partial^2}{\partial q^\lambda \partial q^\rho} = g_0^{\lambda\rho} \frac{\partial^2}{\partial x^\lambda \partial x^\rho}$$
$$= \nabla^2 - \frac{1}{c^2} \frac{\partial^2}{\partial t^2}$$
$$= \Box \qquad ; \text{ d'alembertian} \qquad (12.13)$$

With these steps, the Ricci tensor of Eq. (12.6) can be rewritten as

$$R_{\mu\nu} = -\frac{1}{2} \Box h_{\mu\nu} + \frac{1}{2} \frac{\partial}{\partial x^\mu} \left(\frac{\partial}{\partial x^\lambda} \psi_\nu{}^\lambda \right) + \frac{1}{2} \frac{\partial}{\partial x^\nu} \left(\frac{\partial}{\partial x^\lambda} \psi_\mu{}^\lambda \right) \qquad (12.14)$$

We observe that

- It is assumed that there is a small dimensionless parameter in h, say ε, characterizing its size;[1]

---

[1]To get some feeling for the magnitude of this effect with a familiar astronomical system, there is a relevant quote from [Wiki (2006)], "the total power radiated by the Earth-Sun system in the form of gravitational waves is about as much as five typical (60 Watt) light bulbs." See, however, Prob. 12.7.

- This relation is then exact through $O(\varepsilon)$.

12.2   Auxiliary (Lorentz) Condition

In E&M one has the freedom to choose the covariant Lorentz gauge $\partial A^\nu/\partial x^\nu = 0$ for the vector field A^μ (the vector potential). As a consequence, Maxwell's equations in free space then reduce to the wave equation for the vector potential. One has this freedom due to the gauge invariance of E&M.

It would be advantageous if one could impose a similar Lorentz condition on the tensor field $\psi_\mu{}^\nu$ in Eq. (12.11) characterizing the perturbation of the metric in Eq. (12.2)

$$\frac{\partial}{\partial x^\lambda}\psi_\nu{}^\lambda = 0 \qquad ; \text{ auxiliary condition}$$
$$\nu = 1, 2, 3, 4 \qquad\qquad (12.15)$$

Einstein's equations in free space, as expressed through the Ricci tensor in Eq. (12.14), would then immediately reduce to the wave equation for the perturbation of the metric (see below).

Can one do this? Is there freedom to do this? Here we return to the beginning of the text, where it was observed that the metric by itself has no intrinsic meaning — the form of the metric depends on the choice of generalized coordinates. Thus

One still has the freedom of going to a different set of generalized coordinates.

It is this freedom of choice of generalized coordinates that corresponds to the gauge freedom in E&M. We proceed to show that it is always possible to choose a set of generalized coordinates so that the auxiliary condition in Eq. (12.15) is satisfied.

Go to a *new set* of generalized coordinates (Fig. 12.4)

$$\xi^\mu = \xi^\mu(q^1, \cdots, q^4) \qquad ; \mu = 1, \cdots, 4 \qquad (12.16)$$

Look for a set of coordinates that *again* differ from the pseudo-euclidian coordinates x^μ in Eq. (12.1) by terms of $O(h)$

$$\xi^\mu = x^\mu + O(h) \qquad\qquad (12.17)$$

Now define the *difference* between ξ^μ and q^μ as $\eta^\mu(q)$

$$\xi^\mu = q^\mu + \eta^\mu(q) \tag{12.18}$$

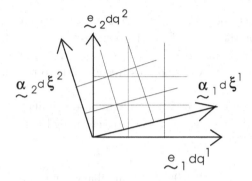

Fig. 12.4   Choice of new coordinates in proof that auxiliary (Lorentz) condition can be satisfied. We seek a new set $\xi^\mu = q^\mu + \eta^\mu(q)$ where both $q^\mu = x^\mu + O(h)$ and $\xi^\mu = x^\mu + O(h)$ so that $\eta^\mu(q) = O(h)$.

Then observe that:

(1) η^μ is again of $O(h)$

$$\xi^\mu = x^\mu + O(h)$$
$$q^\mu = x^\mu + O(h)$$
$$\Rightarrow \quad \xi^\mu - q^\mu = O(h) \tag{12.19}$$

(2) $\eta^\mu(q)$ and $\eta^\mu(x)$ thus differ by terms of $O(h^2)$

$$\eta^\mu(q) = \eta^\mu(x) + O(h^2) \tag{12.20}$$

(3) The derivatives therefore also differ by terms of $O(h^2)$

$$\frac{\partial \eta^\mu}{\partial q^\lambda} = \frac{\partial \eta^\mu}{\partial x^\lambda} + O(h^2) \tag{12.21}$$

Under the coordinate transformation of Eq. (12.16) the metric will be changed to

$$\bar{g}_{\mu\nu} = g^0_{\mu\nu} + \gamma_{\mu\nu} \tag{12.22}$$

Here $g^0_{\mu\nu}$ is the Lorentz metric of Eq. (12.1) and $\gamma_{\mu\nu}$ is again of $O(h)$. Work to first order in $\gamma_{\mu\nu}$, as was done above. Then all the above equations also hold in terms of $\gamma_{\mu\nu}$ and $\bar{\psi}_\mu{}^\nu$ (it is the same derivation!)

$$\bar{R}_{\mu\nu} = -\frac{1}{2}\Box\,\gamma_{\mu\nu} + \frac{1}{2}\frac{\partial}{\partial x^\mu}\left(\frac{\partial}{\partial x^\lambda}\bar{\psi}_\nu{}^\lambda\right) + \frac{1}{2}\frac{\partial}{\partial x^\nu}\left(\frac{\partial}{\partial x^\lambda}\bar{\psi}_\mu{}^\lambda\right)$$

$$\bar{\psi}_\mu{}^\nu = \gamma_\mu{}^\nu - \frac{1}{2}\gamma\,\delta_\mu{}^\nu \tag{12.23}$$

In addition, from the general tensor law of Eq. (3.29) and Eq. (12.18),

$$\bar{a}^\mu{}_\nu = \frac{\partial \xi^\mu}{\partial q^\nu}$$

$$= \delta^\mu{}_\nu + \frac{\partial \eta^\mu}{\partial q^\nu}$$

$$= \delta^\mu{}_\nu + \frac{\partial \eta^\mu}{\partial x^\nu} + O(h^2) \tag{12.24}$$

From the general tensor transformation law of Eq. (3.50), the metric of Eq. (12.2) is correspondingly transformed according to

$$\bar{g}^{\mu\nu} = \bar{a}^\mu{}_{\mu'}\bar{a}^\nu{}_{\nu'}\,g^{\mu'\nu'}$$

$$= \left(\delta^\mu{}_{\mu'} + \frac{\partial \eta^\mu}{\partial x^{\mu'}}\right)\left(\delta^\nu{}_{\nu'} + \frac{\partial \eta^\nu}{\partial x^{\nu'}}\right)\left(g_0^{\mu'\nu'} + h^{\mu'\nu'}\right)$$

$$= g_0^{\mu\nu} + h^{\mu\nu} + g_0^{\mu'\nu}\frac{\partial \eta^\mu}{\partial x^{\mu'}} + g_0^{\mu\nu'}\frac{\partial \eta^\nu}{\partial x^{\nu'}} + O(h^2) \tag{12.25}$$

Hence $\gamma^{\mu\nu}$ can be identified as

$$\gamma^{\mu\nu} = h^{\mu\nu} + \frac{\partial \eta^\mu}{\partial x_\nu} + \frac{\partial \eta^\nu}{\partial x_\mu} + O(h^2) \tag{12.26}$$

This is the *metric with the new coordinates*.[2]

It follows that the tensor $\bar{\psi}_\mu{}^\nu$ in Eq. (12.23) takes the following form after this transformation

$$\bar{\psi}_\mu{}^\nu = \left(h_\mu{}^\nu + \frac{\partial \eta_\mu}{\partial x_\nu} + \frac{\partial \eta^\nu}{\partial x^\mu}\right) - \frac{1}{2}\left(h + \frac{\partial \eta^\lambda}{\partial x^\lambda} + \frac{\partial \eta_\lambda}{\partial x_\lambda}\right)\delta_\mu{}^\nu \tag{12.27}$$

Since it is only the contracted pair of indices that are interchanged in the last two terms in the second parentheses, they are identical. The divergence

---

[2]Here we have used the fact that the metric g^0 raises and lowers the indices on x, and we recall that $\nabla_\mu = \partial/\partial x^\mu$ while $\nabla^\mu = \partial/\partial x_\mu$.

of this expression then follows as

$$\frac{\partial}{\partial x^\nu}\bar{\psi}_\mu{}^\nu = \frac{\partial}{\partial x^\nu}\left(h_\mu{}^\nu - \frac{1}{2}h\,\delta_\mu{}^\nu\right) + \frac{\partial^2}{\partial x^\nu \partial x_\nu}\eta_\mu + \frac{\partial^2}{\partial x^\nu \partial x^\mu}\eta^\nu - \frac{\partial^2}{\partial x^\mu \partial x^\lambda}\eta^\lambda$$

(12.28)

Since the order of partial derivatives can be interchanged in this flat space, the last two terms cancel. We again note that

$$\frac{\partial^2}{\partial x^\nu \partial x_\nu} = g_0^{\mu\nu}\frac{\partial^2}{\partial x^\nu \partial x^\mu}$$

$$= \nabla^2 - \frac{1}{c^2}\frac{\partial^2}{\partial t^2} = \Box \qquad (12.29)$$

Hence Eq. (12.28) takes the form

$$\frac{\partial}{\partial x^\nu}\bar{\psi}_\mu{}^\nu = \frac{\partial}{\partial x^\nu}\left(h_\mu{}^\nu - \frac{1}{2}h\,\delta_\mu{}^\nu\right) + \Box\,\eta_\mu \qquad (12.30)$$

Thus the auxiliary (Lorentz) condition is satisfied with the new choice of coordinates provided the r.h.s. of this expression vanishes

$$\text{If} \qquad \Box\,\eta_\mu = -\frac{\partial}{\partial x^\nu}\left(h_\mu{}^\nu - \frac{1}{2}h\,\delta_\mu{}^\nu\right)$$

$$\Rightarrow \qquad \frac{\partial}{\partial x^\nu}\bar{\psi}_\mu{}^\nu = 0 \qquad (12.31)$$

We observe the following:

- It is always possible to solve the first equation for a given $h_\mu{}^\nu$ and h, since this is simply the *inhomogeneous wave equation*;
- Thus one can always *find a new coordinate system on the deformed surface so that $\bar{\psi}_\mu{}^\nu$ satisfies the auxiliary condition.*[3]
- Note that the auxiliary condition *remains* satisfied under further coordinate transformations satisfying

$$\xi^\mu \to \xi^\mu + (\eta')^\mu$$

$$\Box\,(\eta')_\mu = 0 \qquad (12.32)$$

Let us *summarize* these results. One starts with cartesian coordinates (x^1, x^2, x^3, ct) in pseudo-euclidian (Minkowski) space and the Lorentz metric of Eq. (12.1). It is assumed there is a small distortion of the space so

---

[3]We explicitly exhibit such a coordinate system in the subsequent analysis.

that this metric is changed to

$$g_{\mu\nu} = g^0_{\mu\nu} + \gamma_{\mu\nu} \tag{12.33}$$

The distortion is characterized by a small dimensionless parameter, say ε, with

$$\gamma_{\mu\nu} = O(\varepsilon) \tag{12.34}$$

and the theory is *linearized* in ε. It is always possible to pick a corresponding set of generalized coordinates in the deformed space so that the following *auxiliary condition* is satisfied[4]

$$\frac{\partial}{\partial x_\nu}\psi_\mu{}^\nu = 0 \qquad\qquad ; \mu = 1, 2, 3, 4$$

$$\psi_\mu{}^\nu = \gamma_\mu{}^\nu - \frac{1}{2}\gamma\,\delta_\mu{}^\nu \tag{12.35}$$

Here $\gamma = \gamma^\lambda{}_\lambda$. The Ricci tensor then takes the form

$$R_{\mu\nu} = -\frac{1}{2}\Box\,\gamma_{\mu\nu} + \frac{1}{2}\frac{\partial}{\partial x^\mu}\left(\frac{\partial}{\partial x^\lambda}\psi_\nu{}^\lambda\right) + \frac{1}{2}\frac{\partial}{\partial x^\nu}\left(\frac{\partial}{\partial x^\lambda}\psi_\mu{}^\lambda\right)$$

$$= -\frac{1}{2}\Box\,\gamma_{\mu\nu} \tag{12.36}$$

These results are exact through $O(\varepsilon)$.

The Einstein Eqs. (11.26) in free space where $T_{\mu\nu} = 0$ then simply take the form of the *wave equation for the metric*

$$\Box\,\gamma_{\mu\nu} = 0 \tag{12.37}$$

It is evident from the form of the d'alembertian in Eq. (12.13) that the gravitational waves travel with velocity c.

We have here analyzed small oscillation about free, spatially flat Minkowski space. More generally, one can look for oscillations of the space by *linearizing about one of the previous solutions* (Schwarzschild, TOV, Robertson-Walker) instead of just flat space.[5]

We proceed to study a simple solution to the wave equation for gravitational radiation.

---

[4]Now that it has served its purpose, we drop the bar over the symbols. We also revert to the familiar notation $x^\mu = (x, y, z, ct)$ for the cartesian components.

[5]We leave these as exercises for the *very* dedicated reader!

12.3   Plane-Wave Solution to the Einstein Field Equations

Suppose that in free pseudo-euclidian (Minkowski) space one has cartesian coordinates and a metric of the form

$$x_0^\mu = (x_0, y_0, z_0, ct_0)$$
$$g_{\mu\nu} = g_{\mu\nu}^0 \tag{12.38}$$

where $g_{\mu\nu}^0$ is the Lorentz metric of Eq. (12.1). We are now careful to explicitly label the free-space cartesian coordinates of the previous discussion by x_0^μ. We seek a solution to the Einstein field Eqs. (12.37) using coordinates $q^\mu = (x, y, z, ct)$ embedded *in* the deformed space (*i.e. on* the deformed surface in our two-dimensional visualization),[6] together with a metric of the following form

$$q^\mu = (x, y, z, ct)$$
$$g_{\mu\nu} = g_{\mu\nu}^0 + \gamma_{\mu\nu} \tag{12.39}$$

We work to first order in $\gamma_{\mu\nu}$. As in the previous discussion, we note that

$$q^\mu = x_0^\mu + O(\gamma) \tag{12.40}$$

Any derivative with respect to x_0^μ in a term already of $O(\gamma)$ is the same as a derivative with respect to q^μ, to this order. As always, the *physics* lies in the interval

$$(ds)^2 = g_{\mu\nu} dq^\mu dq^\nu = g_{\mu\nu} dx^\mu dx^\nu \qquad ; \text{ interval} \tag{12.41}$$

Let us try to find a solution where

$$\gamma_{xx}, \gamma_{xy}, \gamma_{yy} \neq 0 \qquad\qquad ; \text{ Try it}$$
$$\text{all others vanish} \tag{12.42}$$

Since only the spatial (x, y) parts of the metric are deformed, there is no modification of the z coordinate, nor of the time t.[7]

---

[6] Just as the coordinates were embedded in the rubber sheet in the previous chapter. It is always essential to carefully define the generalized coordinates that go along with the metric — here they must be defined to the same order. The tensor transformation law then relates other coordinate systems and other metrics.

[7] Thus $(z_0, ct_0) \equiv (z, ct)$; all of the "action" is in the transverse (x, y) plane!

We look for a gravitational plane-wave moving with velocity c in the z-direction (Fig. 12-5)

$$\gamma_{ij} = h_{ij}e^{ik(z - ct)} \qquad ; (i, j) = (x, y)$$
$$h_{ij} = \text{constant} \qquad (12.43)$$

One is dealing with a linear wave equation, and physics is always the *real part* (Re) of this expression.

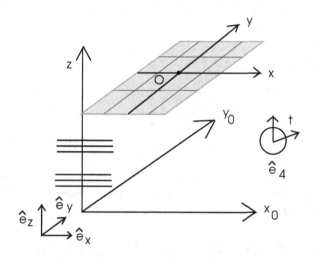

Fig. 12.5 Plane-wave solution to the gravitational wave equation traveling in the z-direction. The coordinates (x, y) are now embedded *in the two-dimensional surface* of the plane wave (just as the coordinates were embedded in the rubber sheet in the previous chapter). Here only the spatial parts of the metric $(\gamma_{xx}, \gamma_{xy}, \gamma_{yy})$ are deformed, and there is no modification of the z coordinate, nor of the time t. The orthonormal unit vectors $(\hat{e}_x, \hat{e}_y, \hat{e}_y)$ are defined in the pseudo-euclidian Minkowski space.

Through terms linear in γ, the indices on $\gamma_{\mu\nu}$ itself are raised and lowered with g^0

$$g^0_{\mu\nu} = g_0^{\mu\nu} = \begin{bmatrix} 1 & 0 & 0 & 0 \\ 0 & 1 & 0 & 0 \\ 0 & 0 & 1 & 0 \\ 0 & 0 & 0 & -1 \end{bmatrix} \qquad (12.44)$$

Thus

$$\gamma_x{}^x = g_0^{xx} \gamma_{xx} = \gamma_{xx} = \gamma^x{}_x$$
$$\gamma_y{}^y = g_0^{yy} \gamma_{yy} = \gamma_{yy} = \gamma^y{}_y$$
$$\gamma_x{}^y = g_0^{yy} \gamma_{xy} = \gamma_{xy} = \gamma^x{}_y \tag{12.45}$$

Hence

$$\gamma = \gamma^x{}_x + \gamma^y{}_y = \gamma_{xx} + \gamma_{yy} \tag{12.46}$$

Since Eq. (12.43) obviously provides a solution to the wave Eq. (12.37), it remains to examine the *auxiliary condition*. From the definition of $\psi_\mu{}^\nu$ in Eq. (12.35), one has

$$\psi_x{}^x = \gamma_{xx} - \frac{1}{2}\left(\gamma_{xx} + \gamma_{yy}\right) = \frac{1}{2}\left(\gamma_{xx} - \gamma_{yy}\right)$$
$$\psi_x{}^y = \gamma_{xy}$$
$$\psi_y{}^y = \gamma_{yy} - \frac{1}{2}\left(\gamma_{xx} + \gamma_{yy}\right) = \frac{1}{2}\left(\gamma_{yy} - \gamma_{xx}\right) = -\psi_x{}^x$$
$$\psi_z{}^z = \psi_4{}^4 = -\frac{1}{2}\left(\gamma_{xx} + \gamma_{yy}\right) \qquad ; \text{all others vanish} \tag{12.47}$$

Since the Ricci tensor in Eqs. (12.36) is already linear in $\gamma_{\mu\nu}$, the additional terms can equally well be written in terms of q^μ rather than x_0^μ through $O(\gamma)$ [see Eq. (12.40)]

$$R_{\mu\nu} + \frac{1}{2}\Box\,\gamma_{\mu\nu} = \frac{1}{2}\frac{\partial}{\partial q^\mu}\left(\frac{\partial}{\partial q^\lambda}\psi_\nu{}^\lambda\right) + \frac{1}{2}\frac{\partial}{\partial q^\nu}\left(\frac{\partial}{\partial q^\lambda}\psi_\mu{}^\lambda\right)$$
$$= \frac{1}{2}\frac{\partial}{\partial x^\mu}\left(\frac{\partial}{\partial x^\lambda}\psi_\nu{}^\lambda\right) + \frac{1}{2}\frac{\partial}{\partial x^\nu}\left(\frac{\partial}{\partial x^\lambda}\psi_\mu{}^\lambda\right) \tag{12.48}$$

If we can show that

$$\frac{\partial}{\partial x^\lambda}\psi_\mu{}^\lambda = 0 \qquad ; \mu = x, y, z, 4 \tag{12.49}$$

then it is clear that we will have constructed a solution to the Einstein field equations in free space. Since γ_{ij} only depends on $(z - ct)$, it is simple to check the relation for $\mu = x, y$

$$\frac{\partial}{\partial x^\lambda}\psi_x{}^\lambda = \frac{\partial}{\partial x}\psi_x{}^x + \frac{\partial}{\partial y}\psi_x{}^y = 0$$
$$\frac{\partial}{\partial x^\lambda}\psi_y{}^\lambda = \frac{\partial}{\partial x}\psi_y{}^x + \frac{\partial}{\partial y}\psi_y{}^y = 0 \qquad ; \mu = x, y \tag{12.50}$$

For $\mu = z, 4$ one has

$$\frac{\partial}{\partial x^\lambda} \psi_z{}^\lambda = \frac{\partial}{\partial z} \psi_z{}^z = -\frac{1}{2} \frac{\partial}{\partial z} \left(\gamma_{xx} + \gamma_{yy} \right)$$

$$\frac{\partial}{\partial x^\lambda} \psi_4{}^\lambda = \frac{\partial}{\partial (ct)} \psi_4{}^4 = -\frac{1}{2} \frac{\partial}{\partial (ct)} \left(\gamma_{xx} + \gamma_{yy} \right) \qquad ; \mu = z, 4 \quad (12.51)$$

These expressions will also vanish, provided we impose the following condition on the metric

$$\gamma_{xx} + \gamma_{yy} = 0 \qquad ; \text{condition on metric} \qquad (12.52)$$

In *summary*, the plane-wave expression for the metric in Eq. (12.43), with the relation between components of Eq. (12.52), satisfies the equations

$$\Box \gamma_{\mu\nu} = 0$$

$$\frac{\partial}{\partial x^\lambda} \psi_\mu{}^\lambda = 0 \qquad ; \mu = x, y, z, 4 \qquad (12.53)$$

It follows from Eqs. (12.48) that the Ricci tensor then also vanishes

$$R_{\mu\nu} = 0 \qquad (12.54)$$

Thus we have indeed constructed a linearized solution to the Einstein field equations representing a plane gravitational wave propagating in free space in the z-direction with velocity c.

To recapitulate the assumptions:

(1) Work to first order in $\gamma_{\mu\nu}$;
(2) The perturbation of the metric has two independent components

$$\gamma_{xx} = -\gamma_{yy}, \quad \gamma_{xy} = \gamma_{yx} \quad ; \text{all others vanish} \qquad (12.55)$$

(3) The solution is a plane wave propagating in the z-direction

$$\gamma_{ij} = h_{ij} e^{ik(z - ct)} \qquad ; h_{ij} = \text{constant} \qquad (12.56)$$

12.4 Interpretation

We now have coordinates $q^\mu = (x, y, z, ct)$ where (z, ct) are the usual Minkowski coordinates and (x, y) are a certain number of tick marks along coordinate axes embedded in the surface perpendicular to the z-axis. Since all agree on (z, t), we can just hold these values constant in order to interpret our results

$$t = \text{constant}$$
$$z = \text{constant}$$
$$dt = dz = 0 \qquad ; \text{ for interpetation} \qquad (12.57)$$

The physics is contained in the *interval*, which then takes the form[8]

$$(ds)^2 = (1 + \gamma_{xx})(dx)^2 + (1 + \gamma_{yy})(dy)^2 + 2\gamma_{xy}\, dx\, dy \qquad (12.58)$$

Suppose two events occur at coordinate points (x, y) and $(x+dx, y+dy)$ at a time t, and these are reported to the record keeper. The record keeper notes the events on his or her screen as in Fig. 12.6.

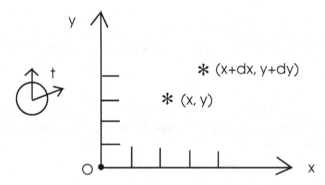

Fig. 12.6   Two events on the record keeper's screen at the points (x, y) and $(y+dy, x+dx)$ and at the time t. The screen remains unchanged for a second pair of events that occur at the same points, but at a later time t'. Only the clock moves.

Now suppose two more events occur *at the same coordinate points* but at a later time t'. Nothing changes on the record keeper's screen. Only the clock moves. But the *physical distance* between the events has now changed, and it is only through the metric that the record keeper can keep track of the physical separation between the simultaneous events.

The metric in Eq. (12.58) describes a two-dimensional surface. To *visualize* this surface, go to the tangent space [which is here the global (x, y) plane]. Imagine that one has a flat elastic sheet as in chap. 12 that can

---

[8]Note that $\gamma_{xy}\, dx\, dy + \gamma_{yx}\, dy\, dx = 2\gamma_{xy}\, dx\, dy$.

both stretch and compress in different directions as illustrated in Fig. 12.7.[9] The x and y coordinate axes, which are embedded in the sheet, now change directions due to the stretching and compressing. The line element on the sheet is given as usual in terms of the basis vectors by

$$d\mathbf{s} = \mathbf{e}_x\,dx + \mathbf{e}_y\,dy \qquad (12.59)$$

where $\mathbf{e}_x\,dx$ is the *physical distance* in tangent space for a change dx at fixed y, and *vice versa*. Suppose that the basis vectors on the deformed sheet in Fig. 12.7 can be written in terms of the original orthonormal unit vectors $(\hat{\mathbf{e}}_x, \hat{\mathbf{e}}_y)$ of Fig. 12.5 as

$$\mathbf{e}_x = (1 + \gamma_{xx})^{1/2}\left[\cos\chi\,\hat{\mathbf{e}}_x + \sin\chi\,\hat{\mathbf{e}}_y\right]$$
$$\mathbf{e}_y = (1 + \gamma_{yy})^{1/2}\left[\sin\chi\,\hat{\mathbf{e}}_x + \cos\chi\,\hat{\mathbf{e}}_y\right] \qquad (12.60)$$

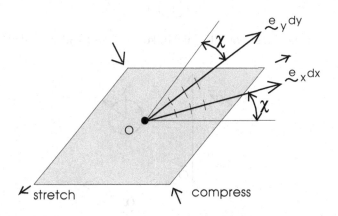

Fig. 12.7   Visualization of the surface in the (x, y) plane as an *elastic sheet* that can both stretch and compress.

It follows that the corresponding *metric* is then given by

$$g_{xx} = \mathbf{e}_x \cdot \mathbf{e}_x = 1 + \gamma_{xx}$$
$$g_{yy} = \mathbf{e}_y \cdot \mathbf{e}_y = 1 + \gamma_{yy}$$
$$g_{xy} = \mathbf{e}_x \cdot \mathbf{e}_y = (1 + \gamma_{xx})^{1/2}(1 + \gamma_{yy})^{1/2}\sin 2\chi \qquad (12.61)$$

---

[9]Our previous discussion of cosmology corresponded to a uniform stretching of this sheet.

If one now defines the angle χ by the relation

$$(1 + \gamma_{xx})^{1/2}(1 + \gamma_{yy})^{1/2} \sin 2\chi \equiv \gamma_{xy} \tag{12.62}$$

then the metric of Eq. (12.58) is reproduced and one has an exact analog of the gravitational plane wave. Note that in this analog the angle $\chi(t)$ is a function of time, so the stretching and compressing oscillates, and note also that χ is of $O(\gamma)$.

Consider one particular frame in the tangent space, the LF^3 (Fig. 12.8). Here the coordinates, interval, and metric are given by

$$\bar{x}^\mu = (\bar{x}, \bar{y}, \bar{z}, c\bar{t})$$
$$(ds)^2 = \bar{g}_{\mu\nu} \, d\bar{x}^\mu d\bar{x}^\nu$$
$$\bar{g}_{\mu\nu} = \begin{bmatrix} 1 & 0 & 0 & 0 \\ 0 & 1 & 0 & 0 \\ 0 & 0 & 1 & 0 \\ 0 & 0 & 0 & -1 \end{bmatrix} \tag{12.63}$$

In the LF^3 one just has spatially flat Minkowski space and special relativity.

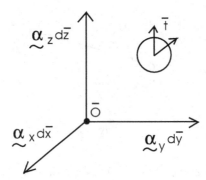

Fig. 12.8   The local freely falling frame LF^3 for the gravitational plane wave.

One can identify the basis vectors in the LF^3 as

$$\boldsymbol{\alpha}_\mu = \hat{\mathbf{e}}_\mu \qquad ; \mu = x, y, z, 4 \tag{12.64}$$

Here the $\hat{\mathbf{e}}_\mu$ are the orthonormal unit vectors of Fig. 12.5, with $\hat{\mathbf{e}}_4 \cdot \hat{\mathbf{e}}_4 = -1$. The coordinate transformation from the LF^3 to the global inertial laboratory frame can then be found from the relation between the basis vectors [see Eq. (3.40)]

$$\mathbf{e}_\mu = \bar{a}^\nu{}_\mu \, \boldsymbol{\alpha}_\nu \qquad (12.65)$$

With the definition of the matrix $[\underline{\bar{a}}]_{\nu\mu} \equiv \bar{a}^\nu{}_\mu$, and the aid of Eqs. (12.60), one then finds

$$[\underline{\bar{a}}]_{\nu\mu} = \bar{a}^\nu{}_\mu$$

$$[\underline{\bar{a}}]_{\nu\mu} = \begin{bmatrix} (1 + \gamma_{xx})^{1/2} \cos\chi & (1 + \gamma_{yy})^{1/2} \sin\chi & 0 & 0 \\ (1 + \gamma_{xx})^{1/2} \sin\chi & (1 + \gamma_{yy})^{1/2} \cos\chi & 0 & 0 \\ 0 & 0 & 1 & 0 \\ 0 & 0 & 0 & 1 \end{bmatrix} \qquad (12.66)$$

We note that

- This gives the transformation from the coordinates (x, y, z, ct) in the global inertial laboratory frame to the coordinates $(\bar{x}, \bar{y}, \bar{z}, c\bar{t})$ in the LF^3;
- There is a Lorentz contraction and expansion in the transverse space;
- The gravitational plane wave involves a *time-dependent stretching and contraction of the transverse space*;
- The transformation is independent of (x, y);
- The LF^3 here is again *global* in space, but *local* in time.

12.5 Detection

How would one detect the gravitational wave? Suppose a test particle of mass m is placed at the coordinate point (x, y, z, ct). The particle's motion will reflect the time-dependent stretching and contracting of the transverse surface caused by the passing wave (Fig. 12.9). The surface oscillates both *in time and in the z-direction* since the additional terms in the metric behave as[10]

$$\gamma_{ij} = h_{ij} \cos k(z - ct) \qquad\qquad ; (i,j) = (x, y)$$
$$= h_{ij} \cos\left[\frac{2\pi z}{\lambda} - 2\pi\nu t \right] \qquad\qquad \nu\lambda = c \qquad (12.67)$$

---

[10]We now explicity work with the real part of the solution in Eq. (12.43), assuming that hij is real. Remember that $\omega = kc$.

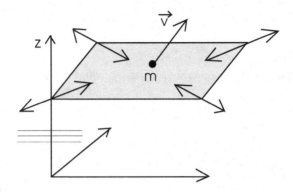

Fig. 12.9 Motion of a point particle of mass m and velocity $\vec{v}$ as affected by the stretching and contracting of the transverse surface caused by a passing gravitational wave of the form $\gamma_{ij} = h_{ij} \cos k(z - ct)$ with $(i,j) = (x,y)$ and h_{ij} real and constant.

One can track the motion through the *particle lagrangian*. The interval follows from the coordinates and metric, which we have just generated

$$(ds)^2 = (d\vec{x})^2 - c^2(dt)^2 + \left[\gamma_{xx}(dx)^2 + \gamma_{yy}(dy)^2 + 2\gamma_{xy}\,dx\,dy\right] \quad (12.68)$$

Here

$$(d\vec{x})^2 = (dx)^2 + (dy)^2 + (dz)^2 \quad (12.69)$$

The proper time is related to the interval by

$$(ds)^2 = -c^2(d\tau)^2 \quad (12.70)$$

The particle lagrangian, which describes motion along the geodesics, is then obtained from the proper time by Eq. (8.10)

$$L = -mc^2 \left(\frac{d\tau}{dt}\right) \quad (12.71)$$

Hence[11]

$$L(x,y,z;\,\dot{x},\dot{y},\dot{z};\,t) = -mc^2 \left[1 - \frac{\vec{v}^2}{c^2} - \frac{1}{c^2}\left(\gamma_{xx}\,\dot{x}^2 + \gamma_{yy}\,\dot{y}^2 + 2\gamma_{xy}\,\dot{x}\dot{y}\right)\right]^{1/2}$$

$$\vec{v}^2 = \dot{x}^2 + \dot{y}^2 + \dot{z}^2 \quad (12.72)$$

---

[11]Note that this lagrangian now has an *explicit* time dependence since the additional terms in the metric in Eq. (12.68) are functions of t.

This lagrangian for the motion of a point test particle is exact under the assumptions stated at the end of the previous section. With this lagrangian, the record keeper can now just do classical particle mechanics to track the motion of the particle on his or her screen (Fig. 12.10).

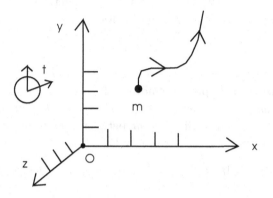

Fig. 12.10   Motion of a point particle of mass m due to a passing gravitational wave as tracked by the record keeper using the lagrangian in Eq. (12.72).

The non-relativistic limit (NRL) for the particle motion, obtained from the limit $c^2 \to \infty$, simplifies L

$$L_{\text{NRL}} = -mc^2 + \frac{m}{2} \left[\dot{z}^2 + (1 + \gamma_{xx})\dot{x}^2 + (1 + \gamma_{yy})\dot{y}^2 + 2\gamma_{xy}\,\dot{x}\dot{y} \right] \quad (12.73)$$

This expression is correct up to $O(1/c^2)$. The equations of motion follow from L_{NRL} through the use of Lagrange's equations for (x, y, z) respectively[12]

$$\frac{d}{dt} \left[(1 + \gamma_{xx})\dot{x} + \gamma_{xy}\,\dot{y} \right] = 0$$

$$\frac{d}{dt} \left[(1 + \gamma_{yy})\dot{y} + \gamma_{xy}\,\dot{x} \right] = 0$$

$$\left[\frac{d}{dt}\dot{z} + \frac{k}{2} \left(\gamma_{xx}\,\dot{x}^2 + \gamma_{yy}\,\dot{y}^2 + 2\gamma_{xy}\,\dot{x}\dot{y} \right) \tan k(z - ct) \right] = 0 \quad (12.74)$$

We note the following:

- One solution is $(x, y, \dot{z}) = $ constant, in which case there is only straight-line motion in the z-direction on the record keeper's screen;

---

[12]The overall factor of m has been canceled.

- The whole effect thus depends on $\vec{v}_\perp = (\dot{x}, \dot{y})$;
- Since it depends on the particle's velocity, the effect of the gravitational wave is like a *viscous drag*.

Additionally, we note that it follows from the wave equation and form of the solution in Eq. (12.43) that, just as with light, the phase and group velocity of the gravitational wave are identical

$$v_{\text{ph}} = v_{\text{gp}} = c \qquad (12.75)$$

Although a single particle at rest at a position (x, y) on the record keeper's screen does not appear to move in the transverse plane, the *physical distance between two such particles does change as the gravitational wave passes by.*[13] The physical distance dl between two points infinitesimally close together at fixed (z, ct) is given by Eq. (12.58)

$$dl = [(ds)^2]^{1/2} = \left[(1 + \gamma_{xx})(dx)^2 + (1 + \gamma_{yy})(dy)^2 + 2\gamma_{xy}\,dx\,dy\right]^{1/2} \qquad (12.76)$$

For simplicity, assume the two particles are on the x-axis so that

$$dl = (1 + \gamma_{xx})^{1/2}dx$$
$$\approx \left(1 + \frac{1}{2}\gamma_{xx}\right)dx \qquad (12.77)$$

where the second line is a first-order expansion in γ. Since the spatial part of the metric is *global*, this relation can be extended to a finite physical separation l corresponding to a finite, fixed coordinate separation d of the particles [just as in Eqs. (11.59) and (11.60)], with the result that

$$\frac{l - d}{d} = \frac{1}{2}\gamma_{xx}$$
$$= \frac{h_{xx}}{2}\cos k(z - ct) \qquad (12.78)$$

Here the explicit time dependence of Eq. (12.67) has been inserted in the second line.

Thus there is a harmonic modulation of the physical distance between particles. If these particles are bound by a spring, or inside of a lattice, this modulation can be expected to act as a driving force in the relative motion that causes the system to oscillate. Then as a function of the driving frequency of the gravitational wave $\nu = c/\lambda$, the response should

---

[13]See Fig. 12.6 and accompanying discussion.

be resonant at the natural frequency of the detector. This is the basis of "bar detectors" for gravitational waves, as pioneered by Weber at Maryland, Fairbank and Michelson at Stanford, and others.[14]

The most ambitious detector is the Laser Interferometer Gravitational-Wave Observatory (LIGO). Here one uses long-baseline interferometry of reflected laser beams to detect minute changes in the relative separation of the mirrors. More detailed information about LIGO, and the latest results from it, are always available from the LIGO website [LIGO (2006)].

---

[14]See Prob. 12.4. For an extended discussion of gravitational-wave detectors, see [Misner, Thorne, and Wheeler (1973)].

Chapter 13

Special Topics

With some confidence in the basic framework, we turn to a few special topics in general relativity.[1]

It is very powerful to be able to treat the problem as one in continuum mechanics and derive the Einstein field equations as the Euler-Lagrange equations following from a given lagrangian density.[2] We introduce the *Einstein-Hilbert lagrangian* and demonstrate that it serves the purpose (see, for example, [Peebles (1993)]). The addition of an arbitrary constant in this lagrangian density, to be determined by experiment, has several interesting consequences, not the least of which is that it can serve to increase the rate of expansion of the universe. This *cosmological constant*, originally included by Einstein, has had an intriguing history.[3] The current evidence is that it is indeed present at a certain level [Sky and Telescope (2006a)]. The lagrangian formulation allows us to easily introduce additional fields as sources, and we study in some detail the role of an *additional scalar field*.

We extend our study of cosmology and examine the consequences of the *Robertson-Walker metric with $k \neq 0$* [Robertson (1935); Walker (1936)]. We shall see through the *Friedmann equation* [Friedmann (1922)] that within this metric, the current observation of spatial flatness and isotropy out to the horizon is very puzzling. The spontaneous symmetry breaking of the additional contribution of a scalar field can lead to a rapid "stretching" of the space in the very early universe, and this theory of *inflation* satisfactorily explains many of the current observations [Guth (2000)].

In order to understand these observations in detail, a new form of *dark matter* must be present, in addition to the baryonic matter, cosmological

---

[1] In preparing this chapter, the author found [Amore (2000)] to be a valuable resource.

[2] See [Fetter and Walecka (2003)] for the basics of continuum mechanics.

[3] Einstein is said to have considered it at one time to be the worst blunder he had ever made.

constant, and new scalar field. Thus, while Einstein's theory of general relativity does provide the framework for understanding the time development of the universe, almost as many questions are raised as answered.

13.1 Einstein-Hilbert Lagrangian

In this section the Einstein field equations are derived from a lagrangian density, an approach that has several meritorious features:

- The equations are derived from Hamilton's variational principle;
- The lagrangian serves to illuminate the structure of the theory;
- It is then a simple matter to include additional terms in the lagrangian for the matter fields.

Choose a set of generalized coordinates $q^\mu = (q^1, q^2, q^3, q^4)$. Then consider variations where the *metric* is varied at each point in space-time according to

$$g^{\mu\nu} \to g^{\mu\nu} + \delta g^{\mu\nu}(q) \qquad ; \text{ variation} \qquad (13.1)$$

We demand that the metric *remain a symmetric second-rank tensor* under the variations.[4]

Equation (5.152) tells us that the physical volume element in the four-dimensional riemannian space is given by[5]

$$dV = \sqrt{-g}\, dq^1 dq^2 dq^3 dq^4 \qquad (13.2)$$

The Einstein-Hilbert lagrangian density for the gravitational field in free space is then simply given by

$$\mathcal{L}_{\mathrm{G}} = \frac{c^4}{16\pi G}(R + 2\bar\Lambda) \qquad ; \text{ Einstein-Hilbert} \qquad (13.3)$$

Here R is the *scalar curvature* obtained from the Ricci tensor through

$$R = R^\mu{}_\mu = R_{\mu\nu}g^{\mu\nu} \qquad ; \text{ scalar curvature} \qquad (13.4)$$

The quantity $\bar\Lambda$ is a possible additional constant in this expression, the *cosmological constant*, whose value is to be determined by experiment.

$$\bar\Lambda = \text{constant} \qquad\qquad ; \text{ cosmological constant} \qquad (13.5)$$

---

[4]The inverse relation $g^{\mu\lambda} g_{\lambda\nu} = \delta^\mu{}_\nu$ must also be maintained (see appendix A).

[5]Here $g = \det \underline{g}$ is the determinant of the metric. Recall the discussion following Eq. (10.24) for the presence of the minus sign in $\sqrt{-g}$.

The factor in front in Eq. (13.3) will be justified later when we discuss the coupling to matter fields.

The claim is that if *Hamilton's principle* is applied to the corresponding lagrangian[6]

$$\delta \int \mathcal{L}_G \sqrt{-g}\, dq^1 dq^2 dq^3 dq^4 = 0 \qquad ; \text{Hamilton's principle}$$

$$\text{fixed endpoints in time} \quad (13.6)$$

then one obtains Einstein's field equations in free space

$$\Rightarrow \qquad G_{\mu\nu} = R_{\mu\nu} - \frac{1}{2} R\, g_{\mu\nu} = \bar{\Lambda}\, g_{\mu\nu} \quad ; \text{Einstein's equations} \qquad (13.7)$$

Here the equations have been augmented by the term on the r.h.s. coming from the cosmological constant, to which we will return. We proceed to prove that Eqs. (13.7) are the Euler-Lagrange equations for the variational principle in Eq. (13.6).

The change in $\sqrt{-g}$ under the variation follows from Eqs. (5.166)[7]

$$\frac{\delta\sqrt{-g}}{\sqrt{-g}} = -\frac{1}{2} g_{\mu\nu}\, \delta g^{\mu\nu} \qquad (13.8)$$

Hence

$$\delta\left(\mathcal{L}_G \sqrt{-g}\right) = \frac{c^4}{16\pi G}\sqrt{-g}\left\{-\frac{1}{2} g_{\mu\nu}\delta g^{\mu\nu}(R + 2\bar{\Lambda}) + (\delta R + 2\delta\bar{\Lambda})\right\} \quad (13.9)$$

The last term in $\delta\bar{\Lambda}$ vanishes since $\bar{\Lambda}$ is a constant

$$\delta\bar{\Lambda} = 0 \qquad ; \text{constant} \qquad (13.10)$$

From Eq. (13.4)

$$\delta R = R_{\mu\nu}\delta g^{\mu\nu} + g^{\mu\nu}\,\delta R_{\mu\nu} \qquad (13.11)$$

Thus

$$\delta\left(\mathcal{L}_G\sqrt{-g}\right) = \frac{c^4}{16\pi G}\sqrt{-g}\left\{-\frac{1}{2}g_{\mu\nu}\delta g^{\mu\nu}(R + 2\bar{\Lambda}) + [R_{\mu\nu}\,\delta g^{\mu\nu} + g^{\mu\nu}\,\delta R_{\mu\nu}]\right\}$$

$$(13.12)$$

---

[6]Recall that we generally choose $q^4 = ct$, so the integral is then the *action*.

[7]Note

$$\frac{\delta g}{g} = \frac{2\,\delta\sqrt{g}}{\sqrt{g}} = \frac{2\,\delta\sqrt{-g}}{\sqrt{-g}} = g^{\mu\nu}\delta g_{\mu\nu} = -g_{\mu\nu}\delta g^{\mu\nu}$$

We now claim that the last term can be reduced to the difference of two *covariant divergences of four-vectors*

$$g^{\mu\nu}\,\delta R_{\mu\nu} = \left[g^{\mu\nu}\,\delta\Gamma^{\lambda}_{\mu\nu}\right]_{;\lambda} - \left[\delta\Gamma^{\lambda}_{\lambda\mu}\,g^{\mu\nu}\right]_{;\nu} \qquad (13.13)$$

Here $\delta\Gamma^{\lambda}_{\mu\nu}$ is the corresponding variation in the affine connection. Although this relation is the essential step in the proof, its derivation is somewhat technical, and in order not to interrupt the flow of the narrative, we have relegated the proof of Eq. (13.13) to appendix A. If we take Eq. (13.13) as established, then one can use Gauss' theorem as derived in Eq. (5.175) to convert the resulting integral to a *surface integral*

$$\int_{\text{vol}} v^{\mu}{}_{;\mu}\,\sqrt{-g}\,dq^1\cdots dq^4 = \oint_{\text{surface}} v^{\mu}dS_{\mu}$$
$$dS_{\mu} = \sqrt{-g}\,dq^1\cdots(dq^{\mu})\cdots dq^4 \quad;\ \text{strike}\,(dq^{\mu})\ (13.14)$$

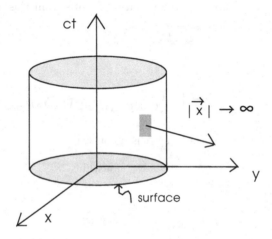

Fig. 13.1 Integration region in Hamilton's principle in Eq. (13.6). The variation is required to vanish on the fixed time surfaces, and the spatial boundary condition is that the integrand must represent a sufficiently localized disturbance that it vanishes on very far-away space-like surfaces.

This surface integral *vanishes* due to (see Fig. 13.1)

- fixed endpoints in time;
- appropriate boundary conditions in space, where the integrand represents a sufficiently localized disturbance that it vanishes on very far-

away space-like surfaces.[8]

The variation of the integrand in Hamilton's principle is thus reduced to

$$\delta\left(\mathcal{L}_G\sqrt{-g}\right) \doteq \frac{c^4}{16\pi G}\sqrt{-g}\left\{R_{\mu\nu} - \frac{1}{2}(R + 2\bar{\Lambda})g_{\mu\nu}\right\}\delta g^{\mu\nu} \quad (13.15)$$

where $\doteq$ implies that the covariant divergences have been discarded. Since $\delta g^{\mu\nu}(q)$ is now arbitrary at each point in space-time, its coefficient must vanish to satisfy the variational principle, and hence

$$R_{\mu\nu} - \frac{1}{2}R\,g_{\mu\nu} = \bar{\Lambda}\,g_{\mu\nu} \qquad ; \text{Einstein's equations} \quad (13.16)$$

These are just the Einstein field Eqs. (13.7), and the result is established. Note that Hamilton's principle can here be simply interpreted as a *minimization of the integrated scalar curvature*.[9]

13.2 Cosmological Constant

The cosmological constant appearing in Eq. (13.16) can be interpreted as a vacuum source term in the Einstein field equations

$$G_{\mu\nu} = \frac{8\pi G}{c^4}T_{\mu\nu}^{\text{vac}} \qquad ; \text{vacuum source term}$$

$$T_{\mu\nu}^{\text{vac}} = \frac{\bar{\Lambda}c^4}{8\pi G}g_{\mu\nu} \quad (13.17)$$

The general form of the stress tensor for a fluid at rest in the LF^3 is

$$\bar{T}_{\mu\nu} = P\bar{g}_{\mu\nu} + \left(P + \rho c^2\right)\frac{\bar{u}_\mu \bar{u}_\nu}{c^2}$$

$$\frac{\bar{u}_\mu}{c} = (0, 0, 0, -1) \quad (13.18)$$

A comparison of the two expressions for the stress tensor then gives the effective *equation of state* (EOS) for the cosmological term

$$\left(P + \rho c^2\right)^{\text{vac}} = 0 \qquad ; \text{effective EOS}$$

$$\frac{\bar{\Lambda}c^4}{8\pi G} = P^{\text{vac}} \quad (13.19)$$

---

[8]Alternatively, one can impose periodic boundary conditions in the spatial directions.
[9]More generally, one is making it stationary.

With a positive $\bar{\Lambda}$, the cosmological constant exerts a positive P^{vac} that counteracts the gravitational attraction of matter and acts to increase the expansion rate of the universe.[10]

It is an experimental question as to whether this term is actually present in the cosmology we observe today, and if so, to what extent. Current values of the observed "deceleration" parameter indicate that the contribution of the cosmological constant is of the same order of magnitude as that of matter in the source term in the Einstein field equations [Sky and Telescope (2006a)]. Understanding the actual observed value of $\bar{\Lambda}$ presents a formidable theoretical challenge.

13.3   Additional Scalar Field

It is now an easy task to include an additional scalar source field ϕ by augmenting the lagrangian density with a term[11]

$$\mathcal{L}_\phi = \left[-\frac{c^2}{2} g^{\mu\nu} \frac{\partial \phi}{\partial q^\mu} \frac{\partial \phi}{\partial q^\nu} - \mathcal{V}(\phi) \right] \tag{13.20}$$

We note that both $\mathcal{L}_\phi$ and $\sqrt{-g}\, dq^1 dq^2 dq^3 dq^4$ are *physical quantities* that must be invariant under coordinate transformations. We also note that this lagrangian density is of the following form in Minkowski space

$$\mathcal{L}_\phi = \mathcal{T}_\phi - \mathcal{V}_\phi$$
$$\mathcal{T}_\phi = \frac{1}{2}\dot{\phi}^2 - \frac{c^2}{2}\left(\vec{\nabla}\phi\right)^2 \tag{13.21}$$

which serves to set the normalization of the kinetic energy term.

The integrand in Hamilton's principle is formed from the total lagrangian density

$$\mathcal{L} = \mathcal{L}_{\text{G}} + \mathcal{L}_\phi \tag{13.22}$$

Now compute the variation of the terms in the integrand under the variation

---

[10]Note that the effective mass density $\rho^{\text{vac}} = -P^{\text{vac}}/c^2$ is then *negative* for the cosmological term.

[11]By definition, a scalar field is a physical quantity whose value is independent of coordinate system. We do not, at this point, attempt to identify the scalar field with any observed particle. It is simply a matter field that serves as a source in the Einstein field equations. For the form of the lagrangian, see Prob. 13.7.

of the metric $\delta g^{\mu\nu}$. First, making use of Eq. (13.8),

$$\delta \left(\mathcal{L}_\phi \sqrt{-g}\right) = \sqrt{-g}\left[-\frac{1}{2}g_{\mu\nu}\,\mathcal{L}_\phi - \frac{c^2}{2}\frac{\partial\phi}{\partial q^\mu}\frac{\partial\phi}{\partial q^\nu}\right]\delta g^{\mu\nu}$$

$$= \sqrt{-g}\left[-\frac{1}{2}T^\phi_{\mu\nu}\right]\delta g^{\mu\nu} \tag{13.23}$$

Here the energy-momentum tensor for the scalar field has been identified in the second line

$$T^\phi_{\mu\nu} = c^2 \frac{\partial\phi}{\partial q^\mu}\frac{\partial\phi}{\partial q^\nu} + g_{\mu\nu}\,\mathcal{L}_\phi \qquad ;\text{ energy-momentum tensor} \tag{13.24}$$

From before

$$\delta \left(\mathcal{L}_G\sqrt{-g}\right) \doteq \sqrt{-g}\left\{\frac{c^4}{16\pi G}\left[R_{\mu\nu} - \frac{1}{2}(R + 2\bar{\Lambda})g_{\mu\nu}\right]\right\}\delta g^{\mu\nu} \tag{13.25}$$

where $\doteq$ again means that the additional covariant divergences have been discarded.

Since the variation $\delta g^{\mu\nu}(q)$ is arbitrary, its coefficient in the integrand in Hamilton's principle must vanish, with the result that[12]

$$R_{\mu\nu} - \frac{1}{2}R\,g_{\mu\nu} = \frac{8\pi G}{c^4}T^\phi_{\mu\nu} + \bar{\Lambda}g_{\mu\nu} \qquad ;\text{ Einstein's equations}$$

$$T^\phi_{\mu\nu} = c^2 \frac{\partial\phi}{\partial q^\mu}\frac{\partial\phi}{\partial q^\nu} + g_{\mu\nu}\,\mathcal{L}_\phi \tag{13.26}$$

Given the energy-momentum tensor, we are now in a position to identify the pressure, energy density, and hence the equation of state (EOS) for the scalar field. Go to the LF^3. In this frame, the general form of the stress tensor for a fluid at rest is

$$\bar{T}^\phi_{\mu\nu} = P\,\bar{g}_{\mu\nu} + (P + \rho c^2)\frac{\bar{u}_\mu \bar{u}_\nu}{c^2}$$

$$\frac{\bar{u}_\mu}{c} = (0, 0, 0, -1)$$

$$\bar{g}_{\mu\nu} = \begin{bmatrix} 1 & 0 & 0 & 0 \\ 0 & 1 & 0 & 0 \\ 0 & 0 & 1 & 0 \\ 0 & 0 & 0 & -1 \end{bmatrix} \tag{13.27}$$

---

[12]Here we finally get an explanation for the factor in front in Eq. (13.3). Note that the source term in $T^\phi_{\mu\nu}$ further complicates the non-linearity of the problem since it, in turn, depends on the metric $g_{\mu\nu}$ in a nonlinear fashion.

One can then use the following relations to identify the mass density and pressure

$$\rho c^2 = \bar{T}^\phi_{44}$$
$$P = \frac{1}{3}\bar{T}^\phi_{ii} \tag{13.28}$$

where the repeated Latin index is again summed from 1 to 3. Thus from the second of Eqs. (13.26), now evaluated in the LF^3,

$$\rho c^2 = \frac{1}{2}\dot{\phi}^2 + \frac{c^2}{2}\left(\vec{\nabla}\phi\right)^2 + \mathcal{V}(\phi) \qquad ; \text{ EOS for scalar field in } LF^3$$
$$P = \frac{1}{2}\dot{\phi}^2 - \frac{c^2}{6}\left(\vec{\nabla}\phi\right)^2 - \mathcal{V}(\phi) \tag{13.29}$$

Note that, as is the case for the cosmological constant, a *constant* scalar field has the following equation of state

$$\left(P + \rho c^2\right)^\phi = 0 \qquad ; \text{ constant } \phi \tag{13.30}$$

The sign of the pressure (and of ρ) then depends on the sign of $\mathcal{V}(\phi)$.

13.4 Robertson-Walker Metric with $k \neq 0$

In chapter 11 we studied the cosmology produced by a cold uniform matter distribution in a space-time that is homogeneous, isotropic, and *spatially flat*. We did this by looking for a solution to the Einstein field equations starting from the Robertson-Walker metric with $k = 0$. There is no *a priori* reason that the riemannian space for such a system should be spatially flat, and so we now generalize that discussion to include the possibility of a *spatial curvature* and repeat the analysis starting from the Robertson-Walker metric with $k \neq 0$. In this case, the interval and corresponding metric are given by

$$(d\mathbf{s})^2 = \Lambda(t)^2\left[\frac{(dr)^2}{1 - kr^2} + r^2(d\theta)^2 + r^2\sin^2\theta(d\phi)^2\right] - c^2(dt)^2$$

$$g_{\mu\nu} = \begin{bmatrix} \Lambda^2(t)/(1 - kr^2) & 0 & 0 & 0 \\ 0 & \Lambda^2(t)r^2 & 0 & 0 \\ 0 & 0 & \Lambda^2(t)r^2\sin^2\theta & 0 \\ 0 & 0 & 0 & -1 \end{bmatrix} \tag{13.31}$$

The procedure is now clear, we have carried it out several times. It is first necessary to compute the affine connection, and then the Ricci tensor.

Again, so as not to interrupt the flow of the narrative, this calculation is relegated to appendix B. The result is that the Ricci tensor for this metric is given by

$$R_{ij} = \frac{1}{c^2}\left[2\left(\frac{\dot{\Lambda}}{\Lambda}\right)^2 + \left(\frac{\ddot{\Lambda}}{\Lambda}\right) + \frac{2kc^2}{\Lambda^2}\right] g_{ij} \qquad ; (i,j) = (r,\theta,\phi)$$

$$R_{44} = -\frac{3}{c^2}\left(\frac{\ddot{\Lambda}}{\Lambda}\right) \qquad\qquad q^4 = ct$$

$$R_{i4} = 0 \qquad\qquad\qquad (13.32)$$

Here $\dot{\Lambda} = d\Lambda/dt$ and $\ddot{\Lambda} = d^2\Lambda/dt^2$. There are some checks one can make:

- This reproduces the result obtained in Eq. (11.19) for $k = 0$ and the cartesian basis where $g_{ij} = \Lambda^2\,\delta_{ij}$
- This reproduces the result obtained in Eq. (7.71) for $\Lambda = 1$, $B = 1$, and $A = (1 - kr^2)^{-1}$.

The source in the Einstein field equations is obtained from the energy-momentum tensor

$$T_{\mu\nu} = P\,g_{\mu\nu} + (\rho c^2 + P)\frac{u_\mu u_\nu}{c^2}$$

$$\frac{u_\mu}{c} = (0,0,0,-1) \qquad\qquad (13.33)$$

where all terms are now retained.[13] The expression for the source is given in Eq. (10.17)

$$T_{\mu\nu} - \frac{1}{2}T\,g_{\mu\nu} = \frac{1}{2}(\rho c^2 - P)\,g_{\mu\nu} + (\rho c^2 + P)\frac{u_\mu u_\nu}{c^2} \qquad (13.34)$$

A combination of Eqs. (13.32) and (13.34) then yields the Einstein field equations for the Robertson-Walker metric with $k \neq 0$ and a source described by the energy-momentum tensor $T_{\mu\nu}$ in Eq. (13.33)

$$-\frac{3}{c^2}\left(\frac{\ddot{\Lambda}}{\Lambda}\right) = \frac{\kappa}{2}(\rho c^2 + 3P) \qquad ; \text{Einstein equations}$$

$$\frac{1}{c^2}\left[\left(\frac{\ddot{\Lambda}}{\Lambda}\right) + 2\left(\frac{\dot{\Lambda}}{\Lambda}\right)^2 + \frac{2kc^2}{\Lambda^2}\right] = \frac{\kappa}{2}(\rho c^2 - P) \qquad (13.35)$$

---

[13]That is, it is no longer assumed that $P \ll \rho c^2$.

In comparison with Eqs. (11.27), we note the addition of the term in $2k/\Lambda^2$ on the l.h.s. of the second equation, and instead of just a common ρc^2 on the r.h.s., one has $(\rho c^2 + 3P)$ in the first equation, and $(\rho c^2 - P)$ in the second.[14]

Equations (13.35) are a pair of coupled, nonlinear, second-order differential equations for the time development of the scale factor $\Lambda(t)$ in the metric in terms of the quantities $[\Lambda(t), \rho(t)]$, where the pressure is given in terms of the mass density by the equation of state $P(\rho)$.

One useful form of the dynamical equations is obtained by simply substituting the first equation in the second. The additional term in the pressure cancels, and one finds

$$2\dot{\Lambda}^2 + 2kc^2 = \frac{2\kappa c^2}{3}\rho c^2 \Lambda^2$$

$$= \frac{16\pi G\rho}{3}\Lambda^2 \qquad (13.36)$$

Here Eq. (10.55) has been used in the second line. This result can be rearranged to yield the *Friedmann equation* [Friedmann (1922)]

$$\dot{\Lambda}^2 - \frac{8\pi G\rho}{3}\Lambda^2 = -kc^2 \qquad ; \text{ Friedmann equation} \qquad (13.37)$$

Note the following:

- The Friedmann equation relates the spatial curvature of the space, as expressed through k, to the mass density $\rho(t)$ and the rate of change of the spatial stretching of the space as expressed through $\dot{\Lambda}(t)$;
- The pressure P has dropped out of this result, and hence it is independent of the specific equation of state of the medium.

We are now in a position to define the *critical density* ρ_c, which will be a function of time

$$\rho_c(t) \equiv \frac{3}{8\pi G}\left(\frac{\dot{\Lambda}}{\Lambda}\right)^2 = \frac{3}{8\pi G}h^2(t) \qquad (13.38)$$

Here Eq. (11.28) has been used in the last equality. The Friedmann Eq. (13.37) can then be written as

$$1 - \frac{\rho(t)}{\rho_c(t)} = -\frac{kc^2}{\left[\dot{\Lambda}(t)\right]^2} \qquad (13.39)$$

---

[14]Compare Prob. 11.2.

We further define the ratio of the current mass density to the critical density as

$$\Omega(t) \equiv \frac{\rho(t)}{\rho_c(t)} \qquad ; \text{ fraction of critical density} \quad (13.40)$$

This allows one to recast the Friedmann equation in the form

$$\Omega(t) - 1 = \frac{kc^2}{\left[\dot{\Lambda}(t)\right]^2} \qquad ; \text{ Friedmann equation} \quad (13.41)$$

This expression relates the spatial curvature of the space to the current ratio of mass density to critical density. If the mass density is at the critical density and $\Omega = 1$, then the space is spatially flat with $k = 0$, and *vice versa*. We shall return to the further implications of these results shortly.

As was the case with the TOV equations, a second useful form of the dynamical relations is obtained from the conservation of the energy-momentum tensor, which serves as the source in the Einstein field equations. In the LF^3, the covariant divergence of the energy-momentum tensor vanishes

$$\bar{T}^{\mu\nu}{}_{;\nu} = 0 \qquad (13.42)$$

This relation is preserved under the coordinate transformation to the global inertial laboratory frame, and hence

$$T^{\mu\nu}{}_{;\nu} = \left[P g^{\mu\nu} + (\rho c^2 + P) \frac{u^\mu u^\nu}{c^2} \right]_{;\nu} = 0$$

$$\frac{u^\mu}{c} = (0,0,0,1) \qquad (13.43)$$

It follows as in Eq. (10.43) that

$$T^{\mu\nu}{}_{;\nu} = [P g^{\mu\nu}]_{;\nu} + \left[(\rho c^2 + P) \frac{u^\mu u^\nu}{c^2} \right]_{;\nu}$$

$$= P g^{\mu\nu}{}_{;\nu} + g^{\mu\nu} P_{;\nu} + \left\{ \frac{1}{\sqrt{-g}} \frac{\partial}{\partial q^\nu} \left[\sqrt{-g} \, (\rho c^2 + P) \frac{u^\mu u^\nu}{c^2} \right] \right\} +$$

$$\Gamma^\mu_{\lambda\nu} (\rho c^2 + P) \frac{u^\lambda u^\nu}{c^2} \qquad (13.44)$$

In the second line, the covariant divergence of the metric tensor in the first term vanishes. The second term in the second line only contributes for

$\mu = 4$, in which case it gives

$$g^{44} \frac{dP}{d(ct)} = -\frac{dP}{d(ct)} \tag{13.45}$$

The last term in Eq. (13.44) vanishes since it is evident from Table B.1 that there is no Γ^μ_{44}

$$\Gamma^\mu_{44}(\rho c^2 + P)\frac{u^4 u^4}{c^2} = 0 \tag{13.46}$$

The term in braces in Eq. (13.44) also only contributes for $\mu = 4$. Since the square root of determinant of the metric is given by $\sqrt{-g} = \Lambda^3(t)\, r^2 \sin\theta/(1 - kr^2)^{1/2}$, this term becomes

$$\frac{1}{\sqrt{-g}} \frac{\partial}{\partial q^4}\left[\sqrt{-g}\,(\rho c^2 + P)\frac{u^4 u^4}{c^2}\right] = \frac{d}{d(ct)}(\rho c^2 + P) + (\rho c^2 + P)\frac{1}{\Lambda^3}\frac{d}{d(ct)}\Lambda^3 \tag{13.47}$$

Hence the conservation of the energy-momentum tensor is satisfied identically for $\mu = (r, \theta, \phi)$, and for $\mu = 4$ it leads to the following relation

$$g^{44}\frac{dP}{d(ct)} + \frac{d}{d(ct)}(\rho c^2 + P) + (\rho c^2 + P)\frac{1}{\Lambda^3}\frac{d}{d(ct)}\Lambda^3 = 0 \quad ; \mu = 4 \tag{13.48}$$

The term in dP/dt cancels, with the result that

$$\frac{d\rho}{dt} + 3\left(\frac{\dot\Lambda}{\Lambda}\right)\left(\rho + \frac{P}{c^2}\right) = 0 \tag{13.49}$$

This is a dynamical equation that relates the rate of change of the mass density and the stretching of the space to the current value of the mass density and the pressure, which is obtained from the mass density through the equation of state. Since k does not appear in this equation, it is the same result as obtained in the case $k = 0$ (see Prob. 13.3).

In *summary*, the Einstein field equations for the Robertson-Walker metric with $k \neq 0$ can be recast in the form

$$\dot\Lambda^2 - \frac{8\pi G\rho}{3}\Lambda^2 = -kc^2 \quad ; \text{Einstein equations for } k \neq 0$$

$$\dot\rho + 3\left(\frac{\dot\Lambda}{\Lambda}\right)\left(\rho + \frac{P}{c^2}\right) = 0 \tag{13.50}$$

Equations (13.50) have the functional form[15]

$$\frac{d\Lambda}{dt} = f[\Lambda(t), \rho(t); k]$$

$$\frac{d\rho}{dt} = g\left[\frac{d\Lambda}{dt}, \Lambda(t), \rho(t)\right] \tag{13.51}$$

Here the pressure has been expressed in terms of the mass density through the equation of state $P(\rho)$. This is now a pair of coupled, nonlinear, first-order differential equations that can be integrated to find $[\rho(t), \Lambda(t)]$ given a set of initial values $[\rho_0, \Lambda_0]$ at the time t_0. The equations depend on the value of the constant parameter k, which determines the scalar curvature of the four-dimensional riemannian space described by the Robertson-Walker metric with $k \neq 0$. This scalar curvature is given in the case $\Lambda = 1$ by (Prob. 13.4)

$$R = R^\mu{}_\mu = 6k \qquad ; \Lambda = 1 \tag{13.52}$$

13.5 Inflation

The cosmology observed experimentally has several characteristic features:

- The universe appears to be spatially flat with $k \approx 0$;
- The mass density appears to be close to the critical density $\Omega(t_p) \approx 1$;[16]
- The temperature, for which COBE and WMAP serve as thermometers, appears to satisfy $T(t_p) \approx$ constant out to the horizon in all directions.

This leads to at least two problems within the framework of the hot big bang theory (HBBT) where the universe started with a hot, dense, uniform medium and then expanded adiabatically:

(1) It follows from Eqs. (13.50), and various traditional equations of state for a matter source, that if one starts with *any non-zero* k, then the quantity $\Omega(t) - 1$ *diverges with time from zero*. For example, if one

---

[15]We choose the positive square root for $\dot{\Lambda}/\Lambda$ in the first of Eqs. (13.50), which certainly provides *a solution*. If $k = 0$, these equations then reproduce the cosmology of chapter 11 (see Prob. 13.5).

[16]The source terms in the Einstein field equations in cosmology arising from the cosmological constant and scalar field are referred to as "dark energy," whereas the additional matter required to achieve $\Omega(t_p) = 1$ is called "dark matter" (compare Prob. 13.12). For the current evidence on the relative contribution of the source terms, see [Sky and Telescope (2006b)].

takes the cosmology of chapter 11 as the zeroth-order solution, then the r.h.s. of Eq. (13.41) takes the form

$$\frac{kc^2}{\left[\dot{\Lambda}(t)\right]^2} \approx \frac{9kc^2(t-t_0)^{2/3}}{4\gamma^{2/3}} \tag{13.53}$$

where Eq. (11.38) has been used in obtaining this result. This clearly grows with time. In order to get $\Omega(t_p) \approx 1$ today, one has to *fine tune* it to be very, very close to one in the far-distant past;

(2) How can the universe have the same temperature $T(t_p)$ out to the horizon in all directions, when a large part of the observed universe is not *causally connected*? Here we refer to that part of the universe whose current spatial separation is greater than the distance to the horizon $D_H = c(t_p - t_0)$, which is the maximum distance that light can have traveled since $t = t_0$.

The cosmology developed in chapter 11, based on a cold, uniform, matter-dominated universe that is spatially flat with $k \approx 0$ describes most of what is observed today. Given the above problems, the question arises as to how one arrives at the *initial conditions* leading to this cosmology. The theory of *inflation* addresses this issue.

The goal is to find a mechanism that leads to the proper initial conditions for the spatially flat ($k \approx 0$), cold ($T \approx 0$), matter dominated ($\rho c^2 \gg P$) cosmology to subsequently hold, as it evidently does.[17]

The basic idea is to include an additional scalar field ϕ with a form of the potential $\mathcal{V}(\phi)$ that gives rise to spontaneous symmetry breaking.

The ordinary mass term for a scalar field of inverse Compton wavelength $m = m_0 c/\hbar$ corresponds to a potential of the form (see Prob. 13.7)

$$\mathcal{V}(\phi) = \frac{1}{2}(mc)^2\phi^2 \quad ; \text{ ordinary mass term} \tag{13.54}$$

This potential has a stable minimum at $\phi = 0$. In particle physics, both to describe the spontaneous breaking of chiral symmetry in the sigma model

---

[17]This is the author's version of inflation, as he understands it. The reader is referred to the extensive literature on inflationary cosmology, for example [Linde (1990); Peebles (1993); Guth (2000)], for further study of this subject. So far, the scalar field has not been associated with any known particle.

and to give the weak vector bosons their mass through the Higgs mechanism, one introduces scalar fields with more general scalar potentials.[18] These potentials have an unstable equilibrium point at $\phi = 0$, and a stable minimum at a non-zero value of the field $\phi = \phi_0$. Suppose the same situation holds here, as illustrated in Fig. 13.2.

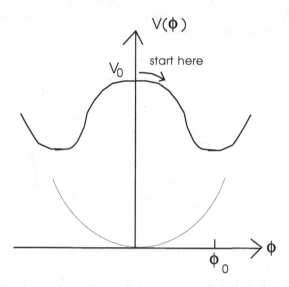

Fig. 13.2 Scalar field potential $\mathcal{V}(\phi)$ that leads to spontaneous symmetry breaking. The origin $\phi = 0$ is a point of equilibrium, but it is unstable, and with this form of the potential, the scalar field develops a vacuum expectation value $\phi = \phi_0$ in stable equilibrium. Here $V(0) \equiv V_0$. The ordinary mass term of Eq. (13.54) is sketched with the broken line.

Suppose the universe originally had $\phi = 0$ everywhere. It is then in unstable equilibrium. In some small region, the scalar field ϕ will start rolling down the potential hill toward the non-zero stable equilibrium value ϕ_0.[19] The energy stored in the scalar field will eventually go into creating a hot plasma of particles and antiparticles (the "hot big bang").[20] This

---

[18] See [Walecka (2004)].

[19] The dynamics of the scalar field follows from the scalar meson field equation derived from the lagrangian in Eq. (13.20). See Prob. 13.8.

[20] The baryon-antibaryon imbalance that we observe in the universe around us is presumably due to the non-conservation of various quantities at the elementary particle level [Dimopoulos and Susskind (1978)].

region will grow with time as more of the scalar field develops (Fig. 13.3). Equations (13.50) describe how this scalar field then stretches the space (see Prob. 13.9).

According to Guth [Guth (2000)]:

> "*The uniformity is created initially on microscopic scales by normal thermal equilibrium processes, and then inflation takes over and stretches the region of uniformity to become large enough to encompass the known universe.*"

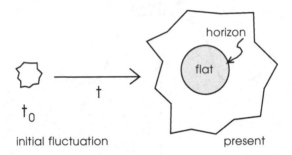

Fig. 13.3   Stretching of space due to the development of the scalar field in inflation.

Why is the resulting region within the current horizon spatially flat? We give a heuristic argument. Consider the two-dimensional surface of a balloon. If the balloon expands rapidly, and one's horizon is restricted to a small region on the surface, the space will appear to be spatially flat.[21] Within that space, one will simply have a flat, stretching rubber sheet (Fig. 13.4).

In this picture, the quantity $1 - \Omega(t)$ *converges with time to zero.* This convergence is illustrated in model problems with the EOS of the cosmological constant in Prob. 13.6, and with that of the scalar field in Prob. 13.9.

Note that if $\Lambda(t) \to \infty$ rapidly enough as $t \to \infty$, as is the case in these models, then a finite physical distance corresponds to a small value of the coordinate change dr in the interval in Eq. (13.31).[22] Hence one never gets far away from $r = 0$. Thus for a given k, the radial part of the metric is unmodified since $(1 - kr^2)^{-1} \approx 1$, and one can just as well start with $k = 0$.

---

[21] Like the tangent space!
[22] A finite physical distance as defined, for example, by the horizon.

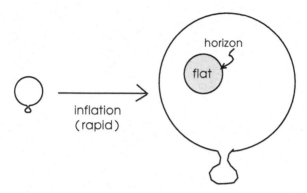

Fig. 13.4 Surface of an inflating balloon where the present horizon confines one to a small flat region on the surface.

It might be considered presumptuous in the extreme to extrapolate the history of the universe back to a time $t - t_0 \approx e^{-xx}$ sec where xx is a very large number, as one does in the theory of inflation. The justification is that the subsequent development follows from the known laws of physics. So far, this seems to be the case.

Chapter 14

Problems

2.1 Consider two parallel flat surfaces interconnected through a smooth circular cylindrical tube to make a single unified surface. Give a qualitative discussion of the particle orbits and geodesics on the unified surface.

3.1 Given the definition of $\delta^i{}_j$ in Eq. (2.13), the fact that the metric raises and lowers indices, and Eq. (2.20), show $\delta_i{}^j = \delta^i{}_j$.

3.2 Suppose a contraction $v_i w^i$ is invariant under coordinate transformations $v_i w^i = \bar{v}_i \bar{w}^i$ for an *arbitrary* vector v_i, which transforms as $\bar{v}_i = \bar{a}_i{}^j v_j$. Show that w^i must then also be a vector transforming as $\bar{w}^i = \bar{a}^i{}_j w^j$.

3.3 Choose cartesian coordinates $(q^1, q^2, q^3) = (x, y, z)$ in three-dimensional euclidian space and write the line element as

$$d\mathbf{s} = \hat{\mathbf{e}}_x \, dx + \hat{\mathbf{e}}_y \, dy + \hat{\mathbf{e}}_z \, dz$$

Here $(\hat{\mathbf{e}}_x, \hat{\mathbf{e}}_y, \hat{\mathbf{e}}_z)$ are the set of global, orthonormal, cartesian unit vectors.

(a) Show the metric is

$$\underline{g} = \begin{pmatrix} g_{xx} & g_{xy} & g_{xz} \\ g_{yx} & g_{yy} & g_{yz} \\ g_{zx} & g_{zy} & g_{zz} \end{pmatrix} = \begin{pmatrix} 1 & 0 & 0 \\ 0 & 1 & 0 \\ 0 & 0 & 1 \end{pmatrix}$$

(b) Show the reciprocal basis is identical to the original basis in this case, and hence there is no need to distinguish upper and lower indices if one sticks to cartesian coordinates in euclidian space.

3.4 The point transformation from cartesian coordinates $(q^1, q^2, q^3) = (x, y, z)$ to spherical coordinates $(\xi^1, \xi^2, \xi^3) = (r, \theta, \phi)$ is given by

$$x = r \sin\theta \cos\phi \qquad ; \quad y = r \sin\theta \sin\phi \qquad ; \quad z = r \cos\theta$$

281

(a) Determine the transformation coefficients $a^i{}_j = \bar{a}_j{}^i$.

(b) Use the results in (a) to relate the components of a vector in spherical coordinates $\bar{v}_i$ to those in cartesian coordinates v_i.

(c) The metric in spherical coordinates is given in Eq. (5.178), and the metric in cartesian coordinates is given in Prob. 3.3. Determine the transformation coefficients $a_i{}^j = \bar{a}^j{}_i$ from the results found in (a).

3.5 Another example of a tensor is the stress tensor for non-viscous fluids, whose components in cartesian coordinates are given by

$$T_{ij} = p\, g_{ij} + \rho\, v_i v_j$$

Here p is the pressure and ρ is the mass density, both quantities being defined in the rest frame of the fluid. The vector **v** is the fluid velocity, and g_{ij} is the metric.[1]

(a) Prove that T_{ij} transforms as a second-rank tensor.

(b) Use the results in Prob. 3.4 to determine the components of the stress tensor in spherical coordinates.

3.6 There are eight elements in the affine connection Γ^i_{jk} in plane polar coordinates. They are given in Eqs. (3.84). Five are computed in the text.

(a) Verify the remaining three.

(b) The change in basis vectors as one moves to a neighboring point $(r + dr, \phi + d\phi)$ is given in terms of the affine connection by Eqs. (3.85). Provide a direct geometrical derivation of these results.

3.7 Write the gradient as $\boldsymbol{\nabla} = \mathbf{e}_i \nabla^i = \mathbf{e}^i \nabla_i$ and show that

$$\nabla^i = g^{ij} \nabla_j$$

3.8 Let $\hat{\mathbf{e}}_k$ with $k = (1, \cdots, n)$ be a global, complete, orthonormal set of cartesian unit vectors in the n-dimensional euclidian space.

(a) Show that any vector field $\mathbf{v}(q^1, \cdots, q^n)$ can be expanded in this basis according to

$$\mathbf{v}(q^1, \cdots, q^n) = v^k(q^1, \cdots, q^n)\, \hat{\mathbf{e}}_k$$

(b) Hence conclude that the rule for interchanging the order of differentiation when the second partial derivative $\partial^2 / \partial q^i \partial q^j$ is applied to this vector field is just that for multivariable functions. What is the rule?

---

[1]See [Fetter and Walecka (2003)] p.297. Note that with cartesian coordinates $g_{ij} = \delta_{ij}$ is the usual Kronecker delta. We use the terms *stress tensor* and *energy-momentum tensor* interchangeably for such a fluid in this book.

4.1 (a) Show from Eq. (4.8) that when the particle motion is constrained to the surface

$$\left(\frac{d\mathbf{v}}{dt}\right)_{\perp} = v^i v^j \Gamma_{ij}^3 \mathbf{e}_3 \qquad ; (i,j) = (1,2)$$

(b) Find the constraint force then implied by Eq. (4.9).

(c) Discuss the implications for Γ_{ij}^3 for the curved surface.

4.2 (a) Start from Eq. (4.22). Write out in detail the three steps given below that equation and show that one indeed arrives at Lagrange's Eqs. (4.12).

(b) Provide all the steps for the derivation of Lagrange's Eqs. (4.12) on the surface from the condition of parallel displacement in the tangent plane given in Eq. (4.25).

(c) Provide all the steps for the derivation of the geodesic Eqs. (4.27) on the surface from the condition of parallel displacement in the tangent plane stated there.

5.1 It is shown in the text that the covariant derivative of a vector is given by

$$v^i_{;j} = \frac{\partial v^i}{\partial q^j} + \Gamma_{jk}^i v^k$$

$$v_{i;j} = \frac{\partial v_i}{\partial q^j} - \Gamma_{ij}^k v_k$$

It is also shown that the metric is constant under covariant differentiation so that it can be moved through the derivative to raise and lower indices. Thus

$$g_{ik}\left(v^k_{;j}\right) = v_{i;j}$$

Use this relation to derive the second of the above equations from the first.

5.2 Prove that the symmetry of the Ricci tensor $R_{ij} = R_{ji}$ follows directly from the symmetry property of the Riemann tensor $R_{jklm} = R_{lmjk}$ derived in Eq. (5.120).

5.3 Show the symmetry properties of the Riemann tensor are preserved under a coordinate transformation.

5.4[2] Choose cartesian coordinates (x, y, z) in three-dimensional euclidian space as in Prob. 3.3.

---

[2]Problems 5.4 and 5.5 are very easy, but they are also very instructive.

(a) What is the affine connection?

(b) What is the Riemann tensor?

(c) What are the Ricci tensor and scalar curvature?

5.5 Generalize the results in Prob. 5.4 to n-dimensional euclidian space. Show the Riemann tensor, Ricci tensor, and scalar curvature all vanish.

5.6 The tensor transformation law in Eq. (3.50) not only provides insight but can also save an incredible amount of work. Start from the cartesian coordinates in n-dimensional euclidian space of Prob. 5.5, and make a point transformation to *any new set of generalized coordinates.*

(a) What is the Riemann tensor in the new coordinate system?

(b) What are the Ricci tensor and scalar curvature in the new coordinate system?

5.7 Consider a flat, two-dimensional euclidian space and the point transformation from cartesian coordinates (x, y) to plane polar coordinates (r, ϕ) (see Fig. 2.5)

$$x = r \cos \phi \qquad ; \ y = r \sin \phi$$

The affine connection in cartesian coordinates vanishes by the argument in Prob. 5.4(a), while the affine connection in plane polar coordinates is given in Eqs. (3.84). Use these specific results to conclude that the affine connection Γ^i_{jk} does *not* transform as a third-rank tensor.

5.8 Cylindrical coordinates are a three-dimensional set (r, z, ϕ) where z is the height above the plane in Fig. 2.5. The line element in cylindrical coordinates in three-dimensional euclidian space can be written

$$ds = \hat{\mathbf{e}}_r \, dr + \hat{\mathbf{e}}_z \, dz + \hat{\mathbf{e}}_\phi \, rd\phi$$

Here $(\hat{\mathbf{e}}_r, \hat{\mathbf{e}}_z, \hat{\mathbf{e}}_\phi)$ form a set of orthonormal unit vectors. The surface of a right circular cylinder is then described with the coordinates (z, ϕ) and the condition $r = $ constant.

(a) Show the metric on the surface of a cylinder with radius r is

$$\underline{g} = \begin{pmatrix} g_{zz} & g_{z\phi} \\ g_{\phi z} & g_{\phi\phi} \end{pmatrix} = \begin{pmatrix} 1 & 0 \\ 0 & r^2 \end{pmatrix}$$

(b) Calculate the affine connection.

(c) Calculate the Riemann tensor, Ricci tensor, and scalar curvature.

(d) Could you have anticipated these results by considering the parallel transport of a vector around a closed loop with two sides along the axis of the cylinder and two sides along the circular circumference?

5.9 The metric in cylindrical coordinates (r, z, ϕ) follows directly from the line element in Prob. 5.8

$$\underline{g} = \begin{pmatrix} g_{rr} & g_{rz} & g_{r\phi} \\ g_{zr} & g_{zz} & g_{z\phi} \\ g_{\phi r} & g_{\phi z} & g_{\phi\phi} \end{pmatrix} = \begin{pmatrix} 1 & 0 & 0 \\ 0 & 1 & 0 \\ 0 & 0 & r^2 \end{pmatrix}$$

(a) What is the covariant divergence in cylindrical coordinates?
(b) What is the volume element?

5.10 Use polar coordinates in the $z = 0$ plane as generalized coordinates $(q^1, q^2) = (r, \phi)$. Consider the problem of a particle of mass m moving without friction, and with only the constraint force, on a surface of revolution $z = z(r)$ about the z-axis. Here $z(r)$ is assumed to be smooth and single-valued.

(a) Show the square of the line element on the surface is given by

$$(d\mathbf{s})^2 = \left\{ 1 + \left[\frac{dz(r)}{dr} \right]^2 \right\} (dr)^2 + (rd\phi)^2$$

(b) What is the lagrangian?
(c) What are Lagrange's equations?
(d) What is the metric?
(e) Sketch how the motion of the particle on the surface is reflected in the space of the generalized coordinates.[3]

5.11 In Prob. 5.10:
(a) What is the affine connection on the surface?
(b) What is the Riemann tensor on the surface?

5.12 The change in coordinate volume under the point transformation in Eqs. (3.26) is given by

$$d\xi^1 \cdots d\xi^n = \frac{\partial(\xi^1, \cdots, \xi^n)}{\partial(q^1, \cdots, q^n)} dq^1 \cdots dq^n$$

---

[3] *Note:* This problem is instructive as to the role of the metric in relating the coordinate motion to the actual physical motion.

where the jacobian determinant is defined by

$$\frac{\partial(\xi^1, \cdots, \xi^n)}{\partial(q^1, \cdots, q^n)} \equiv \begin{vmatrix} \partial\xi^1/\partial q^1 & \cdots & \partial\xi^1/\partial q^n \\ \vdots & & \vdots \\ \partial\xi^n/\partial q^1 & \cdots & \partial\xi^n/\partial q^n \end{vmatrix}$$

(a) Define a matrix $\left(\underline{A}^{-1}\right)_{ij} \equiv \bar{a}^i{}_j$. Show

$$d\xi^1 \cdots d\xi^n = \left(\det \underline{A}^{-1}\right) dq^1 \cdots dq^n$$

(b) Show the corresponding change in metric can be written as a matrix relation

$$\bar{g} = \left(\underline{A}\right)^T \underline{g}\, \underline{A}$$

Hence conclude the new determinant of the metric is

$$\det \bar{g} = \left(\det \underline{A}\right)^2 \det \underline{g}$$

(c) Show

$$\left(\det \underline{A}\right)\left(\det \underline{A}^{-1}\right) = 1$$

and hence conclude that the physical n-dimensional volume element in Eq. (5.152) is a *scalar* under coordinate transformations.[4]

6.1 Given cartesian coordinates $(x^1, \cdots, x^4) = (x^1, x^2, x^3, ct)$ and the Lorentz metric of Eq. (6.5), show the affine connection and Riemann curvature tensor both vanish. One says that such a Minkowski space is "flat."

6.2 Suppose one were to augment the lagrangian L in Eq. (6.36) with a static potential about the origin $V(\vec{x})$ so that

$$L = -mc^2 \left[1 - \frac{1}{c^2}\left(\frac{d\vec{x}}{dt}\right)^2\right]^{1/2} - V(\vec{x})$$

Use Lagrange's equations and the hamiltonian to determine how Eqs. (6.30) are modified. Discuss.

6.3 Consider a particle of energy $E_L^2 = (mc^2)^2 + (c\vec{p}_L)^2$ incident on a target of mass M in the laboratory frame. Let $\mathbf{p} = (\vec{p}_L, E_L/c)$ be the four-momentum of the incident particle in that frame and $\mathbf{P} = (\vec{0}, Mc)$ that of the target. Express the Lorentz invariant $s \equiv -(\mathbf{p} + \mathbf{P})^2$ in both the

---

[4]The more astute reader can probably convert this into a proof that, to within an overall constant, the n-dimensional volume element *must* have the form in Eq. (5.152).

laboratory and center-of-momentum (C-M) frames, and hence determine the total energy available in the C-M system in terms of E_L. The C-M frame is that one where the four-momenta are $\mathbf{p} = (\vec{p}, \ E_1/c)$ and $\mathbf{P} = (-\vec{p}, \ E_2/c)$, respectively.

6.4 (a) Construct a logical argument to show that if an observer in f sees the frame f' moving with velocity $\vec{v}$, then an observer in f' will see f moving with $-\vec{v}$.

(b) Use the Lorentz transformation of Eqs. (6.93), corresponding to the configuration in Fig. 6.10, to compute the motion of the origin O as viewed from the moving frame f'. Show $\bar{z} = -v\bar{t}$ and hence conclude that the frame f indeed appears to move with a velocity $-\vec{v}$ when viewed from the frame f'.

(c) Construct a logical argument to show that transverse spatial vectors should be unchanged under a Lorentz transformation.

6.5 It was Lorentz who first showed that the coordinate transformation from (z, t) to $(\bar{z}, \bar{t})$ in Eq. (6.93) leaves the electromagnetic wave operator invariant (same c)

$$\frac{\partial^2}{\partial z^2} - \frac{1}{c^2}\frac{\partial^2}{\partial t^2} = \frac{\partial^2}{\partial \bar{z}^2} - \frac{1}{c^2}\frac{\partial^2}{\partial \bar{t}^2}$$

Verify this.

It was Einstein who gave this result a physical interpretation as the actual transformation of these quantities between inertial frames, which revolutionized our understanding of space-time.

6.6 Consider all inertial frames whose origins coincide, but which are moving at different velocities $\vec{v}$. An event occurs in one frame at a space-time point $(\vec{x}, \ ct)$.

(a) Show that in all such frames, the event lies on a hyperboloid defined by $\vec{x}^2 - (ct)^2 = $ constant [see Fig. (14.1)].

(b) Show that all events connected by a light signal to the origin will give $\vec{x}^2 - (ct)^2 = 0$; they lie on the *light cone*. Hence conclude that all observers measure the same speed of light.

(c) Show that all events which stand in a causal relationship in one frame will preserve this relationship in all frames.

6.7 A light signal, and a neutrino with energy $E = 2\,\text{MeV}$ and rest mass $mc^2 = 2\,\text{eV}$, are emitted simultaneously from a supernova which is at a distance 10^4 light-years from earth.

(a) What is the difference in arrival times at the earth?

(b) How long does the trip take in the neutrino's rest frame?

(c) What is the distance to earth as viewed in the neutrino's rest frame?

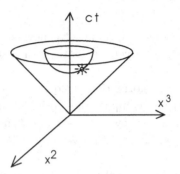

Fig. 14.1   Forward light-cone and event hyperboloid.

6.8 An electron of energy 50 GeV and rest mass $mc^2 = 0.5$ MeV travels down an accelerator pipe of length 1 km. How long does the pipe appear to be in the rest frame of the electron (in cm)?

6.9 Consider the local laboratory current of Eq. (6.113).

(a) Show that the number of particles in the spatial volume d^3x in the laboratory frame is given by $n(x)u^4(x)\, d^3x/c$. Then show that the particle flux through the surface $d\vec{S}$ is $n(x)\vec{u}\,(x) \cdot d\vec{S}$.

(b) The number of particles is conserved. Show this implies

$$\frac{1}{c}\frac{d}{dt}\int_V n(x)u^4(x)\, d^3x = -\oint_S n(x)\vec{u}\,(x) \cdot d\vec{S}$$

where V is a spatial laboratory volume and S is the surface surrounding V.

(c) Use Gauss' theorem and the fact that the relation in (b) must hold for any V to re-derive the laboratory continuity equation

$$\frac{\partial}{\partial x^\mu}\left[\frac{1}{c}n(x)u^\mu(x)\right] = 0$$

6.10 This problem retains the $O(1/c^2)$ corrections in taking the NRL of the relativistic expression for energy conservation in the fluid in Eq. (6.134), and leads to the result quoted in Eq. (6.138).

(a) Justify the following relation between the proper energy density ρc^2 and the laboratory internal energy density defined as $\rho\epsilon$ [compare

Prob. 6.9(a)]

$$\rho c^2 = \rho \epsilon \, (1 - \beta^2)^{1/2} = \rho \epsilon \left(1 - \frac{1}{2} \beta^2 + \cdots \right)$$

(b) Multiply Eq. (6.134) by c, and then expand in $1/c^2$ retaining all terms of overall $O(c^0)$ to obtain

$$\frac{\partial}{\partial t} \left(\rho c^2 + \rho v^2 \right) + \vec{\nabla} \cdot \left[\vec{v} \left(\rho c^2 + \rho v^2 + P \right) \right] = 0$$

or ; $\qquad \dfrac{\partial}{\partial t} \left(\rho \epsilon + \dfrac{1}{2} \rho v^2 \right) + \vec{\nabla} \cdot \left[\vec{v} \left(\rho \epsilon + \dfrac{1}{2} \rho v^2 + P \right) \right] = 0$

If Eq. (6.123) is expanded to the same order, multiplied by mc^2, and subtracted from this expression, the result is to replace $\rho \epsilon \to \rho \epsilon'$ where $\rho \epsilon' = (\rho \epsilon - mc^2 \, n_{\mathrm{lab}})$. The term $mc^2 \, n_{\mathrm{lab}}$ is the rest mass contribution to the internal energy density in the laboratory frame. This difference, although still formally of $O(c^2)$, now makes the term $\rho \epsilon'$ comparable to the other terms, and this is the way the result is generally employed in the NRL.

(c) Verify that this is the result given in [Fetter and Walecka (2003)] p. 300 (where the possibility of an additional source term $\rho \vec{f}_{\mathrm{app}} \cdot \vec{v}$ doing work on the system is also included). Verify also that Eq. (6.137) is the result given on p. 298 of that reference.

6.11 The energy-momentum tensor for a fluid without any shear forces is given by Eq. (6.108), where the four-velocity is obtained from Eq. (6.104). How is the stress tensor modified by the inclusion of shear forces? (*Hint:* see [Fetter and Walecka (2003)] chap. 12 for a discussion of viscosity, and [Weinberg (1972)] for the relativistic generalization.)

6.12 There is another representation of flat Minkowski space, which, while algebraically attractive, greatly obscures the transition from special to general relativity. In this approach, the basis vectors form an ordinary four-dimensional orthonormal set with $\mathbf{e}_\mu \cdot \mathbf{e}_\nu = \delta_{\mu\nu}$, where $\delta_{\mu\nu}$ is the usual Kronecker delta. The metric is then simply the identity matrix, and there is no distinction between up and down indices. The fourth coordinate now becomes *imaginary* so that $x_\mu = (x_1, \cdots, x_4)$ with $x_4 = ix_0 = ict$. The square of this four-vector is then $\mathbf{x}^2 = x_1^2 + x_2^2 + x_3^2 + x_4^2 = \vec{x}^2 - c^2 t^2$ and all of the effects of the negative fourth component in the metric are reproduced. In this approach, a Lorentz transformation is one that preserves the square of the four-vector.

(a) Show that a Lorentz transformation now forms an orthogonal matrix with $\underline{a}^T = \underline{a}^{-1}$.

(b) The Lorentz transformation corresponding to the configuration in Fig. 6.10 is now a rotation

$$\bar{x}_3 = x_3 \cos\psi + x_4 \sin\psi$$
$$\bar{x}_4 = -x_3 \sin\psi + x_4 \cos\psi$$

From the motion of O' as observed in the frame f, show $\tan\psi = iv/c$.

(c) Re-derive the results for $(\bar{z}, \bar{t})$ in Eqs. (6.93).

(d) Exhibit the new Lorentz transformation matrix $\underline{\bar{a}}$ corresponding to that in Eqs. (6.93).

(e) Rewrite part (b) as a relation between real components

$$\bar{x}_3 = x_3 \cosh\chi + x_4' \sinh\chi$$
$$\bar{x}_4' = x_3 \sinh\chi + x_4' \cosh\chi$$

where $x_4' = -ix_4$ and $\tanh\chi = -\beta$.

6.13 With the knowledge that the Lorentz transformation in Prob. 6.12(b) is a rotation, analyze the consequences of two successive Lorentz transformations. Show the velocities add according to

$$\beta_{\text{lab}} = \tanh\left[\tanh^{-1}(\beta_1) + \tanh^{-1}(\beta_2)\right]$$

and conclude that the laboratory velocity always satisfies $|\beta_{\text{lab}}| < 1$.

6.14 Consider the configuration in Fig. 6.10 where a first event marks the coincidence of the origins. A light signal is emitted at that first event and received later at a second event at (z, t), with $z/t = c$. The coordinates of the second event in the second frame are $(\bar{z}, \bar{t})$, which are related to (z, t) by the Lorentz transformation in Eq. (6.93).

(a) Express $\bar{z}$ in terms of z and note the relation between them;

(b) Express $\bar{t}$ in terms of t and note the relation between them;

(c) Show the speed of light given by $\bar{z}/\bar{t}$ in the second frame is again c.

6.15 This problem concerns the Doppler shift in special relativity. A light signal is sent out from a source at $\bar{O}$ moving with velocity V in the z-direction. In its rest frame the source undergoes dN oscillations in a proper time $d\tau$. All observes can agree on this number dN. The proper frequency in the rest frame of the source is

$$\bar{\nu} = \frac{dN}{d\tau} \qquad\qquad ; \, dN = \bar{\nu}\, d\tau$$

Now Lorentz transform to the laboratory frame where the source is moving with velocity V.

(a) Show the corresponding time interval dt in the lab is given by the relation

$$d\tau = \left(1 - \frac{V^2}{c^2}\right)^{1/2} dt$$

(b) During the time dt, the light front has traveled a distance $dl = cdt$ in the laboratory frame and hence arrives at an origin O a distance $dl = cdt$ away as shown in Fig. 14.2(a).

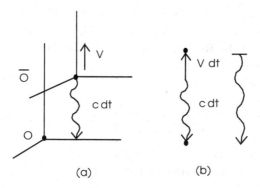

(a) (b)

Fig. 14.2   Configuration for calculation of the Doppler shift in special relativity.

In the time dt, the source has moved to a new position a distance $dl' = cdt + Vdt$ away from O. The wavelength of the radiation received at O is thus increased as indicated in Fig. 14.2(b). Since λ is the actual distance the signal has traveled divided by the number of oscillations, show that

$$\lambda = \frac{(c+V)dt}{dN} = \bar{\lambda}\frac{(1+V/c)}{(1-V^2/c^2)^{1/2}}$$

Here $\bar{\lambda} = c/\bar{\nu}$ is the proper wavelength of the radiation for a source at rest.

(c) Show that in the limit $V/c \to 0$, this reduces to the familiar freshman physics result. Note that $V = \pm|V|$ can have either sign in these arguments.

6.16 A more elegant derivation of Eq. (6.105) is obtained as follows.[5]

(a) Show that Eqs. (6.102) and (6.103) imply that the combination $a^\mu{}_i a^\nu{}_i$ should transform as a symmetric second-rank Lorentz tensor. Then write

---

[5]Courtesy of Paolo Amore.

this expression in terms of the only two such tensors available

$$a^\mu{}_i a^\nu{}_i = c_1 g^{\mu\nu} + c_2 u^\mu u^\nu$$

where (c_1, c_2) are Lorentz scalars.

(b) Use this relation to obtain two equations for (c_1, c_2), and hence rederive Eq. (6.105).

7.1 Given a four-dimensional riemannian space with an arbitrary metric. Take all the symmetry properties into account, and the first Bianchi identity, and determine the number of independent components of the Riemann curvature tensor R_{ijkl}.

7.2 Even though the Ricci tensor vanishes everywhere outside of a spherically symmetric source with the Schwarzschild solution, various individual components of the Riemann curvature tensor do not, and the space is curved. For example,

(a) Show that

$$R^4{}_{\theta 4\theta} = -\frac{rB'}{2AB} = -\frac{R_s}{2r}$$

(b) Find a small loop in a two-dimensional surface in the space where the enclosed surface manifests this curvature.

7.3 Make use of the results in Table 7.1 to write out the geodesic Eqs. (7.8) with coordinates $q^\mu = (r, \theta, \phi, ct)$ and a spherically symmetric source. Show that if one changes variable from proper time to time $(\tau \to t)$ the result is

$$\frac{1}{\rho}\frac{d}{dt}(\rho\dot{r}) + \left[\frac{A'}{2A}(\dot{r})^2 - \frac{r}{A}(\dot{\theta})^2 - \frac{r\sin^2\theta}{A}(\dot{\phi})^2 + \frac{B'}{2A}c^2\right] = 0$$

$$\frac{1}{\rho}\frac{d}{dt}(\rho\dot{\theta}) + \left[\frac{2}{r}(\dot{r}\dot{\theta}) - \sin\theta\cos\theta(\dot{\phi})^2\right] = 0$$

$$\frac{1}{\rho}\frac{d}{dt}(\rho\dot{\phi}) + \left[\frac{2}{r}(\dot{r}\dot{\phi}) + 2\frac{\cos\theta}{\sin\theta}(\dot{\theta}\dot{\phi})\right] = 0$$

$$\frac{1}{\rho}\frac{d}{dt}(\rho c) + \left[\frac{B'}{B}(c\dot{r})\right] = 0$$

Here the dot indicates a time derivative ($\dot{r} = dr/dt$, etc.), and ρ is given by

$$\rho = \left\{ B - \frac{1}{c^2}\left[(\dot{r})^2 A + (r\dot{\theta})^2 + (r\dot{\phi}\sin\theta)^2\right] \right\}^{-1/2} = \frac{dt}{d\tau}$$

Outside of a spherically symmetric source, the Schwarzschild solution for A and B is given in Eqs. (7.88).

7.4 (a) Take the newtonian limit of the equations in Prob. 7.3 by letting $c^2 \to \infty$.

(b) The lagrangian in spherical coordinates for non-relativistic motion outside a spherically symmetric gravitational source is

$$L = T - V = \frac{m}{2}\vec{v}^2 + \frac{mMG}{r}$$

$$= \frac{m}{2}\left[(\dot{r})^2 + (r\dot{\theta})^2 + (r\dot{\phi}\sin\theta)^2\right] + \frac{mMG}{r}$$

Write Lagrange's equations for (r, θ, ϕ) and compare with the results for the first three equations in part (a).

(c) What is the content of the fourth equation in (a)?

7.5 (a) Look for a solution to the equations in Prob. 7.3 with constant $(r = a, \theta = \pi/2, \dot{\phi} = \omega)$ corresponding to circular motion in the Schwarzschild metric in Eqs. (7.88). Show the angular frequency is given by[6]

$$\omega^2 = \frac{MG}{a^3}$$

Compare with the newtonian result.

(b) Show the solution with constant $(r = a, \dot{\theta} = \omega, \phi = \phi_0)$ gives the same result as in (a).

7.6 (a) Look for a solution to the equations in Prob. 7.3 with constant (θ, ϕ) corresponding to purely radial motion. Show the equation of motion for the radial coordinate can be reduced to the form

$$\ddot{r} + \dot{r}^2\left(\frac{A'}{2A} - \frac{B'}{B}\right) + \frac{B'}{2A}c^2 = 0$$

(b) Use the Schwarzschild solution of Eqs. (7.88), expand through first order in $1/c^2$, and show that the result in part (a) becomes

$$\ddot{r} + \frac{MG}{r^2}\left[1 - \frac{2MG}{c^2 r} - 3\frac{\dot{r}^2}{c^2}\right] = 0$$

7.7 (a) Convert the equations in Prob. 7.3 in the Schwarzschild metric to dimensionless form using R_s as the unit of length and R_s/c as the unit of time.

---

[6]The period is $p = 2\pi/\omega$.

(b) Write, or obtain, a program to solve the resulting four, coupled, second-order, nonlinear differential equations.

(c) What is the role of the fourth equation?

(d) Compute some representative particle trajectories. Compare with the newtonian limit.

7.8 The interior of a satellite orbiting outside of a spherically symmetric gravitational source also forms a local freely falling frame (LF^3) since the inertial centrifugal force just balances the gravitational attraction. Suppose one has a clock on a satellite which is in a circular orbit at a distance of $r = 10^8 R_s$ from the center of force, and which orbits it for 2000 days. What is the time difference (in sec) coming from general relativity between that clock and one at rest in the inertial laboratory frame very far away from the source. Work through $O(1/c^2)$.

7.9 A standard meter stick, very small on that scale, lies in the surface in Fig. 7.9 and is oriented in the radial direction. It is then slid inward toward the symmetry axis. The location of its two ends at a given instant in t are reported to the record keeper, who plots the two points in Fig. 7.10.

(a) What does the record keeper actually see?

(b) How is the record keeper able to keep track of what is happening physically?

8.1 This problem constructs the hamiltonian corresponding to the lagrangian in Eq. (8.22).[7]

(a) Show through $O(1/c^2)$ that

$$\dot{\theta} = \frac{p_\theta}{mr^2}\left(1 - \frac{v^2}{2c^2} + \frac{\Phi}{c^2}\right)$$

$$\dot{r} = \frac{p_r}{m}\left(1 - \frac{v^2}{2c^2} + \frac{\Phi}{c^2}\right) + \frac{2p_r}{m}\frac{\Phi}{c^2}$$

(b) Show through $O(1/c^2)$ that the hamiltonian $H = p_r\dot{r} + p_\theta\dot{\theta} - L$ is given by

$$H = T_0\left(1 - \frac{T_0}{2mc^2}\right) + V_0\left(1 + \frac{T_0}{mc^2} - \frac{V_0}{2mc^2} + \frac{1}{mc^2}\frac{p_r^2}{m}\right) + mc^2$$

Here

$$T_0 \equiv \frac{1}{2m}\left(p_r^2 + \frac{p_\theta^2}{r^2}\right) \qquad ; V_0 \equiv -\frac{mMG}{r}$$

----

[7] *Hint:* It is simplest to start with Eqs. (8.14) and (8.21).

(c) The lagrangian in Eq. (8.22) has no explicit dependence on time so that $\partial L/\partial t = 0$, and it is cyclic in θ. Use these facts to show that the result in (b) produces a first integral of the radial equation of motion in the form

$$H(p_r, r) = \text{constant}$$

8.2 The newtonian limit $c^2 \to \infty$ in Prob. 8.1 (NRL) produces the familiar statement of conservation of energy

$$T_0 + V_0 = \frac{1}{2m}\left(p_r^2 + \frac{l^2}{r^2}\right) - \frac{mMG}{r} = E = \text{constant}$$

where $l = \text{constant}$, and $E = H - mc^2$. Define the effective potential by

$$\frac{1}{2m}p_r^2 + V_{\text{eff}}(r) = E$$

(a) Show that for circular orbits of radius a at the minimum of the effective potential

$$\frac{l^2}{m^2 a^3} = \frac{MG}{a^2} \qquad ; E_0 = -\frac{mMG}{2a}$$

(b) Assume $\delta E = E - E_0$ and $\delta r = r - a$ are small and expand about the minimum of the effective potential. Show the phase space orbit is an ellipse given by

$$\frac{p_r^2}{A^2} + \frac{(\delta r)^2}{B^2} = 1$$

$$A^2 = 2m\,\delta E \qquad ; B^2 = \frac{2a^3}{mMG}\delta E$$

(c) Show the area of the ellipse in (b) is $\mathcal{A} = 2\pi\,\delta E/\dot{\theta}_0$ where $\dot{\theta}_0$ is defined in Eq. (8.37).

8.3 Consider the radial Eq. (8.32):
(a) Show that purely radial motion is characterized by $l = 0$;
(b) Show the result obtained with $l = 0$ reproduces that in Prob. 7.6(b).

8.4 The reconciliation of the angular geodesic equations in Prob. 7.3 with the conservation of the angular momentum l of Eq. (8.25) in the lagrangian approach is, in fact, quite subtle. This problem leads the reader through that reconciliation. As in the text, assume a constant ϕ so that $\dot{\phi} = 0$.

(a) Show that the geodesic equation for θ in Prob. 7.3 can be written in the form

$$\frac{d}{dt}(r^2\dot{\theta}) = \frac{2MG}{c^2}\dot{r}\dot{\theta} + O\left(\frac{1}{c^4}\right)$$

(b) Show that the conservation of the angular momentum l in Eq. (8.25) can be recast in the form

$$\frac{d}{dt}(r^2\dot{\theta}) = \frac{2MG}{c^2}\dot{r}\dot{\theta} - \left(\frac{E}{mc^2} + \frac{2MG}{rc^2}\right)\frac{d}{dt}(r^2\dot{\theta}) + O\left(\frac{1}{c^4}\right)$$

where energy conservation in the NRL, that is through $O(c^0)$, has been invoked in the evaluation of the coefficient of the $O(1/c^2)$ term

$$E = \frac{m}{2}(\dot{r}^2 + r^2\dot{\theta}^2) - \frac{mMG}{r} = \text{constant} \qquad ; \text{NRL}$$

(c) Now iterate the result in (b) through $O(1/c^2)$ to reproduce the result in (a).

8.5 Use the following numbers for the earth

$$M_e = 5.98 \times 10^{24}\,\text{kg}$$

$$R_e = 6.38 \times 10^3\,\text{km}$$

Show that the GR contribution to the advance of the perihelion of a satellite in an orbit with very small eccentricity at the surface of the earth is

$$\Omega_{\text{perihelion}} = 16.8''/\text{year} \qquad ; \text{surface of earth}$$

8.6 Consider the *deflection of light in the Schwarzschild metric* (Probs. 8.6–8.7). Work in the plane $\theta = \pi/2$. We are interested in the configuration illustrated in Fig. 14.3, where the light ray moves through the gravitational field in an essentially undeviated straight-line path at an impact parameter b. Define $\Lambda \equiv MG/c^2b$ and assume $\Lambda \ll 1$.

(a) Start in the LF^3, and show that the interval vanishes for a light signal. Hence, conclude that light follows the null interval $(ds)^2 = 0$ in the global laboratory frame where the Schwarzschild metric applies. Now work in this frame;

(b) Consider first the case where $\Lambda = 0$. Show the vanishing of the interval implies that $v^2/c^2 = 1$ for the light signal (here $v = \dot{x}$);

(c) Consider then the case of small Λ. Show that the vanishing of the interval $(ds)^2 = 0$ implies the existence of an *effective index of refraction*

$(v/c)_{\text{eff}} \equiv 1/n$ for light in the gravitational field given by[8]

$$n = 1 + \Lambda \left(\frac{1}{r/b} \right) \left(1 + \frac{\dot{r}^2}{c^2} \right) \qquad ; \text{ effective index of refraction}$$

Thus the deflection of light amounts to tracing a ray through a "medium" with this varying effective index of refraction.

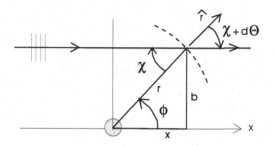

Fig. 14.3   Configuration for deflection of a light ray from an essentially straight-line path at an impact parameter b through the Schwarzschild metric. There is a small change $d\Theta$ in angle of the incident ray relative to the normal $\hat{r}$ due to the change dn in effective index of refraction. The trajectory extends to infinity in both directions.

8.7 Since $r \approx x$ for most of the trajectory in Fig. 14.3, with $\dot{r}^2/c^2 \approx \dot{x}^2/c^2 \approx 1$, assume the contribution of the second term in the final parentheses in Prob. 8.6(c) is approximately equal to that of the first. More generally, write its contribution as γ times that of the first (with $\gamma \approx 1$). The effective index of refraction then takes the approximate form

$$n \approx 1 + 2\Lambda \left(\frac{1}{r/b} \right) \left(\frac{1+\gamma}{2} \right) \qquad ; \text{ effective index of refraction}$$

The advantage of making this approximation is that one now knows that the surfaces of constant n are spheres with the normal to the surface given by $\hat{r}$. The problem is then reduced to an application of *Snell's law* from freshman physics. Snell's law states that at the interface between two media with different indices of refraction, a light ray is refracted according to the relation $n_i \sin \chi_i = n_r \sin \chi_r$, where the angles are measured relative to the normal to the surface.

---

[8]Here $v_{\text{eff}}^2 = \dot{r}^2 + r^2\dot{\theta}^2 + r^2 \sin^2\theta\, \dot{\phi}^2$ is the square of the velocity that the record keeper observes on his or her screen.

(a) Show the differential form of Snell's law as $n \to n + dn$ and $\chi \to \chi + d\Theta$ is (see Fig. 14.3)[9]

$$d\Theta = -\frac{dn}{n} \tan \chi$$

(b) With the observation that $r^2 = x^2 + b^2$, and the definition $u \equiv x^2/b^2$, show that the relation in (a) becomes

$$d\Theta = -\left[\frac{1}{n}\frac{dn}{du} \tan \chi\right] du \qquad\qquad ; u \equiv \frac{x^2}{b^2}$$

$$= \Lambda \frac{(1+\gamma)}{2} \left[\frac{1}{(1+u)^{3/2}} \tan \chi\right] du + O(\Lambda^2)$$

(c) Show from Fig. 14.3 that

$$\tan \chi = \tan \phi = \frac{b}{x} = \frac{1}{\sqrt{u}}$$

(d) Now integrate the relation in (b) from 0 to ∞ to get the net change Θ_0 from the outgoing leg of the trajectory

$$\Theta_0 = 2\Lambda \left(\frac{1+\gamma}{2}\right)$$

(e) Show there is an equivalent deflection from the incoming leg, and hence obtain the total deflection of the light ray through $O(\Lambda)$

$$\Theta_{\text{deflect}} = 2\Theta_0 = 4\Lambda \left(\frac{1+\gamma}{2}\right) \qquad\qquad ; \Lambda \equiv \frac{MG}{c^2 b}$$

This is the result given in [Weinberg (1972)] for the deflection of light in the Schwarzchild metric through $O(\Lambda)$, where it is stated that $\gamma = 1$ is actually an exact result.[10]

8.8 This problem reviews some results for orbital motion in the newtonian limit. Start from $L = L_{\text{NRL}}$ in Eq. (8.15) and use the spherical unit vectors defined in Fig. 5.4.

---

[9]When entering a medium with higher effective index of refraction, the ray is bent *toward* the normal; when entering one with lower effective index of refraction, it is bent *away* from the normal.

[10]For some accurate analytical formulas for gravitational lensing that follow from the result derived in [Weinberg (1972)], see [Amore and Arceo (2006)].

(a) Show the canonical angular momenta are given by

$$p_\phi = \frac{\partial L}{\partial \dot{\phi}} = mr^2 \dot{\phi} \sin^2 \theta \qquad ; \; \dot{p}_\phi = 0$$

$$p_\theta = \frac{\partial L}{\partial \dot{\theta}} = mr^2 \dot{\theta} \qquad ; \; \dot{p}_\theta = p_\phi \dot{\phi} \cot \theta$$

(b) Define the angular momentum in this limit by $\vec{l} = \vec{r} \times (m\vec{v})$ where $\vec{v} = \dot{r}\,\hat{\mathbf{e}}_r + r\dot{\theta}\,\hat{\mathbf{e}}_\theta + r\dot{\phi}\sin\theta\,\hat{\mathbf{e}}_\phi$. Show

$$\hat{\mathbf{e}}_\phi \cdot \vec{l} = p_\theta \qquad ; \; \hat{\mathbf{e}}_z \cdot \vec{l} = p_\phi$$

where $\hat{\mathbf{e}}_z$ is a unit vector in the z-direction. Hence verify the claims made in Figs. 8.3 and 8.4.

8.9 (a) Expand the lagrangian in Eq. (8.9) as a power series in $2\Phi/c^2$ for *any* $v^2 = \dot{r}^2 + r^2\dot{\theta}^2 + r^2\sin^2\theta\,\dot{\phi}^2$ with $v^2/c^2 < 1$. Show that through first order, L takes the form

$$L = -mc^2(1 - \beta^2)^{1/2} - m\Phi(r)\frac{1 + \dot{r}^2/c^2}{(1 - \beta^2)^{1/2}} \qquad ; \; \vec{\beta} = \vec{v}/c$$

This provides one consistent extension of the free relativistic lagrangian in Eq. (6.36), which includes an interaction.

(b) To simplify things, neglect the $\dot{r}^2/c^2$ in the numerator of the second term (this does preserve the NRL). Then write $\vec{\beta} = d\vec{x}/d(ct)$ and use cartesian coordinates. What is the contribution of the interaction to the relativistic form of Newton's second law in the first of Eqs. (6.30)?

(c) Compare with the result in Prob. 6.2.

8.10 Suppose the lagrangian L for the motion of a point particle of mass m in the Schwarzschild metric in Eq. (8.9) is augmented with a potential $V(r)$ as in Prob. 6.2. Expand through $O(1/c^2)$. How are the subsequent equations of motion in the text modified?

9.1 (a) Show through $O(1/c^2)$ that the shift in frequency of a photon traveling up a small distance h at the earth's surface can be written as

$$\frac{\Delta\nu}{\nu} = -\frac{gh}{c^2}$$

where g is the acceleration due to gravity.

(b) Take $g = 9.807\,\text{m/sec}^2$ and assume $h = 10\,\text{m}$. Compute the fractional shift in frequency of a spectral line.

9.2 The lifetime of a free muon is $\tau = 2.197 \times 10^{-6}$ sec. How close to the Schwarzschild radius, $r/R_s = 1 + \epsilon$, must a muon be to have its laboratory lifetime extended to 1 sec?

9.3 A few spectral lines thought to come from the surface of a distant body are observed to have a gravitational redshift of $\Delta\nu/\nu \approx -10^{-6}$. Calculate the ratio of mass to radius for that body.

9.4 Problems 9.4–9.8 examine the radial propagation of light in the Schwarzschild metric.[11] Consider a *classical electromagnetic wave*. We first make some observations concerning the wavelength of this wave in the gravitational potential.

(a) From the fact that light follows the null interval $(d\mathbf{s})^2 = 0$ in the LF^3 [frame III], conclude that for propagation in the radial direction, the light signal must satisfy the following differential relation in the global inertial laboratory frame I

$$dr = \left(1 + \frac{2\Phi}{c^2}\right) c\, dt$$

(b) Define dN as a given number of oscillations of the electric field at the origin in the LF^3 at a given point. Further define the proper time element $d\tau$ and proper frequency $\bar\nu = dN/d\tau$ in that frame. Let $d\bar l$ be an element of proper radial distance traversed by the light signal in the time $d\tau$ in the LF^3. Substitute the relations in Eqs. (7.109) and (7.113) between laboratory quantities and proper quantities into the expression in (a), and verify the following

$$d\bar l = c\, d\tau$$
$$\bar\lambda = \frac{d\bar l}{dN} = \frac{c}{\bar\nu}$$

Hence, one indeed recovers the correct relation between wavelength and frequency in the LF^3 from (a).

(c) In direct analogy to the analysis in (b), define the wavelength for a light signal propagating in the radial direction in the Schwarzschild metric in the global inertial laboratory frame I by $\lambda \equiv dr/dN$. Show from (a) that

$$\lambda = \frac{dr}{dN} = \bar\lambda \left(1 + \frac{2\Phi}{c^2}\right)^{1/2}$$

---

[11] In these problems, λ represents the wavelength.

9.5 In exactly the same manner as in Prob. 9.4(c), derive the following relations for the frequency defined by $\nu \equiv dN/dt$

$$\nu\lambda = c\left(1 + \frac{2\Phi}{c^2}\right)$$

$$\nu = \bar{\nu}\left(1 + \frac{2\Phi}{c^2}\right)^{1/2}$$

Now note that the fractional change in this frequency when the gravitational potential is changed by $\Delta\Phi$ is given through $O(1/c^2)$ by

$$\frac{\Delta\nu}{\nu} = \frac{\Delta\Phi}{c^2} \qquad ; \text{oscillator (1)} \to \text{(2)}$$

Note the sign. This expression is identical to that derived in the text for the frequency of an oscillator in the gravitational field. The difficulty in attempting to apply the analysis in this problem (and in the preceding one) to the *propagation* of a light signal in the Schwarzschild metric is that it *does not fully take into account the effect of the gravitational potential on the frequency of the photon.* This is addressed in the text, and the correct answer for the frequency shift through $O(1/c^2)$ is that given in Eq. (9.21).

9.6 Let us examine what the *record keeper sees* with a radially propagating light signal in the Schwarzschild metric.

(a) Start in the LF^3 where the velocity of light is c, and demonstrate that the light signal follows the null interval $(ds)^2 = 0$ in the global inertial laboratory frame I in the Schwarzschild metric.

(b) The record keeper logs the coordinates (r, ct) for the light signal. Show that what he or she observes for all $r \geq R_s$ is that

$$\frac{dr}{d(ct)} = \left(1 + \frac{2\Phi}{c^2}\right)$$

(c) Note that $\dot{r} \leq c$. What does the record keeper observe for $\dot{r}$ at the Schwarzschild radius?

9.7 Problems 9.4 and 9.5 *do* provide the wavelength and frequency for a classical electromagnetic wave at a given point in the global inertial laboratory frame I in the Schwarzschild metric in terms of the corresponding quantities in the LF^3.

(a) Show

$$\nu\lambda = \dot{r}$$

where $\dot{r}$ is the radial velocity observed by the record keeper in Prob. 9.6.

Discuss.

(b) Show that when combined with the correct expression in Eq. (9.21) for $\Delta\nu/\nu$ for a propagating light signal, this gives the corresponding wavelength shift as

$$\frac{\Delta\lambda}{\lambda} = 3\,\frac{\Delta\Phi}{c^2} \qquad ;\text{ wavelength } (1) \to (2)$$

Again, to this order, one can use either λ_1 or λ_2 in the denominator on the l.h.s.[12] As the light wave moves to a region of higher potential $\Delta\Phi > 0$, the wavelength increases $\Delta\lambda > 0$. This is the *gravitational redshift*.

9.8 Suppose there were to be an (uncalculated) energy shift $\Delta\varepsilon/\varepsilon = O(1/c^2)$ of the atom or nucleus between points (1) and (2), or between the lab and LF[3], in Fig. 9.3. Show the results in Eqs. (9.20) and (9.21) still hold through the stated order.

10.1 Work in the newtonian limit, and assume an equation of state $P(\rho)$.

(a) Use Gauss' law to obtained the gravitational field as a function of r inside a spherically symmetric mass distribution;

(b) Balance the pressure force and gravitational attraction on a small mass element to obtain the newtonian expression for stellar structure in Eq. (10.60).

10.2 Given the fact that the mass density $\rho(r)$ is a positive quantity which decreases with r, and that $\rho(0)$ is finite, show that the inequality in Eq. (10.71) implies that the denominator in the integral in Eq. (10.68) will not vanish.

10.3 (a) Write, or obtain, a program to solve the TOV Eqs. (10.76) numerically.

(b) Choose the "stiff" equation of state $P = \rho c^2$. Find (R, M) for a few neutron stars.[13]

(c) What is the maximum mass of a neutron star in units of $M_\odot$ with this equation of state?

10.4 (a) Derive the parametric equation of state of an ideal non-relativistic degenerate Fermi gas as a limiting form of Eqs. (10.98) and

---

[12]Or even $\bar{\lambda}$.

[13]This equation of state is as stiff as possible, while still consistent with causal propagation of sound signals — see text.

$(10.102)^{14}$

$$\rho c^2 = m_b c^2 \rho_B + \frac{3}{5}\frac{\hbar^2}{2m_b}\left(\frac{6\pi^2}{\gamma}\right)^{2/3}\rho_B^{5/3}$$

$$P = \frac{2}{3}(\rho c^2 - m_b c^2 \rho_B)$$

(b) Repeat Prob. 10.3 with this equation of state for neutron matter.

(c) At what densities does this non-relativistic calculation become inconsistent?

10.5 Repeat Prob. 10.3 with the equation of state of an ideal ultra-relativistic gas[15]

$$P = \frac{1}{3}\rho c^2$$

Compare with the results in Prob. 10.3.

10.6 Use the first law of thermodynamics in Eq. (10.101) to derive the RMFT expression for the pressure in Eq. (10.102).

10.7 Show that with the equation of state $P = \rho c^2$, the thermodynamic speed of sound is equal to the speed of light. (*Hint:* see chap. 49 of [Fetter and Walecka (2003)]; note that there $c^2 \equiv c_{\text{sound}}^2$.)

10.8 Consider the (p, e^-) component present at equilibrium in a cold, neutral neutron star in RMFT. Assume massless electrons.

(a) Show the Fermi wavenumber of the protons is related to that of the neutrons by

$$k_{\text{Fp}} + \left(k_{\text{Fp}}^2 + M_b^{\star 2}\right)^{1/2} = \left(k_{\text{Fn}}^2 + M_b^{\star 2}\right)^{1/2}$$

(b) What is the ratio of proton to neutron densities as $M_b^\star \to \infty$? As $M_b^\star \to 0$?

(c) Estimate from Figs. 10.10 and 10.11 the ratio of proton to neutron densities actually present in the neutron star.

10.9 Problems 10.9-10.12 concern the *thermodynamics* of black holes. We make no attempt to derive these relations from first principles. They are

---

[14]See chap. 2 of [Fetter and Walecka (2003a)] regarding the equations of state used in Probs. 10.4 and 10.5; recall that ρc^2 is the proper energy density.

[15]This is the equation of state of a high-pressure, asymptotically-free, collection of massless quarks and gluons (see [Walecka (2004)]). More exotic phases of this system may exist at high baryon density — it may become a "color superconductor" (see [Alford, Bowers, and Rajagopal (2001)]).

included only so that the reader has some acquaintance with these concepts when approaching the literature.[16]

A black hole can be assigned a *temperature*

$$k_B T = \frac{\hbar}{2\pi c} g_s$$

Here $k_B = 1.381 \times 10^{-16}\,\mathrm{erg/^\circ K}$ is Boltzmann's constant, and g_s is the acceleration of gravity at the Schwarzschild radius

$$g_s = \frac{MG}{R_s^2}$$

(a) Show that the relation for the temperature reduces to[17]

$$k_B T = \frac{\hbar c^3}{8\pi G M}$$

Note the $\hbar$ in this relation — it is a quantum effect. Note also the characteristic energy $(\hbar c^3/GM)$.

(b) Use the numbers in Eqs. (10.87), and $\hbar = h/2\pi = 1.055 \times 10^{-27}$ erg-sec, to compute the temperature of a black hole with mass equal to the solar mass $M = M_\odot$.

10.10 The Stefan-Boltzmann law gives the energy flux of electromagnetic radiation at the surface of a *black body* at a temperature T as [Ohanian (1995)]

$$\mathcal{S} = \sigma T^4 \qquad ; \text{ Stefan-Boltzmann law}$$
$$\sigma = \frac{\pi^2 k_B^4}{60 \hbar^3 c^2}$$

Note that the Stefan-Boltzmann constant σ does not depend on the emission or absorption *mechanism*, but only on $(k_B, \hbar, c)$.

A black hole loses energy by *Hawking radiation*—essentially pair production in the strong gravitational gradient at the Schwarzschild radius, where one member of the pair escapes.

(a) To estimate the lifetime of a black hole against "evaporation," assume that in the global inertial laboratory frame I the rate of loss of mass

---

[16]See, for example, [Ohanian and Ruffini (1994)], [Taylor and Wheeler (2000)], [Weinberg (1972)].

[17]This, and many other topics, are conveniently referenced on the internet at [Wiki (2006)]. See also [Batiz (2000)].

due to Hawking radiation is given by

$$\frac{d}{dt}(Mc^2) = -(4\pi R_s^2)\mathcal{S}$$

Integrate this relation to find the time τ for disappearance of the black hole

$$\tau = \frac{1}{3}\frac{Mc^2}{(4\pi R_s^2)\mathcal{S}}$$

where the quantities of the r.h.s. are initial values.

(b) Rewrite this last result as

$$\tau = 5120\pi \left(\frac{G^2 M^3}{\hbar c^4}\right)$$

Note the dependence on the characteristic time $(G^2 M^3/\hbar c^4)$.[18]

(c) Evaluate τ (in years) for a black hole of one solar mass, $M = M_\odot$.

(d) Evaluate τ (in years) for a black hole of mass $M = 1\,\mathrm{kg}$.

(e) Repeat for $M = m_p$, where $m_p = 1.673 \times 10^{-24}$ gm is the mass of the proton.

10.11 An *entropy* can be associated with a black hole according to[19]

$$TS = \frac{1}{2}Mc^2$$

where T is the temperature introduced in Prob. 10.9(a).

(a) Show this relation can be rewritten as

$$\frac{S}{k_{\mathrm{B}}} = \frac{4\pi M^2 G}{\hbar c} = \frac{c^3}{4\hbar G}A_s$$

where A_s is the *area* of a sphere with the Schwarzschild radius

$$A_s = 4\pi R_s^2 \qquad ; \text{area}$$

(b) Discuss how the second law of thermodynamics might now be applied to black holes.

---

[18]The *Hawking lifetime* of a black hole is indeed given by $\tau_{\mathrm{H}} = 5120\pi G^2 M^3/\hbar c^4$ [Wiki (2006)].

[19]Recall that the Helmholtz free energy of a system is $F = E - TS$, so the product TS should scale like the energy, which in the case of a black hole is Mc^2. We leave it to the dedicated reader to supply an explanation for the factor of 1/2.

10.12 Use the *equivalence principle* to generalize the result in Prob. 10.9 and associate a temperature (and acompanying Planck distribution of photons) with a particle undergoing an acceleration a_s according to

$$k_{\mathrm{B}}T = \frac{\hbar}{2\pi c}a_s$$

It has been proposed that this will be an additional source of "Unruh" radiation for strongly accelerated electrons [Unruh (1976)], [Chen and Tajima (1999)].

11.1 Show that for the cosmology considered in this chapter, $\Lambda(t)$ and $\rho(t)$ develop in time according to

$$\Lambda^3(t) \propto (t - t_0)^2 \qquad\qquad ;\ \rho(t) \propto (t - t_0)^{-2}$$

Hence conclude that

$$\rho(t)\,\Lambda^3(t) = \text{constant}$$

Why might one have anticipated this result?

11.2 Do not assume $P \ll \rho c^2$, and retain the pressure P in the source terms in the Einstein field equations for the cosmology studied in this chapter.

(a) Show that Eqs. (11.27) are modified as follows

$$-\frac{3}{c^2}\frac{\ddot\Lambda}{\Lambda} = \frac{\kappa}{2}(\rho c^2 + 3P)$$

$$\frac{1}{c^2}\left[\frac{\ddot\Lambda}{\Lambda} + 2\left(\frac{\dot\Lambda}{\Lambda}\right)^2\right] = \frac{\kappa}{2}(\rho c^2 - P)$$

(b) Show that Eq. (11.34) becomes

$$h^2 = \left(\frac{\dot\Lambda}{\Lambda}\right)^2 = \frac{1}{3}\kappa\rho c^4$$

Thus this relation and its consequences are *independent* of P. What are those consequences?

11.3 The LF^3 in this cosmology is an unchanging spatially global frame in which there is neither gravity nor inertial forces. In the LF^3, one has only the laws of special relativity. The time t in the LF^3 is identical to that in the global inertial laboratory frame I, and at any instant in time, the LF^3 is tangent to the laboratory space. The *proper distance* is the distance at any instant in the LF^3. The coordinate transformation between the frames is given by Eq. (11.53).

(a) Show that if the physical spatial distance in the global laboratory frame at any instant is l, then l is also the proper distance;

(b) Show that if a light signal is emitted at a time t from a star that is a proper distance l away from the origin O, then the time it takes to arrive at the origin in the LF^3 is $\Delta t = l/c$;

(c) Show that if the star is at a physical distance $l + \Delta l$ in the global inertial laboratory frame when the light signal arrives at the origin at the time $t + \Delta t$, then $l + \Delta l$ is the proper distance at that time.

11.4 One can define the proper distance as

$$(d\mathbf{l})^2 \equiv (d\mathbf{s})^2 + (cdt)^2 \quad \text{; proper distance}$$
$$\equiv (dl)^2$$

(a) Show this quantity is *invariant* under transformations between the global inertial laboratory frame I and the LF^3.

(b) Show that at a given instant in time, when the global inertial laboratory frame and the global LF^3 coincide, this becomes the ordinary spatial separation

$$(d\mathbf{l})^2 = (d\vec{l})^2 \quad \text{; given instant}$$

(c) For a propagating light signal the interval vanishes $(d\mathbf{s})^2 = 0$. Show that the proper distance dl traveled by the light is related to the time interval dt by

$$dl = cdt \quad \text{; light signal}$$

(d) Now imagine that the LF^3 has its origin at the observer and that at the instant a star emits a light signal, the star is a proper distance l away from the origin. Show the time Δt that the light signal takes to reach the origin is obtained from the integration of the result is (c) as

$$\Delta t = \frac{l}{c} \quad \text{; time to go proper distance } l$$

11.5 (a) Let t_e be the time of emission of the radiation and the present time $t_p = t_e + \Delta t$ that of observation. Show Eq. (11.68) is equivalent to the following expression for the redshift in the present cosmology

$$\frac{\lambda(t_p, t_e)}{\lambda_0} = \left(\frac{t_p - t_0}{t_e - t_0}\right)^{2/3} = \left(\frac{t_p - t_0}{t_p - \Delta t - t_0}\right)^{2/3}$$

$$\Delta t = \frac{l}{c} = t_p - t_e$$

Here l is the physical spatial distance of the source at the time of emission.

(b) Hence conclude that there is an infinite redshift for light that comes from the horizon where $t_e \to t_0$ and $l \to D_H$.

11.6 An object on the horizon emitted light that gets to us now, and when the light was emitted the object was a physical spatial distance D_H away. What is the physical distance of the object now?

(a) Let t_e be the time of emission, and t_p the present time. Make use of Eqs. (11.60) and (11.67) to show that the physical spatial distance of an object at a fixed coordinate point (q^1, q^2) satisfies

$$\frac{l(t_p)}{l(t_e)} = \left(\frac{t_p - t_0}{t_e - t_0}\right)^{2/3}$$

(b) Consider an object a small distance εc inside the horizon so that $l(t_e) = D_H - \varepsilon c$. If the light gets to us at the present time, then the light was emitted at a time $t_e = t_0 + \varepsilon$. Show from part (a) that these quantities are related by

$$\frac{l(t_p)}{D_H - \varepsilon c} = \left(\frac{t_p - t_0}{\varepsilon}\right)^{2/3}$$

$$\text{or ;} \qquad l(t_p) \to D_H \left(\frac{t_p - t_0}{\varepsilon}\right)^{2/3} \qquad ; \text{ as } \varepsilon \to 0$$

Hence conclude that an object on the horizon will be an *infinite distance away* at the time the light gets to us.[20]

11.7 The measured value of the Hubble constant in Eq. (11.83) corresponds to what value of the mass density ρ for the cosmology studied in this chapter?

11.8 There exists a cosmic microwave background (CMB) that fills all of space. It remains from the early hot era of the universe after the radiation decoupled and adiabatically expanded with time. The CMB exhibits an

---

[20]This correlates with the infinite redshift of the light from the horizon found in Prob. 11.5(b).

almost perfect *black-body spectrum* with an energy density

$$d\varepsilon_\gamma = \frac{8\pi\nu^2\, d\nu}{c^3}\frac{h\nu}{e^{h\nu/k_B T} - 1}$$

and a current temperature $T(t_p) = 2.73\,^\circ\text{K}$ [Ohanian (1995)].

(a) Show that when integrated over frequencies, this gives rise to a radiation density of

$$\varepsilon_\gamma = \frac{4\sigma}{c}T^4$$

Here σ is the Stefan-Boltzmann constant of Prob. 10.10, with the value $\sigma = 5.670 \times 10^{-5}\,\text{erg/sec-cm}^2\text{-}^\circ\text{K}^4$.

(b) Define an effective radiation mass density by $\varepsilon_\gamma \equiv \rho_\gamma c^2$ and compute ρ_γ. Compare with the value of the matter (baryon) ρ appearing in Eqs. (11.48) and Prob. 11.7, and show that ρ_γ is a negligible source of gravity in the Einstein field equations.

11.9 One can define a *photon number density* spectrum dn_γ by dividing the energy density spectrum $d\varepsilon_\gamma$ of Prob. 11.8 by the energy/photon of $h\nu$. Thus

$$dn_\gamma = \frac{8\pi\nu^2\, d\nu}{c^3}\frac{1}{e^{h\nu/k_B T} - 1}$$

(a) Integrate this expression and show the photon number density is then given by

$$n_\gamma = \left[\frac{2.404\, k_B^3}{\pi^2(\hbar c)^3}\right]T^3$$

Note that this quantity goes as T^3.

The mass density in the present cosmology comes from baryons (protons) and can be written $\rho = m_b n_b$ where n_b is the baryon density. After "decoupling" in the era where the medium is so diffuse that photon reactions are no longer important, one can imagine that the mixture maintains a constant ratio of n_γ/n_b and merely adiabatically expands as the space stretches.

(b) Combine the result in part (a) with that in Prob. 11.1 to conclude that the time development of the temperature $T(t)$ and the scale factor in the metric $\Lambda(t)$ are thus related by

$$T^3(t)\Lambda^3(t) = \text{constant}$$

(c) Given the present value of $t_p - t_0$ in Eqs. (11.48), and $T(t_p)$ from Prob. 11.8, at what time in the past was $k_B T$ at 1 MeV? (Note $k_B =$ $8.620 \times 10^{-5} \, \text{eV}/^\circ\text{K}$).[21] What was the corresponding mass density ρ at that time?

12.1 (a) Expand the coordinate transformation matrix for the gravitational wave in Eq. (12.66) through first order in γ_{ij} [recall Eq. (12.62)], and then insert the time dependence of the metric in Eq. (12.67).

(b) Now explicitly carry out the transformation from the coordinates (x, y, z, ct) in the global inertial laboratory frame to the coordinates $(\bar{x}, \bar{y}, \bar{z}, c\bar{t})$ in the LF^3. Discuss.

12.2 Consider the lagrangian $L_{\text{NLR}}(\dot{x}, \dot{y}, \dot{z}; x, y, z; t)$ of Eq. (12.73) where the perturbation of the metric is given in Eq. (12.67).

(a) What are the consequences of the fact that the coordinates (x, y) are cyclic?

(b) What are the consequences of the fact that this lagrangian has an *explicit* dependence on the time?

12.3 Write, or obtain, a program to solve numerically the three coupled non-linear differential Eqs. (12.74) for particle motion in the presence of a gravitational wave. Use dimensionless variables $(kx, ky, kz, \omega t)$ with $\omega = kc$, and assume small values for the constants (h_{xx}, h_{xy}). Investigate the subsequent motion for various initial conditions.

12.4 The physical separation l of two particles a fixed coordinate distance d from each other on the x-axis, for a gravitational plane wave moving in the z-direction, is given by Eq. (12.78) as

$$l = d + \frac{dh_{xx}}{2} \cos k(z - ct)$$

(a) Use $\mu \ddot{l}$ to show that the equivalent gravitational interparticle *force* on the pair f_g is given by

$$f_g = -\mu \omega^2 \frac{dh_{xx}}{2} \cos k(z - ct)$$

where μ is the reduced mass.

(b) Suppose the two particles in (a) are connected by a spring of equilibrium length l_0 and spring constant $k = \mu \omega_0^2$. Assume $z = 0$ for simplicity,

---

[21] The "freezing out" of various reactions with falling temperature plays a central role in the study of element formation in the early universe (see, for example, [Kolb and Turner (1990)]).

and set $d \approx l_0$ in the driving term. Write Newton's second law for the relative motion of the pair, and show that the general solution to that equation is given by

$$\zeta(t) = A\cos\omega_0 t + B\sin\omega_0 t + \frac{1}{2}\frac{l_0 h_{xx}\omega^2}{(\omega^2 - \omega_0^2)}\cos\omega t \qquad ; \zeta \equiv l - l_0$$

Here (A, B) are constants.

(c) What is it that prevents the resonant contribution from actually becoming infinite?

12.5 (a) Consider the *source* of the gravitational radiation. Suppose one has a given, localized energy-momentum tensor $T_{\mu\nu}(\vec{x}, t)$. Show that if the source is weak enough, Einstein's linearized field equations for gravitational radiation can be written *everywhere* as

$$\Box \gamma_{\mu\nu}(\vec{x}, t) = -\frac{16\pi G}{c^4} S_{\mu\nu}(\vec{x}, t)$$

where the source tensor $S_{\mu\nu}$ is given by

$$S_{\mu\nu} = T_{\mu\nu} - \frac{1}{2}T g_{\mu\nu}^0 \qquad ; T = T^\lambda_{\ \lambda}$$

What does "weak enough" mean?

(b) Insert a Fourier transform in the time

$$f(\vec{x}, t) = \frac{1}{2\pi}\int_{-\infty}^{\infty} \tilde{f}(\vec{x}, \omega)e^{-i\omega t}\, d\omega$$

and show that the inhomogeneous wave equation in (a) reduces to the inhomogeneous Helmholtz equation

$$(\nabla^2 + k^2)\tilde{\gamma}_{\mu\nu}(\vec{x}, \omega) = -\frac{16\pi G}{c^4}\tilde{S}_{\mu\nu}(\vec{x}, \omega)$$

Here $k = \omega/c$.

(c) Show that the driven solution to this equation can be written with the aid of the Green's function for the Helmholtz equation as[22]

$$\tilde{\gamma}_{\mu\nu}(\vec{x}, \omega) = -\frac{16\pi G}{c^4}\int \frac{e^{ik|\vec{x} - \vec{y}|}}{4\pi|\vec{x} - \vec{y}|}\tilde{S}_{\mu\nu}(\vec{y}, \omega)d^3 y$$

(d) Explore the consequences of the solution in (c).

---

[22]See [Fetter and Walecka (2003)] pp. 314-317. Here one takes Re at the end.

12.6 (a) Start from the expressions in Prob. 12.5(a), and show that the quantity $\psi_\mu{}^\nu$ defined in Eq. (12.35) satisfies the following equation

$$\Box\, \psi_\mu{}^\nu = -\frac{16\pi G}{c^4} T_\mu{}^\nu$$

(b) Use the conservation of the energy-momentum tensor in Minkowski space to show that

$$\Box\left(\frac{\partial}{\partial x^\nu}\psi_\mu{}^\nu\right) = 0$$

Hence conclude that if the auxiliary condition is satisfied initially, it will continue to be satisfied.[23]

12.7 In the next chapter we will relate the source to the motion of massive bodies. Here we simply make some dimensional arguments on the source. The power $\mathcal{P}$ is the total energy radiated per unit time. The dimensions of power are $[ml^2t^{-3}]$. The dimensions of Newton's constant G are $[m^{-1}l^3t^{-2}]$. The angular frequency of the source ω has dimensions of $[t^{-1}]$, and the ratio ω/c provides an inverse length $[l^{-1}]$.

(a) The power radiated must involve a factor of G, which enters into Einstein's field equations and governs the strength of coupling of gravity to mass. Show that to obtain the correct dimensions, one must then have

$$\mathcal{P} \propto GM^2\frac{\omega^2}{c}$$

Here M is a characteristic moving mass in the source.

(b) There is no *dipole* radiation since there will never be any net motion of the mass of an isolated system relative to its *center* of mass. Thus the leading multipole for radiation whose wavelength is long compared to the size of the system will be *quadrupole*.[24] The quadrupole moment depends on a mean value of the square of the radial size of the system, and thus the radiated power will also depend on the square of a (now dimensionless)

---

[23] This is the direct analog of the demonstration in E&M, through the use of Maxwell's equation $\Box A^\nu = -j^\nu$ and current conservation, that if the condition for the Lorentz gauge $\partial A^\nu/\partial x^\nu = 0$ is satisfied originally, it will continue to be satisfied.

[24] Although we will not demonstrate it here, the quantum of gravitational radiation, the *graviton*, is a spin-2 particle corresponding to the tensor nature of the gravitational field, and since it is massless, it can have only the maximum and minimum values of the helicity $\lambda = \pm 2$ (just as a photon, the massless spin-1 quantum of the vector electromagnetic field, has helicity $\lambda = \pm 1$). There is no monopole gravitational radiation in this theory. The factors in the estimate view graviton emission in analogy with photon emission.

radiating quadrupole moment

$$\left(\ddot{Q}\right)^2 \propto \left[\left(\frac{\omega}{c}\right)^2 \langle r^2 \rangle\right]^2$$

Thus the total power radiated goes as

$$P_{\text{rad}} \propto GM^2 c \left(\frac{\omega}{c}\right)^6 \left(\langle r^2 \rangle\right)^2$$

(c) The shrinking of the orbit with time of the Hulse-Taylor pulsar, evidently a binary neutron star, is a beautiful example of a star system *emitting* gravitational radiation [Wiki (2006)]. Use the following relevant values, as well as those of Eqs. (10.87), to obtain a number for the above estimate for P_{rad}

$$M \approx 2M_\odot \qquad ; \langle r^2 \rangle \approx (2R_\odot)^2 \qquad ; \tau = 2\pi/\omega \approx 8\,\text{hrs}$$

Compare with the measured value for the Hulse-Taylor pulsar of $P_{\text{rad}} \approx 10^{33}\,\text{erg/sec}$.

(d) Verify that indeed $2R_\odot \ll \lambda$ in part (c).[25]

The direct *detection* of gravitational radiation is the goal of LIGO.

13.1 Obtain a symbolic manipulation program, and verify the results in Table B.1 for the affine connection in the case of the Robertson-Walker metric with $k \neq 0$. Start from the elements of the metric in Eqs. (B.2).

13.2 Obtain a symbolic manipulation program, and verify the results in Eqs. (B.6) for the Ricci tensor in the case of the Robertson-Walker metric with $k \neq 0$. Start from the elements of the affine connection in Table B.1.

13.3 Carry the proof through from the beginning, and verify that the relation in Eq. (13.49) also holds for the Robertson-Walker metric with $k = 0$ as studied in chapter 11.[26]

13.4 (a) Use the Ricci tensor of Eqs. (7.71) and the metric with $A(r) = 1/(1-kr^2)$ and $B(r) = 1$ to evaluate the *scalar curvature* of the riemannian space described by the Robertson-Walker metric with $k \neq 0$ and $\Lambda = 1$. Show that it is given by

$$R = R^\mu{}_\mu = 6k \qquad ; \text{scalar curvature with } \Lambda = 1$$

Hence conclude that it is a space of *constant curvature*.

---

[25] It is also interesting to compare with Prob. 7.5(a).

[26] While the result may be obvious, it is a useful exercise to think through each step again from the beginning.

(b) What is the answer when $\Lambda(t) \neq 1$?

13.5 Consider the solution to the Einstein Eqs. (13.50) in the case where $k = 0$ and $\rho c^2 \ll P$. Take the positive root, and show that the previous results for a flat, cold, matter-dominated cosmology are reproduced:

$$\frac{\dot{\Lambda}}{\Lambda} = \frac{2}{3(t - t_0)}$$

$$\rho(t) = \frac{1}{6\pi G} \frac{1}{(t - t_0)^2}$$

$$\rho(t_p) = \frac{3}{8\pi G} H_0^2 \qquad ; H_0 = \text{Hubble's constant}$$

13.6 Consider the solution to the Einstein Eqs. (13.50) in the case where the source is simply that of the cosmological constant, with the effective equation of state in Eq. (13.19).

(a) Show that $\rho = \rho_0$ where ρ_0 is a constant. Assume for the purposes of this problem that ρ_0 is positive.

(b) Assume that $k/\dot{\Lambda}^2$ is negligible. Take the positive root, and show the solution is

$$\Lambda \propto e^{\mathcal{H}_0 t} \qquad ; \mathcal{H}_0 \equiv \left(\frac{8\pi G \rho_0}{3} \right)^{1/2}$$

where $\mathcal{H}_0$ is a constant. Hence conclude that in this case Λ exhibits *exponential growth*.

(c) Show that now

$$\frac{k}{(\dot{\Lambda})^2} \propto \frac{k}{\mathcal{H}_0^2} e^{-2\mathcal{H}_0 t}$$

Thus verify that the assumption in (b), which is equivalent to setting $k = 0$ at the outset, is consistent. In this case one *converges with time to the flat solution with* $\Omega = 1$.

13.7 Go to the LF^3 and introduce cartesian coordinates $x^\mu = (\vec{x}, ct)$. The metric $\bar{g}^{\mu\nu}$ is just that of Minkowski space in Eq. (13.27), so the volume element is $d\tau = d^3x\, d(ct)$. Assume the usual form of the potential in Eq. (13.54), and write the lagrangian density for a scalar field ϕ with inverse Compton wavelength $m = m_0 c/\hbar$ as

$$\mathcal{L}_\phi \left(\phi, \frac{\partial \phi}{\partial x^\mu} \right) = -\frac{c^2}{2} \left[\bar{g}^{\mu\nu} \frac{\partial \phi}{\partial x^\mu} \frac{\partial \phi}{\partial x^\nu} + m^2 \phi^2 \right]$$

(a) Start from Hamilton's principle in Eq. (13.6) and derive the continuum form of Lagrange's equation[27]

$$\frac{\partial}{\partial x^\mu} \frac{\partial \mathcal{L}_\phi}{\partial (\partial \phi / \partial x^\mu)} - \frac{\partial \mathcal{L}_\phi}{\partial \phi} = 0$$

Here $\partial/\partial x^\mu$ implies that the other coordinates are kept fixed in the differentiation.

(b) Show that the equation of motion of the scalar field is then

$$\left(\Box - m^2 \right) \phi = 0$$

which is the Klein-Gordon equation [compare the second of Eqs. (10.91)].

(c) Now make a coordinate transformation to coordinates q^μ in the global, inertial laboratory frame, assume a general potential $\mathcal{V}(\phi)$, and remember that the scalar field ϕ is invariant. Show the lagrangian density takes the form in Eq. (13.20).

13.8 Start from Hamilton's principle in Eq. (13.6) and the scalar field lagrangian of Eq. (13.20).

(a) Show Lagrange's equation for the scalar field takes the form

$$\frac{c^2}{\sqrt{-g}} \frac{\partial}{\partial q^\mu} \left(\sqrt{-g} \, g^{\mu\nu} \frac{\partial \phi}{\partial q^\nu} \right) = \frac{\partial \mathcal{V}(\phi)}{\partial \phi}$$

or;
$$c^2 (\nabla^\mu \phi)_{;\mu} = \frac{\partial \mathcal{V}(\phi)}{\partial \phi}$$

b) Assume a spatially constant scalar field which is a function of time $\phi(t)$. Show the equation of motion of the scalar field in the LF^3 is

$$\ddot{\phi} = -\frac{\partial \mathcal{V}(\phi)}{\partial \phi}$$

Here $\ddot{\phi} = d^2\phi/dt^2$.

(c) Assume a spatially constant scalar field which is a function of time $\phi(t)$ and the Robertson-Walker metric with $k \neq 0$ of Eq. (13.31). Show the equation of motion of the scalar field in the global inertial laboratory frame takes the form

$$\ddot{\phi} + 3 \left(\frac{\dot{\Lambda}}{\Lambda} \right) \dot{\phi} = -\frac{\partial \mathcal{V}(\phi)}{\partial \phi}$$

Here $\dot{\phi} = d\phi/dt$ and $\ddot{\phi} = d^2\phi/dt^2$.

---
[27] See [Fetter and Walecka (2003)] p. 129.

(d) Use the equation of state for the scalar field in Eqs. (13.29) to show that under the stated conditions, the result in part (c) is identical to the second of Eqs. (13.50).

13.9 Try to find an approximate solution to Eqs. (13.50) in the case where there is only an additional spatially constant scalar field $\phi(t)$ that satisfies the results in Prob. 13.8. Assume a potential of the form in Fig. 13.2, and use the equation of state of Eq. (13.29).

(a) Start in the LF^3 where ρ is defined. Differentiate the first of Eqs. (13.29) with respect to time and show that

$$\dot{\rho}c^2 = \left[\ddot{\phi} + \frac{\partial \mathcal{V}(\phi)}{\partial \phi}\right]\dot{\phi} = 0$$

Hence conclude that ρ must be constant everywhere in space as the scalar field develops with time

$$\rho = \frac{1}{2c^2}\dot{\phi}^2 + \frac{1}{c^2}\mathcal{V}(\phi) = \rho_0 \qquad ; \text{ constant}$$

Here $\rho_0 = \mathcal{V}(0)/c^2$, which we define to be $\rho_0 \equiv \mathcal{V}_0/c^2$. Since ϕ is a scalar, and the times in the two frames are identical, observe that this constant mass density also fills all of space in the global inertial laboratory frame.

(b) Substitute the result in (a) into the first of Eqs. (13.50) and show

$$\dot{\Lambda}^2 - \mathcal{H}_0^2\Lambda^2 = -kc^2 \qquad\qquad ; \mathcal{H}_0 \equiv \left(\frac{8\pi G\rho_0}{3}\right)^{1/2}$$

This expression is now identical to that appearing in Prob. 13.6. As done there, assume that $k/\dot{\Lambda}^2$ is negligible. Take the positive root, and show the solution to this differential equation is $\Lambda \propto e^{\mathcal{H}_0 t}$. Thus

$$\frac{k}{(\dot{\Lambda})^2} \propto \frac{k}{\mathcal{H}_0^2}e^{-2\mathcal{H}_0 t} \qquad\qquad ; \Lambda \propto e^{\mathcal{H}_0 t}$$

Hence verify the assumption, which is equivalent to setting $k = 0$ at the outset, and conclude that in this case one again *converges with time to the flat solution with $\Omega = 1$*.

(c) It remains to satisfy the second of Eqs. (13.50), which, because of the result in part (a), can only be done approximately. From Eqs. (13.29) one has

$$\left(\rho + \frac{P}{c^2}\right) = \frac{1}{c^2}\dot{\phi}^2$$

If one demands that the last term in the second of Eqs. (13.50) be negligible with respect to, say, the second term in $\dot{\rho}$ in part (a), one should have a consistent approximation. Show this condition becomes (note that $\dot{\phi}^2 = 2[\mathcal{V}_0 - \mathcal{V}(\phi)]$)

$$\left| 6 \left(\frac{\dot{\Lambda}}{\Lambda}\right) \left[\frac{\mathcal{V}_0 - \mathcal{V}(\phi)}{\mathcal{V}'(\phi)\,\dot{\phi}}\right] \right| \ll 1$$

Show that this may be rewritten approximately as

$$\left[\frac{48\pi G\,\mathcal{V}_0[\mathcal{V}_0 - \mathcal{V}(\phi)]}{c^2\,[\mathcal{V}'(\phi)]^2}\right]^{1/2} \ll 1$$

Verify that the expression on the l.h.s. is dimensionless.[28] Conclude that this condition is always satisfied in the limit $c^2 \to \infty$. Since the l.h.s. is homogeneous in $\mathcal{V}$, conclude also that this condition is always satisfied in the limit $c^2 \to \infty$ *at fixed* $\mathcal{V}/c^2$ (that is, at fixed ρ_0).

13.10 Show through a rescaling of coordinates that to extract the physics of the Robertson-Walker metric it is sufficient to examine the cases with $k = 0$ and $k = \pm 1$.

13.11 (a) Work in the LF^3. Consider a scalar field $\phi(x)$ with lagrangian density and energy-momentum tensor

$$\mathcal{L}_\phi = -\frac{c^2}{2}\left[\bar{g}^{\mu\nu}\frac{\partial\phi}{\partial x^\mu}\frac{\partial\phi}{\partial x^\nu} + m^2\phi^2\right]$$

$$\bar{T}_\phi^{\mu\nu} = c^2\frac{\partial\phi}{\partial x_\mu}\frac{\partial\phi}{\partial x_\nu} + \bar{g}^{\mu\nu}\mathcal{L}_\phi$$

Use the results in Prob. 13.7, and show that the energy-momentum tensor is conserved along the dynamical path

$$\frac{\partial\bar{T}^{\mu\nu}}{\partial x^\nu} = 0 \qquad ; \text{ along dynamical path}$$

(b) Consider a more general form of the lagrangian density $\mathcal{L}(\phi, \partial\phi/\partial x^\mu)$. Show that if the energy-momentum tensor is defined by

$$\bar{T}_\phi^{\mu\nu} = \bar{g}^{\mu\nu}\,\mathcal{L}_\phi - \frac{\partial\phi}{\partial x_\mu}\frac{\partial\mathcal{L}_\phi}{\partial(\partial\phi/\partial x^\nu)}$$

---

[28]*Hint*: Recall from Prob. 12.7 that the dimensions of G are $[m^{-1}l^3t^{-2}]$, and verify that the dimensions of ϕ^2 are $[ml^{-1}]$. What are the dimensions of $\mathcal{V}$ and $\mathcal{L}$?

then it is again conserved along the dynamical path. Discuss the relevance to Eqs. (13.20) and (13.24).

13.12 Suppose one has the Robertson-Walker metric with $k \neq 0$ and all three sources are simultaneously present: (1) uniform matter mass density with its own equation of state; (2) cosmological constant; (3) uniform, time-dependent scalar field $\phi(t)$ with dynamics governed as in Prob. 13.8.

(a) Write the full set of field equations for this system;

(b) Note how the different source terms contribute;

(c) Discuss how one would go about solving these equations.

13.13 (a) Specialize the energy-momentum tensor of a fluid in Eq. (6.131) to describe the motion of a point mass m located at the postion $\vec{x}_p(t)$ through the replacements

$$P \to 0$$
$$\rho(x) \to m\,\delta^{(3)}[\vec{x} - \vec{x}_p(t)]$$

Show the non-relativistic energy-momentum tensor then takes the following form

$$T_p^{ij} = m\,\delta^{(3)}[\vec{x} - \vec{x}_p(t)]\frac{dx_p^i}{dt}\frac{dx_p^j}{dt}$$
$$T_p^{i4} = mc\,\delta^{(3)}[\vec{x} - \vec{x}_p(t)]\frac{dx_p^i}{dt}$$
$$T_p^{44} = mc^2\,\delta^{(3)}[\vec{x} - \vec{x}_p(t)]$$

Assume that the combination $(8\pi G/c^4)T_p^{\mu\nu}$ is characterized by a small quantity of $O(h)$. Discuss how these relations might be used to model a source in the discussion of gravitational radiation in chapter 12 and Probs. 12.5-12.7.

(c) How would you improve this model?

13.14 Show that one can construct a quantity with the dimension of mass from the fundamental constants $(\hbar, c, G)$, and a corresponding length, in the following manner

$$M_{\text{Planck}} = \left(\frac{\hbar c}{G}\right)^{1/2} \qquad ; \text{ Planck mass}$$

$$\frac{\hbar}{M_{\text{Planck}}\,c} = \left(\frac{\hbar G}{c^3}\right)^{1/2} \qquad \text{Planck length}$$

It is widely believed that the effects of quantum gravity must become important at this distance and corresponding energy scale. Find the numerical values of these quantities.[29]

---

[29] Current experiments probe our understanding of the strong, electromagnetic, and weak interactions to a distance scale of the order of 10^{-16} cm. One should take note of the magnitude of the extrapolation to the Planck length.

Appendix A

Reduction of $g^{\mu\nu}\,\delta R_{\mu\nu}$ to covariant divergences

The goal of this appendix is to show that the term $g^{\mu\nu}\,\delta R_{\mu\nu}$ appearing in the application of Hamilton's principle to the Einstein-Hilbert lagrangian can be written as the difference of two covariant divergences of four-vectors

$$g^{\mu\nu}\,\delta R_{\mu\nu} = \left[g^{\mu\nu}\,\delta\Gamma^\lambda_{\mu\nu}\right]_{;\lambda} - \left[\delta\Gamma^\lambda_{\lambda\mu}\,g^{\mu\nu}\right]_{;\nu} \tag{A.1}$$

where $\delta\Gamma^\lambda_{\mu\nu}$ is the corresponding variation in the affine connection. Gauss' theorem then allows one to convert this term to a surface integral, where it vanishes due to the boundary conditions in the problem. The proof is after [Blau (2006)].

The Ricci tensor is related to the affine connection by

$$R_{\mu\nu} = \frac{\partial}{\partial q^\lambda}\Gamma^\lambda_{\mu\nu} - \frac{\partial}{\partial q^\nu}\Gamma^\lambda_{\mu\lambda} + \Gamma^\lambda_{\lambda\rho}\Gamma^\rho_{\mu\nu} - \Gamma^\lambda_{\nu\rho}\Gamma^\rho_{\mu\lambda} \tag{A.2}$$

We are interested in the change when the metric $g^{\mu\nu}$ is varied locally, subject to the conditions that the metric remain a symmetric second-rank tensor and that the inverse relation also be maintained

$$\delta\left(g_{\mu\lambda}\,g^{\lambda\rho}\right) = 0 \tag{A.3}$$

The variation of the Ricci tensor follows directly from Eq. (A.2) as

$$\delta R_{\mu\nu} = \frac{\partial}{\partial q^\lambda}\delta\Gamma^\lambda_{\mu\nu} - \frac{\partial}{\partial q^\nu}\delta\Gamma^\lambda_{\mu\lambda} + \delta\Gamma^\lambda_{\lambda\rho}\Gamma^\rho_{\mu\nu} + \Gamma^\lambda_{\lambda\rho}\delta\Gamma^\rho_{\mu\nu} - \delta\Gamma^\lambda_{\nu\rho}\Gamma^\rho_{\mu\lambda} - \Gamma^\lambda_{\nu\rho}\delta\Gamma^\rho_{\mu\lambda}$$

$$\tag{A.4}$$

It is then necessary to examine the variation of the affine connection.

The affine connection is defined as

$$\Gamma^\lambda_{\mu\nu} = \frac{1}{2}g^{\lambda\alpha}\left[\frac{\partial}{\partial q^\mu}g_{\alpha\nu} + \frac{\partial}{\partial q^\nu}g_{\alpha\mu} - \frac{\partial}{\partial q^\alpha}g_{\mu\nu}\right] \tag{A.5}$$

The variation of this quantity follows as

$$\delta\Gamma^{\lambda}_{\mu\nu} = \frac{1}{2}\delta g^{\lambda\alpha}\left[\frac{\partial}{\partial q^{\mu}}g_{\alpha\nu} + \frac{\partial}{\partial q^{\nu}}g_{\alpha\mu} - \frac{\partial}{\partial q^{\alpha}}g_{\mu\nu}\right] +$$
$$\frac{1}{2}g^{\lambda\alpha}\left[\frac{\partial}{\partial q^{\mu}}\delta g_{\alpha\nu} + \frac{\partial}{\partial q^{\nu}}\delta g_{\alpha\mu} - \frac{\partial}{\partial q^{\alpha}}\delta g_{\mu\nu}\right] \quad \text{(A.6)}$$

Now recall from our discussion of the covariant derivative in chap. 5 that the covariant derivative of a tensor is given by

$$T^{ab\cdots}_{lm\cdots;j} = \frac{\partial T^{ab\cdots}_{lm\cdots}}{\partial q^{j}} + \Gamma^{a}_{a'j}T^{a'b\cdots}_{lm\cdots} + \cdots - \Gamma^{l'}_{lj}T^{ab\cdots}_{l'm\cdots} - \cdots \quad \text{(A.7)}$$

Thus, for example,

$$g_{\alpha\nu;\mu} = \frac{\partial g_{\alpha\nu}}{\partial q^{\mu}} - \Gamma^{\alpha'}_{\alpha\mu}g_{\alpha'\nu} - \Gamma^{\nu'}_{\nu\mu}g_{\alpha\nu'} \quad \text{(A.8)}$$

Or, solving for the derivative,

$$\frac{\partial g_{\alpha\nu}}{\partial q^{\mu}} = g_{\alpha\nu;\mu} + \Gamma^{\alpha'}_{\alpha\mu}g_{\alpha'\nu} + \Gamma^{\nu'}_{\nu\mu}g_{\alpha\nu'} \quad ; \text{etc.} \quad \text{(A.9)}$$

We observe that this relation is true for *any* second-rank tensor, in particular for $\delta g_{\alpha\nu}$.

Consider the second line of Eq. (A.6). Substitution of the above for the derivatives of δg leads to

$$\frac{1}{2}g^{\lambda\alpha}\left[\frac{\partial}{\partial q^{\mu}}\delta g_{\alpha\nu} + \frac{\partial}{\partial q^{\nu}}\delta g_{\alpha\mu} - \frac{\partial}{\partial q^{\alpha}}\delta g_{\mu\nu}\right] =$$
$$\frac{1}{2}g^{\lambda\alpha}\left\{\delta g_{\alpha\nu;\mu} + \delta g_{\alpha\mu;\nu} - \delta g_{\mu\nu;\alpha} + \Gamma^{\alpha'}_{\alpha\mu}\delta g_{\alpha'\nu} + \Gamma^{\nu'}_{\nu\mu}g_{\alpha\nu'}\right.$$
$$\left. + \Gamma^{\alpha'}_{\alpha\nu}\delta g_{\alpha'\mu} + \Gamma^{\mu'}_{\nu\mu}\delta g_{\alpha\mu'} - \Gamma^{\mu'}_{\mu\alpha}\delta g_{\mu'\nu} - \Gamma^{\nu'}_{\nu\alpha}\delta g_{\mu\nu'}\right\} \quad \text{(A.10)}$$

A cancelation and combination of identical terms then leads to

$$\frac{1}{2}g^{\lambda\alpha}\left[\frac{\partial}{\partial q^{\mu}}\delta g_{\alpha\nu} + \frac{\partial}{\partial q^{\nu}}\delta g_{\alpha\mu} - \frac{\partial}{\partial q^{\alpha}}\delta g_{\mu\nu}\right] =$$
$$\frac{1}{2}g^{\lambda\alpha}\left(\delta g_{\alpha\nu;\mu} + \delta g_{\alpha\mu;\nu} - \delta g_{\mu\nu;\alpha}\right) + g^{\lambda\alpha}\Gamma^{\nu'}_{\nu\mu}\delta g_{\alpha\nu'} \quad \text{(A.11)}$$

Consider now the first term on the r.h.s. of Eq. (A.6). Recall that the covariant derivative of the metric tensor vanishes, and in this case Eq. (A.9) reduces to the form[1]

$$\frac{\partial g_{\alpha\nu}}{\partial q^\mu} = \Gamma^{\alpha'}_{\alpha\mu}\, g_{\alpha'\nu} + \Gamma^{\nu'}_{\nu\mu}\, g_{\alpha\nu'} \qquad ; \text{ etc.} \tag{A.12}$$

Hence this first term takes the form

$$\frac{1}{2}\delta g^{\lambda\alpha}\left[\frac{\partial}{\partial q^\mu}g_{\alpha\nu} + \frac{\partial}{\partial q^\nu}g_{\alpha\mu} - \frac{\partial}{\partial q^\alpha}g_{\mu\nu}\right] =$$

$$\frac{1}{2}\delta g^{\lambda\alpha}\left\{\Gamma^{\alpha'}_{\alpha\mu}\, g_{\alpha'\nu} + \Gamma^{\nu'}_{\nu\mu}\, g_{\alpha\nu'} + \Gamma^{\alpha'}_{\alpha\nu}\, g_{\alpha'\mu} + \Gamma^{\mu'}_{\nu\mu}\, g_{\alpha\mu'} - \Gamma^{\mu'}_{\mu\alpha}\, g_{\mu'\nu} - \Gamma^{\nu'}_{\nu\alpha}\, g_{\mu\nu'}\right\} \tag{A.13}$$

Cancelation and combination of identical terms reduces this to

$$\frac{1}{2}\delta g^{\lambda\alpha}\left[\frac{\partial}{\partial q^\mu}g_{\alpha\nu} + \frac{\partial}{\partial q^\nu}g_{\alpha\mu} - \frac{\partial}{\partial q^\alpha}g_{\mu\nu}\right] = \delta g^{\lambda\alpha}\,\Gamma^{\nu'}_{\mu\nu}\, g_{\nu'\alpha} \tag{A.14}$$

This term now cancels the last term in Eq. (A.11) since by Eq. (A.3)

$$\delta g^{\lambda\alpha}\, g_{\alpha\nu'} = -g^{\lambda\alpha}\,\delta g_{\alpha\nu'} \tag{A.15}$$

Hence the variation of the affine connection in Eq. (A.6) reduces to

$$\delta\Gamma^\lambda_{\mu\nu} = \frac{1}{2}g^{\lambda\alpha}\left[\delta g_{\alpha\nu;\mu} + \delta g_{\alpha\mu;\nu} - \delta g_{\mu\nu;\alpha}\right] \tag{A.16}$$

We are now in a position to make the important observation that while the affine connection $\Gamma^\lambda_{\mu\nu}$ is *not* by itself a third-rank tensor, the *variation* of the affine connection *is* a third-rank tensor.[2]

Thus the covariant derivative of the rank-one tensor $\delta\Gamma^\lambda_{\lambda\mu}$ is given by the general relation in Eq. (A.7) as

$$\delta\Gamma^\lambda_{\lambda\mu;\nu} = \frac{\partial}{\partial q^\nu}\delta\Gamma^\lambda_{\lambda\mu} - \Gamma^{\mu'}_{\mu\nu}\,\delta\Gamma^\lambda_{\lambda\mu'} \tag{A.17}$$

This is just the covariant derivative of a four-vector.

Furthermore, the covariant divergence of the third-rank tensor $\delta\Gamma^\lambda_{\mu\nu}$ is given by the general relation as

$$\delta\Gamma^\lambda_{\mu\nu;\lambda} = \frac{\partial}{\partial q^\lambda}\delta\Gamma^\lambda_{\mu\nu} + \Gamma^\lambda_{\lambda'\lambda}\,\delta\Gamma^{\lambda'}_{\mu\nu} - \Gamma^{\mu'}_{\mu\lambda}\,\delta\Gamma^\lambda_{\mu'\nu} - \Gamma^{\nu'}_{\lambda\nu}\,\delta\Gamma^\lambda_{\mu\nu'} \tag{A.18}$$

---

[1] While it is true that the covariant derivative of the metric vanishes so that $g_{\alpha\nu;\mu} = 0$, this is not necessarily true for the covariant derivative of the *variation* of the metric $\delta g_{\alpha\nu;\mu}$, so in that case one must retain all the terms in Eq. (A.9).

[2] Note that while the affine connection and Ricci tensor are intrinsically nonlinear in the metric, they both become linear expressions in the *variation* of the metric.

These results can now be combined to write the variation of the Ricci tensor in Eq. (A.4) as

$$\delta R_{\mu\nu} = \delta\Gamma^\lambda_{\mu\nu;\lambda} - \delta\Gamma^\lambda_{\lambda\mu;\nu} \qquad (A.19)$$

Since the covariant derivative of the metric tensor vanishes, it can be taken right through the derivatives to give

$$g^{\mu\nu}\,\delta R_{\mu\nu} = \left[g^{\mu\nu}\,\delta\Gamma^\lambda_{\mu\nu}\right]_{;\lambda} - \left[\delta\Gamma^\lambda_{\lambda\mu}\,g^{\mu\nu}\right]_{;\nu} \qquad (A.20)$$

This is the stated result.

Appendix B

Robertson-Walker Metric with $k \neq 0$

Introduce spatial spherical coordinates

$$q^\mu = (r, \theta, \phi, ct) \qquad ; \mu = r, \theta, \phi, 4 \qquad (B.1)$$

The Robertson-Walker metric with $k \neq 0$ then generalizes the metric with $k = 0$ to the following, and its inverse, [Robertson (1935); Walker (1936)]

$$g_{\mu\nu} = \begin{bmatrix} \Lambda^2(t)/(1 - kr^2) & 0 & 0 & 0 \\ 0 & \Lambda^2(t)r^2 & 0 & 0 \\ 0 & 0 & \Lambda^2(t)r^2 \sin^2\theta & 0 \\ 0 & 0 & 0 & -1 \end{bmatrix}$$

$$g^{\mu\nu} = \begin{bmatrix} (1 - kr^2)/\Lambda^2(t) & 0 & 0 & 0 \\ 0 & 1/\left[\Lambda^2(t)r^2\right] & 0 & 0 \\ 0 & 0 & 1/\left[\Lambda^2(t)r^2 \sin^2\theta\right] & 0 \\ 0 & 0 & 0 & -1 \end{bmatrix} \quad (B.2)$$

The procedure is clear. The affine connection must first be computed from the metric

$$\Gamma^\lambda_{\mu\nu} = \frac{1}{2} g^{\lambda\sigma} \left[\frac{\partial g_{\sigma\nu}}{\partial q^\mu} + \frac{\partial g_{\sigma\mu}}{\partial q^\nu} - \frac{\partial g_{\mu\nu}}{\partial q^\sigma} \right] \qquad (B.3)$$

The Ricci tensor must then be obtained from the affine connection

$$R_{\mu\nu} = \frac{\partial}{\partial q^\lambda} \Gamma^\lambda_{\mu\nu} + \Gamma^\lambda_{\lambda\sigma} \Gamma^\sigma_{\mu\nu} - \left[\frac{\partial}{\partial q^\nu} \Gamma^\lambda_{\mu\lambda} + \Gamma^\lambda_{\nu\sigma} \Gamma^\sigma_{\mu\lambda} \right] \qquad (B.4)$$

We have carried this procedure through in great detail in the text for several examples — Schwarzschild solution, TOV equations, $k = 0$ cosmology, and gravitational radiation. The calculation now simply involves, long, tedious, repetitive algebra. But this is exactly where a computer excels! As a learning experience, the author carried out the required calculations on his PC using the symbolic manipulation program in Mathcad11. The results for the affine connection are given in Table B.1. The exercise was so informative that it is assigned as a problem for the dedicated reader.

Table B.1   SUMMARY —Affine connection for Robertson-Walker metric with $k \neq 0$ and coordinates $(q^1, \cdots, q^4) = (r, \theta, \phi, ct)$. It is symmetric in its lower two indices so that $\Gamma^\lambda_{\mu\nu} = \Gamma^\lambda_{\nu\mu}$. In this table $\dot{\Lambda} = d\Lambda(t)/dt$.

$\Gamma^r_{rr} = kr/(1 - kr^2)$	$\Gamma^r_{\theta\theta} = -r(1 - kr^2)$	$\Gamma^r_{\phi\phi} = -r(1 - kr^2)\sin^2\theta$
$\Gamma^r_{r4} = \dot{\Lambda}/\Lambda c$	$\Gamma^\theta_{\phi\phi} = -\sin\theta\cos\theta$	$\Gamma^\theta_{r\theta} = 1/r$
$\Gamma^\theta_{\theta 4} = \dot{\Lambda}/\Lambda c$	$\Gamma^\phi_{\phi 4} = \dot{\Lambda}/\Lambda c$	$\Gamma^\phi_{\theta\phi} = \cos\theta/\sin\theta$
$\Gamma^\phi_{r\phi} = 1/r$	$\Gamma^4_{rr} = \Lambda\dot{\Lambda}/(1 - kr^2)c$	$\Gamma^4_{\theta\theta} = r^2\Lambda\dot{\Lambda}/c$
$\Gamma^4_{\phi\phi} = r^2\Lambda\dot{\Lambda}\sin^2\theta/c$	All others vanish	

In calculating the Ricci tensor, it is useful to remember just where the indices on it go, for example

$$R_{r\theta} = \frac{\partial}{\partial q^\lambda}\Gamma^\lambda_{r\theta} - \frac{\partial}{\partial \theta}\Gamma^\lambda_{r\lambda} + \Gamma^\lambda_{\lambda\sigma}\Gamma^\sigma_{r\theta} - \Gamma^\lambda_{\theta\sigma}\Gamma^\sigma_{r\lambda} \qquad (B.5)$$

The actual evaluation is again left for the reader as an exercise in symbolic manipulation. This is evidently the approach of choice in problems with far more complicated metrics.

The final expression for the Ricci tensor is

$$R_{ij} = \frac{1}{c^2}\left[2\left(\frac{\dot{\Lambda}}{\Lambda}\right)^2 + \left(\frac{\ddot{\Lambda}}{\Lambda}\right) + \frac{2kc^2}{\Lambda^2}\right]g_{ij} \qquad ; (i,j) = (r, \theta, \phi)$$

$$R_{44} = -\frac{3}{c^2}\left(\frac{\ddot{\Lambda}}{\Lambda}\right)$$

$$R_{i4} = 0 \qquad\qquad\qquad (B.6)$$

Here $\dot{\Lambda} = d\Lambda/dt$ and $\ddot{\Lambda} = d^2\Lambda/dt^2$. Note the presence of g_{ij} on the r.h.s. of the first equation. This is the principal result of this appendix. As discussed in the text, there are several checks one can perform on it.

Bibliography

Abraham, R., Marsden, J. E., and Ratiu, J. (1993). *Manifolds, Tensor Analysis, and Applications, 2nd ed.*, Springer, New York

Adler, R., Bazin, M., and Schiffer, M. (1965). *Introduction to General Relativity*, McGraw-Hill, New York

Alford, M., Bowers, J. A., and Rajagopal, K. (2001). *Lecture Notes in Physics* **578**, Springer, Berlin, p. 235

Amore, P. (2000). "Introduction to Inflation," term paper in Physics 786, College of William and Mary, Williamsburg, VA (unpublished)

Amore, P. and Arceo, S. (2006). *Phys. Rev.* **D73**, 083004

Batiz, Z. (2000). "Black Hole Thermodynamics," term paper in Physics 786, College of William and Mary, Williamsburg, VA (unpublished)

Blau, M. (1994). "Lecture Notes on General Relativity," available at http://www.unine.ch/phys/ string/lecturesGR.pdf

Brown, G. E. (1994). *Nucl. Phys.* **A574**, 217c

Chandra (2006). *The Chandra X-ray Observatory*, http://chandra.nasa.gov/

Chen, P., and Tajima, T. (1999). "Testing Unruh Radiation with Intense Lasers," *Phys. Rev. Lett.* **83**, 256

COBE (2006). *The Cosmic Background Explorer*, http://lambda.gsfc.nasa.gov/ product/cobe/

Dimopoulos, S., and Susskind, L. (1978). *Phys. Rev.* **D18**, 4500

Einstein, A. (1905). "Zur Electrodynamik bewegter Körper," *Annalen der Physik* **17**, 891

Einstein, A. (1916). "Die Grundlage der Allgemeinen Relativitäts Theorie," *Annalen der Physik* **49**, 50

Einstein, A. (1936). *Science* **84**, 506

Einstein (2006). *Gravity Probe B*, http://einstein.stanford.edu/

Falco, E. E. (1994). *New Journal of Physics* **7**, 200

Foster, J., and Nightengale, J. D. (2006). *A Short Course on General Relativity, 3rd ed.*, Springer, New York

Fetter, A. L. and Walecka, J. D. (2003). *Theoretical Mechanics of Particles and Continua*, McGraw-Hill, New York (1980); reissued by Dover Publications, Mineola, New York

Fetter, A. L. and Walecka, J. D. (2003a). *Quantum Theory of Many-Particle Systems*, McGraw-Hill, New York (1971); reissued by Dover Publications, Mineola, New York

Freedman, W. L., *et al.* (2001). *The Astrophysical Journal* **553**, 47 (2001)

Friedmann, A. (1922). *Z. Phys.* **10**, 377; also **21**, 326 (1924)

Glendenning, N. K. (2000). *Compact Stars: Nuclear Physics, Particle Physics and General Relativity, 2nd ed.*, Springer, New York

Guth, A. H. (2000). "Inflation and Eternal Inflation," *Physics Reports* **333**, 555

Hartle, J. (2002). *Gravity: An Introduction to Einstein's General Relativity*, Addison Wesley, Reading, MA

Hubble (2006). *The Hubble Space Telescope*, http://hubble.nasa.gov/ ; for images, see http://hubblesite.org/gallery/

Hughston, L. P., and Tod, K. P. (1990). *An Introduction to General Relativity*, Cambridge U. Press, New York

Kerr, R. (1963). *Phys. Rev. Lett.* **11**, 237

Kolb, E. W., and Turner, M. S. (1990). *The Early Universe*, Addison-Wesley, Reading, MA

Landau, L. D., and Lifshitz, E. M. (1975). *Classical Theory of Fields*, Pergamon Press, New York

LIGO (2006). *The Laser Interferometer Gravitational-Wave Observatory (LIGO)*, www.ligo.caltech.edu/

Linde, A. (1990). *Particle Physics and Inflationary Cosmology*, Harwood, New York

Misner, C. W., Thorne, K. S., and Wheeler, J. A. (1973). *Gravitation*, W. H. Freeman, San Francisco

Müller, H., and B. D. Serot, B.D. (1996). *Nucl. Phys.* **A606**, 508

Ohanian, H. C., and Ruffini, R. (1994). *Gravitation and Spacetime, 2nd ed.*, W. W. Norton and Company, New York

Ohanian, H. C. (1995). *Modern Physics, 2nd ed.*, Prentice-Hall, Upper Saddle River, NJ

Oppenheimer, J. R., and Volkoff, G. M. (1939). *Phys. Rev.* **55**, 374

Peacock, J. A. (1999). *Cosmological Physics*, Cambridge U. Press, Cambridge, UK

Peebles, P. J. E. (1993). *Principles of Physical Cosmology*, Princeton U. Press, Princeton, NJ

Pound, R. V., and Rebka, G. A. (1960). *Phys. Rev. Lett.* **4**, 337

Robertson, H. P. (1935). *Ap. J.* **82**, 284; also **83**, 187, 257 (1936)

Schwarzschild, K. (1916). "Über das Gravitationsfeld eines Massenpunktes nach der Einsteinschen Theorie," *Sitzungsberichte der Königlich Preussischen*

Akademie der Wissenschaften **1**, 189

Serot, B. D., and Walecka, J. D. (1986). "The Relativistic Nuclear Many-Body Problem," *Advances in Nuclear Physics* **16**, eds. J. W. Negele and E. Vogt, Plenum Press, New York

Shutz, B. F. (1985). *A First Course in General Relativity*, Cambridge U. Press, New York

Sky and Telescope (2000). "A Black Hole for Nearly Every Galaxy," May, 2000, p. 22

Sky and Telescope (2006). "Putting Einstein to the Test," July, 2006, p. 32

Sky and Telescope (2006a). "Einsteinian Energy Bolstered," April, 2006, p. 23

Sky and Telescope (2006b). "Case Strengthened for Inflation," June, 2006, p. 22

Taylor, E. F., and Wheeler, J. A. (2000). *Exploring Black Holes: Introduction to General Relativity*, Addison-Wesley, Reading, MA

Tolman, R. C. (1939). *Phys. Rev.* **55**, 364

Unruh, W. (1976). *Phys. Rev.* **D14**, 870

Wald, R. M. (1984). *General Relativity*, U. Chicago Press, Chicago

Walecka, J. D. (2004). *Theoretical Nuclear and Subnuclear Physics, 2nd ed.*, World Scientific, Singapore

Walker, A. G. (1936). *Proc. Lond. Math. Soc.* **42**, 90

Weinberg, S. (1972). *Gravitation and Cosmology: Principles and Applications of the General theory of Relativity*, John Wiley and Sons, New York

Wiki (2006). *The Wikipedia*, http://en.wikipedia.org/wiki/(topic)

WMAP (2006). *The Wilkinson Microwave Anisotropy Probe (WMAP)*, http://map.gsfc.nasa.gov/

Index